T0189384

Lecture Notes in Information Systems and Organisation

Volume 52

Lecture Notes in Information Systems and Organization—LNISO—is a series of scientific books that explore the current scenario of information systems, in particular IS and organization. The focus on the relationship between IT, IS and organization is the common thread of this collection, which aspires to provide scholars across the world with a point of reference and comparison in the study and research of information systems and organization. LNISO is the publication forum for the community of scholars investigating behavioral and design aspects of IS and organization. The series offers an integrated publication platform for high-quality conferences, symposia and workshops in this field. Materials are published upon a strictly controlled double blind peer review evaluation made by selected reviewers.

LNISO is abstracted/indexed in Scopus

More information about this series at http://www.springer.com/series/11237

Fred D. Davis · René Riedl ·
Jan vom Brocke · Pierre-Majorique Léger ·
Adriane B. Randolph · Gernot Müller-Putz
Editors

Information Systems and Neuroscience

NeuroIS Retreat 2021

 Springer

Editors
Fred D. Davis
Texas Tech University
Lubbock, TX, USA

René Riedl ⓘ
University of Applied Sciences Upper
Austria and University of Linz
Steyr/Linz, Austria

Jan vom Brocke
University of Liechtenstein
Vaduz, Liechtenstein

Pierre-Majorique Léger
HEC Montréal
Montréal, QC, Canada

Adriane B. Randolph
Kennesaw State University
Kennesaw, GA, USA

Gernot Müller-Putz
Graz University of Technology
Graz, Austria

ISSN 2195-4968   ISSN 2195-4976   (electronic)
Lecture Notes in Information Systems and Organisation
ISBN 978-3-030-88899-2   ISBN 978-3-030-88900-5   (eBook)
https://doi.org/10.1007/978-3-030-88900-5

This Springer imprint is published by the registered company Springer Nature Switzerland AG
The registered company address is: Gewerbestrasse 11, 6330 Cham, Switzerland

Preface

The proceedings contain papers presented at the 13th annual NeuroIS Retreat held June 1–3, 2021. NeuroIS is a field in Information Systems (IS) that uses neuroscience and neurophysiological tools and knowledge to better understand the development, adoption, and impact of information and communication technologies (www.neurois.org).

The NeuroIS Retreat is a leading academic conference for presenting research and development projects at the nexus of IS and neurobiology. This annual conference promotes the development of the NeuroIS field with activities primarily delivered by and for academics, though works often have a professional orientation.

In 2009, the inaugural NeuroIS Retreat was held in Gmunden, Austria. Since then, the NeuroIS community has grown steadily, with subsequent annual Retreats in Gmunden from 2010 to 2017. Beginning in 2018, the conference is taking place in Vienna, Austria.

Due to the corona crisis, the organizers decided to host a virtual NeuroIS Retreat in 2021.

The NeuroIS Retreat provides a platform for scholars to discuss their studies and exchange ideas. A major goal is to provide feedback for scholars to advance their research papers toward high-quality journal publications. The organizing committee welcomes not only completed research, but also work in progress. The NeuroIS Retreat is known for its informal and constructive workshop atmosphere. Many NeuroIS presentations have evolved into publications in highly regarded academic journals.

This year is the seventh time that we publish the proceedings in the form of an edited volume. A total of 27 research papers were accepted and are published in this volume, and we observe diversity in topics, theories, methods, and tools of the contributions in this book. The 2021 keynote presentation entitled "Decision Neuroscience: How it started and where we are today" was given by Antoine Bechara, professor of neuroscience and psychology at the University of Southern California (USC) in Los Angeles, USA. Moreover, Moritz Grosse-Wentrup, professor and head of the Research Group Neuroinformatics at the University of

Vienna, Austria, gave a hot topic talk entitled "How (not) to interpret Multivariate Decoding Models in Neuroimaging".

Altogether, we are happy to see the ongoing progress in the NeuroIS field. Also, we can report that the NeuroIS Society, established in 2018 as a non-profit organization, has been developing well. We foresee a prosperous development of NeuroIS.

June 2021

Fred D. Davis
René Riedl
Jan vom Brocke
Pierre-Majorique Léger
Adriane B. Randolph
Gernot Müller-Putz

Organization

Conference Co-chairs

Fred D. Davis — Texas Tech University, Texas, USA
René Riedl — University of Applied Sciences Upper Austria, Steyr, Austria & Johannes Kepler University Linz, Linz, Austria

Program Co-chairs

Jan vom Brocke — University of Liechtenstein, Vaduz, Liechtenstein
Pierre-Majorique Léger — HEC Montréal, Montréal, Canada
Adriane B. Randolph — Kennesaw State University, Kennesaw, USA
Gernot Müller-Putz — Graz University of Technology, Graz, Austria

Program Committee

Marc Adam — University of Newcastle, Callaghan, Australia
Bonnie Anderson — Brigham Young University, Utah, USA
Ricardo Buettner — Aalen University, Aalen, Germany
Colin Conrad — Dalhousie University, Halifax, Canada
Constantinos Coursaris — HEC Montréal, Montréal, Canada
Alan Dennis — Indiana University, Indiana, USA
Thomas Fischer — Johannes Kepler University Linz, Linz, Austria
Rob Gleasure — Copenhagen Business School, Frederiksberg, Denmark
Jacek Gwizdka — University of Texas at Austin, Austin, Texas
Alan Hevner — Muma College of Business, Florida, USA
Marco Hubert — University of Aarhus, Aarhus, Denmark
Peter Kenning — Heinrich-Heine-University Düsseldorf, Düsseldorf, Germany

Brock Kirwan	Brigham Young University, Utah, USA
Élise Labonté-LeMoyne	HEC Montréal, Montréal, Canada
Ting-Peng Liang	National Sun Yat-sen University, Kaohsiung, Taiwan
Aleck Lin	National Dong Hwa University, Taiwan
Jan Mendling	Vienna University of Economics and Business, Vienna, Austria
Fiona Nah	Missouri University of Science and Technology, Missouri, USA
Aaron Newman	Dalhousie University, Halifax, Canada
Jella Pfeiffer	Justus-Liebig-University Giessen, Giessen, Germany
Sylvain Sénécal	HEC Montréal, Montréal, Canada
Stefan Tams	HEC Montréal, Montréal, Canada
Lars Taxén	Linköping University, Linköping, Sweden
Ofir Turel	California State University, California, USA
Anthony Vance	Fox School of Business, Pennsylvania, USA
Karin VanMeter	University of Applied Sciences Upper Austria, Hagenberg, Austria
Eric Walden	Texas Tech University, Texas, USA
Barbara Weber	University of St. Gallen, St. Gallen, Switzerland
Robert West	DePauw University, Indiana, USA
Eoin Whelan	National University of Ireland Galway, Galway, Ireland
Selina Wriessnegger	Graz University of Technology, Graz, Austria

Organization Support

Thomas Kalischko	University of Applied Sciences Upper Austria, Steyr, Austria
Fabian Stangl	University of Applied Sciences Upper Austria, Steyr, Austria

Sponsors

We thank the sponsors of the NeuroIS Retreat 2021:

Main Sponsors

Further Sponsors

Decision Neuroscience: How It Started and Where We Are Today (Keynote)

Antoine Bechara

Decision neuroscience is an emerging area of research whose goal is to integrate research in neuroscience and behavioral decision making. Neuroeconomics is a more specialized field of study that seeks to bridge neuroscience research on human choice with economic theory, whereas neuromarketing addresses the neuroscience behind consumers' choices, including product branding, preference, and purchase decisions. More recent research seeks to include the field of information science by examining the impact of social media and other technology use on the human brain. All these areas capitalize on knowledge from the fields of neuroscience, behavioral economics, finances, marketing, and information science to explore the neural "road map" for the physiological processes intervening between knowledge and behavior, and the potential interruptions that lead to a disconnection between what one knows and what one decides to do. Thus, decision neuroscience is the domain that captures the interests of scientists who are attempting to understand the neural basis of judgment and decision making in health as well as social behavior.

How (not) to Interpret Multivariate Decoding Models in Neuroimaging (Hot Topic Talk)

Moritz Grosse-Wentrup

Multivariate decoding models are replacing traditional univariate statistical tests in the analysis of neuroimaging data. Their interpretation, however, is far from trivial. In this presentation, I outline various pitfalls and discuss under which conditions they can provide insights into the (causal) question of how neuronal activity gives rise to cognition and behavior.

Success Factors in Publishing NeuroIS Research in Top IS Journals: Experiences of MIS Quarterly Editors and Reviewers (Panel Discussion)

Moderator: Fred D. Davis

Panelists: Ofir Turel, Anthony Vance, Adriane B. Randolph, Eric Walden

Contents

Where NeuroIS Helps to Understand Human Processing of Text: A Taxonomy for Research Questions Based on Textual Data

Florian Popp[✉], Bernhard Lutz, and Dirk Neumann

University of Freiburg, Freiburg, Germany
{bernhard.lutz,dirk.neumann}@is.uni-freiburg.de

Abstract. Several research questions from information systems (IS) are based on textual data, such as product reviews and fake news. In this paper, we investigate in which areas NeuroIS is best suited to better understand human processing of text and subsequent human behavior or decision making. To evaluate this question, we propose a taxonomy to distinguish these research questions depending on how users' corresponding response is formed. We first review all publications about textual data in the IS basket journals from 2010–2020. Then, we distinguish text-based research questions along two dimensions, namely, if a user's response is influenced by subjectivity and if additional information is required to make an objective assessment. We find that NeuroIS research on textual data is still in its infancy. Existing NeuroIS studies focus on texts, where users' responses are subject to a higher need for additional data, which is not part of the text.

Keywords: Textual data · Taxonomy · Information processing · Decision-making · NeuroIS

1 Introduction

Social media and other forms of computer-mediated communication provide users with large amounts of textual data [1]. Text provides an unstructured and high-dimensional source of information as English texts can easily comprise vocabularies with tens of thousands of words [2]. So far, IS research has analyzed how users respond to texts of many types, such as fake news, online reviews, financial reports, spam emails, and computer-mediated communication. While the factors of a helpful product review are well understood (e.g., [3–5]), little is known of why users fall for deception in the form of fake news. Deciding if a news article is real or fake may be subjective and depends on prior beliefs and attitudes [6–8]. Making an objective veracity assessment of a news article may also require additional information from external sources, which is not part of the actual text [9]. Whether or not users invest cognitive effort into a critical assessment also depends on their mindset. Users in a utilitarian mindset are used to spending cognitive effort in reading the text as part of their daily (working) routines, whereas users in a hedonic mindset (e.g., when being active on social media) are less

© The Author(s), under exclusive license to Springer Nature Switzerland AG 2021
F. D. Davis et al. (Eds.): NeuroIS 2021, LNISO 52, pp. 1–8, 2021.
https://doi.org/10.1007/978-3-030-88900-5_1

likely to carefully reflect on their actions [6]. As a consequence, their behavior may be driven by unconscious processes, which can be studied by employing tools and theories from NeuroIS [10–13].

One possible reason for the limited understanding of specific IS phenomena based on textual data is that users' responses may exhibit higher degrees of subjectivity or that making an objective assessment requires additional information, which is not part of the text. In this study, we propose a taxonomy that distinguishes research questions based on textual data according to whether a user's response is driven by a low or high degree of subjectivity and the need for additional information in order to make an objective judgement. We specifically build our taxonomy for the research questions instead of the type of text themselves. The underlying reasoning is that the same type of text can be subject to different research questions. For instance, a news article can be categorized as positive or negative, or as real or fake. However, categorizing the article as positive or negative is less affected by political opinions and unconscious processes than categorizing an article as real or fake. The proposed taxonomy was developed based on all articles about text-related research questions from the IS basket journals published between 2010 and 2020. We categorized each of these studies along the dimensions "degree of subjectivity" and "need for additional information". In addition, we distinguish all studies according to (1) whether they collected user-specific data to account for individual beliefs and attitudes, and (2) studies that can be attributed to the field of NeuroIS. Taken together, we address the following research question:

"For which research questions based on textual data should researchers employ methods from NeuroIS?".

2   Literature Review

For our literature review about research on textual data, we consider all articles published in the IS basket journals between 2010 and 2020. We first read the title of all issues to determine whether an article is about textual data or not. Subsequently, we read the abstract of the remaining studies to validate our initial assessment. Thereby, we encountered 47 studies about text. We excluded nine of these studies [14–22] as they did not specifically analyze human behavior as a direct response to reading the text. Regarding the publication dates, we find that the number of studies about textual data and human responses has increased over the last few years. While 10 out of 38 papers were published in 2010–2015, the majority of 28 were published in the years 2016–2020. Finally, we scan the remaining 38 papers in regard to their collected data. We distinguish between (1) studies without user-specific data (e.g., political attitude, disposition to trust), (2) studies with user-specific data, and (3) studies with NeuroIS measurements.

Table 1 shows the result of our literature review. We identified four studies from NeuroIS, from which three [6, 23, 24] are published in MIS Quarterly (MISQ) and one [25] in the European Journal of Information Systems (EJIS). Moravec et al. [6] analyzed the influence of IS design modifications on users' cognition when processing fake news. In their experiment, subjects were shown several online news articles with and without warning messages, while their cognitive activity was measured with electroencephalography (EEG). The authors found that adding warning messages increases cognitive activity, but this has no impact on the final veracity assessment. Bera et al. [23] used eye tracking to explain cognitive processing of users in reading conceptual modeling scripts to perform problem solving tasks. Their results suggest that high task performance can be explained and detected by attention-based eye tracking metrics. Meservy et al. [24] studied how individuals evaluate and filter posts in online forums, while seeking the solution to a problem. For this purpose, the authors employed functional magnetic resonance imaging (fMRI) to measure the neural correlates that are involved in evaluating both solution content and contextual cues. Their results indicate that both content and contextual cues impact users' final filtering decisions. Finally, Brinton Anderson et al. [25] used text mining and eye tracking to study how users perceive security messages and found that habituation to security warnings significantly reduces the effect of such warnings.

Table 1. Studies about textual data in IS basket journals from 2010 to 2020.

Journal	No user-specific data	User-specific data	NeuroIS	Total
JMIS	[26–38]	[8]		14
MISQ	[39–44]	[7]	[6, 23, 24]	10
JAIS	[45–49]	[50]		6
ISR	[5, 51–53]			4
JIT	[54, 55]			2
JSIS	[56]			1
EJIS			[25]	1
ISJ				0
Total	31	3	4	38

We identified three studies which collected user-specific data, but without employing tools and theories from NeuroIS. For instance, Hong et al. [50] used a combination of surveys and text mining in order to explain and predict deviations in restaurant reviews based on cultural differences. Their findings suggest that users from a collectivist culture tend to agree with prior ratings. Kim and Dennis [8] collected users' political attitude and studied whether fake news can be mitigated by adding user and expert ratings to the presentation of a news article. Their results suggest that expert ratings have a stronger impact in steering users to a correct veracity assessment. However, this only holds when the rating indicates low reliability, whereas ratings of high reliability have no effect. In another study on fake news, Kim et al. [7] studied the effects of alternating the presentation format. Their results suggest that presenting the source before the headline (instead of showing the headline before the source) makes users less likely to perceive fake news as real. In addition, the authors find that source reputation ratings have a stronger effect on the perceived believability. However, this only holds when the rating is low (i.e., indicating fake content).

All other studies did not collect user-specific data as part of their research design. Instead, they rely on statistical methods from text mining to empirically analyze how users respond to textual data. These studies covered consumer behavior [30, 33, 39, 40, 48, 49, 51, 53, 55, 56], financial decision-making [29, 31, 32, 35, 36, 43, 46, 47, 54], perception of online reviews [5, 26–28, 42, 45], deception detection [37, 38], and others [41, 52].

3   Taxonomy for Research Questions Based on Text

We now present our taxonomy to distinguish research questions based on textual data depending on whether users' response is influenced by subjectivity and if there is a need for additional information in order to make a correct assessment. Hereby, "degree of subjectivity" refers to the questions: How subjectively can the probands answer the (research) question? Would people with different prior experience or belief answer differently? "Need for additional information" on the other hand, refers to the questions: Would a proband's answer become better or more correct if they had additional information or additional knowledge? Is all relevant information that is needed for the proband to complete the task or answer the (research) question given in the text? Or is additional information, knowledge, or background required? We propose this taxonomy in order to address our research question. Based on the following cluster analysis, we identify areas in which NeuroIs seems best suited in order to understand human processing of text and their subsequent behavior or decision making? Fig. 1 presents the proposed taxonomy along these two dimensions. All studies including user specific data from surveys are highlighted with a square, while all NeuroIS studies are highlighted with a circle. For the purpose of this study, we distinguish all studies from our literature based on our own reading. Overall, the NeuroIS studies were located in the upper part of Fig. 1, which contains those research questions, where users have a higher need for additional information. The upper right corner of Fig. 1 contains all studies analyzing research questions, where users' responses are driven by a high degree of subjectivity with a high need for additional information to make an objective assessment. Prominent examples are studies that analyze why users fall for fake news [6, 7]. The decision if a news article is real of fake depends on users' individual political attitudes, while making an objective assessment generally requires users to research additional information.

The studies in the bottom right area of Fig. 1 consider research questions, where users' responses involve high degree of subjectivity without a need for additional information. One example is to explain why social media posts receive more likes (e.g., [40]). The decision of whether to like a social media post or not is subjective as it depends on personal tastes, while all relevant information based on which users decide to like a post is generally included in the text. The upper left corner of Fig. 1 contains all studies analyzing research questions, where users' responses are driven by low degree of subjectivity, but with a high need for additional information. We classified all studies about fraud or deception detection as such [32, 34, 37, 38, 47]. Deciding if a message originates from a deceiver is generally not driven by personal attitudes. However, the decision is often made based on incomplete information. Ultimately, the studies in the bottom left area of Fig. 1 consider research questions, where users' responses are not influenced by

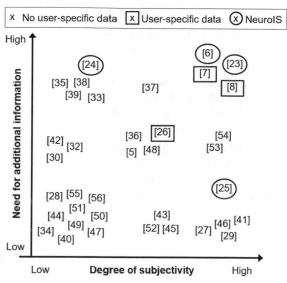

Fig.1. Taxonomy to distinguish research questions based on textual data depending on the characteristics of users' responses.

subjectivity without a need for additional information. For instance, this applies to the question of how daily deals impact a restaurant's online reputation [51]. The review text generally stands for itself, while the factors of a positive or negative review (length, number of arguments, high readability) are usually not influenced by subjectivity.

4 Conclusion and Future Research

Understanding how humans respond to textual data provides novel challenges for IS research and practice. We proposed a taxonomy to distinguish different forms of texts after reviewing recent studies from the IS literature according to their research design. In this taxonomy, we distinguish research questions based on textual data with respect to the characteristics of users' responses. Specifically, we consider two dimensions, namely how much a user's response is driven by subjectivity and the need for additional information in order to make an objective assessment. We find that NeuroIS is used for research questions, where users arrive at conclusions based on incomplete information. All corresponding studies [6, 23–25] focused on users' cognitive processing. It appears that this is the area, where NeuroIS is best equipped to help understand human processing of text and help explain human behavior and decision making. However, the vast majority of existing studies in the IS literature considered texts where users' responses can be explained without user-specific data.

IS research on textual data should hence select their study design depending on the characteristics of users' responses. If users' conclusions exhibit only little subjectivity and the text contains all relevant information, there is no need for experiments and employing measurements from neuroscience. Conversely, if users' responses are driven by a higher degree of subjectivity and if additional information is required for an objective assessment, users may rely on unconscious and/or heuristic processes. For example, affective processes may not be fully reflected through self-reports, which requires IS researchers to employ neurophysiological measurements such as ECG or startle reflex modulation [13].

We plan to extend this study as follows. First, we need to evaluate our taxonomy by gathering feedback from external domain experts. Presenting our work at the NeuroIS retreat would be one important step in this direction. After this, we can perform a sequence of evaluation-improvement iterations until we reach an acceptable state. Second, we aim to extend our literature review to all studies in the existing IS literature to provide a full overview of IS studies on textual data. Third, we are planning to elaborate more on the specific tools from NeuroIS, in particular, how specific measurements can be applied to study users' responses to different types of text.

References

1. Debortoli, S., Müller, O., Junglas, I., Vom Brocke, J.: Text mining for information systems researchers: an annotated topic modeling tutorial. Commun. Assoc. Inf. Syst. **39**, 110–135 (2016)
2. Gentzkow, M., Kelly, B., Taddy, M.: Text as data. J. Econ. Lit. **57**(3), 535–574 (2019)
3. Mudambi, S.M., Schuff, D.: What makes a helpful online review? A study of customer reviews on Amazon.com. MIS Q. **34**(1), 185–200 (2010)
4. Korfiatis, N., García-Bariocanal, E., Sánchez-Alonso, S.: Evaluating content quality and helpfulness of online product reviews: the interplay of review helpfulness vs. review content. Electron. Commer. Res. Appl. **11**(3), 205–217 (2012). https://doi.org/10.1016/j.elerap.2011.10.003
5. Yin, D., Mitra, S., Zhang, H.: Research note—when do consumers value positive vs. negative reviews? An empirical investigation of confirmation bias in online word of mouth. Inf. Syst. Res. **27**(1), 131–144 (2016). https://doi.org/10.1287/isre.2015.0617
6. Moravec, P., Kim, A., Dennis, A., Minas, R.: Fake news on social media: people believe what they want to believe when it makes no sense at all. MIS Q. **43**(4), 1343–1360 (2019)
7. Kim, A., Moravec, P.L., Dennis, A.R.: Combating fake news on social media with source ratings: the effects of user and expert reputation ratings. J. Manag. Inf. Syst. **36**(3), 931–968 (2019)
8. Kim, A., Dennis, A.: Says who? The effects of presentation format and source rating on fake news in social media. MIS Q. **43**(3), 1025–1039 (2019)
9. Graves, L.: Boundaries not drawn: mapping the institutional roots of the global fact-checking movement. J. Stud. **19**(5), 613–631 (2016)
10. Dimoka, A., Pavlou, P.A., Davis, F.D.: NeuroIS: the potential of cognitive neuroscience for information systems research. Inf. Syst. Res. **22**(4), 687–702 (2011)
11. Dimoka, D., et al.: On the use of neurophysiological tools in IS research: developing a research agenda for NeuroIS. MIS Q. **36**(3), 679 (2012). https://doi.org/10.2307/41703475
12. Riedl, R., Fischer, T., Léger, P.M., Davis, F.D.: A decade of NeuroIS research: progress, challenges, and future directions. DATA BASE Adv. Inf. Syst. 1–51 (2019)
13. vom Brocke, J., Hevner, A., Léger, P.M., Walla, P., Riedl, R.: Advancing a neurois research agenda with four areas of societal contributions. Eur. J. Inf. Syst. **29**(1), 9–24 (2020). https://doi.org/10.1080/0960085X.2019.1708218
14. Safi, R., Yu, Y.: Online product review as an indicator of users' degree of innovativeness and product adoption time: a longitudinal analysis of text reviews. Eur. J. Inf. Syst. **26**(4), 414–431 (2017)

15. Abbasi, A., Li, J., Adjeroh, D., Abate, M., Zheng, W.: Don't mention it? Analyzing user-generated content signals for early adverse event warnings. Inf. Syst. Res. **30**(3), 1007–1028 (2019)
16. Mejia, J., Mankad, S., Gopal, A.: A for effort? Using the crowd to identify moral hazard in New York City restaurant hygiene inspections. Inf. Syst. Res. **30**(4), 1363–1386 (2019)
17. Liu, Y., Pant, G., Sheng, O.R.: Predicting labor market competition: leveraging interfirm network and employee skills. Inf. Syst. Res. **31**(4), 1443–1466 (2020)
18. Shi, D., Guan, J., Zurada, J., Manikas, A.: A data-mining approach to identification of risk factors in safety management systems. J. Manag. Inf. Syst. **34**(4), 1054–1081 (2017)
19. Vlas, R.E., Robinson, W.N.: Two rule-based natural language strategies for requirements discovery and classification in open source software development projects. J. Manag. Inf. Syst. **28**(4), 11–38 (2012)
20. Hu, Y., Xu, A., Hong, Y., Gal, D., Sinha, V., Akkiraju, R.: Generating business intelligence through social media analytics: measuring brand personality with consumer-, employee-, and firm-generated content. J. Manag. Inf. Syst. **36**(3), 893–930 (2019)
21. Zhou, S., Qiao, Z., Du, Q., Wang, G.A., Fan, W., Yan, X.: Measuring customer agility from online reviews using big data text analytics. J. Manag. Inf. Syst. **35**(2), 510–539 (2018)
22. Wang, Z., Jiang, C., Zhao, H., Ding, Y.: Mining semantic soft factors for credit risk evaluation in peer-to-peer lending. J. Manag. Inf. Syst. **37**(1), 282–308 (2020)
23. Bera, P., Soffer, P., Parsons, J.: Using eye tracking to expose cognitive processes in understanding conceptual models. MIS Q. **43**(4), 1105–1126 (2019)
24. Meservy, T.O., Fadel, K.J., Kirwan, C.B., Meservy, R.D.: An fMRI exploration of information processing in electronic networks of practice. MIS Q. **43**(3), 851–872 (2019)
25. Brinton Anderson, B., Vance, A., Kirwan, C.B., Eargle, D., Jenkins, J.L.: How users perceive and respond to security messages: a NeuroIS research agenda and empirical study. Eur. J. Inf. Syst. **25**(4), 364–390 (2016)
26. Huang, L., Tan, C.H., Ke, W., Wei, K.K.: Comprehension and assessment of product reviews: a review-product congruity proposition. J. Manag. Inf. Syst. **30**(3), 311–343 (2013)
27. Ma, X., Khansa, L., Deng, Y., Kim, S.S.: Impact of prior reviews on the subsequent review process in reputation systems. J. Manag. Inf. Syst. **30**(3), 279–310 (2013)
28. Jensen, M.L., Averbeck, J.M., Zhang, Z., Wright, K.B.: Credibility of anonymous online product reviews: a language expectancy perspective. J. Manag. Inf. Syst. **30**(1), 293–324 (2013)
29. Luo, X., Zhang, J.: How do consumer buzz and traffic in social media marketing predict the value of the firm? J. Manag. Inf. Syst. **30**(2), 213–238 (2013)
30. Ghiassi, M., Zimbra, D., Lee, S.: Targeted twitter sentiment analysis for brands using supervised feature engineering and the dynamic architecture for artificial neural networks. J. Manag. Inf. Syst. **33**(4), 1034–1058 (2016)
31. Lash, M.T., Zhao, K.: Early predictions of movie success: the who, what, and when of profitability. J. Manag. Inf. Syst. **33**(3), 874–903 (2016)
32. Siering, M., Koch, J.A., Deokar, A.V.: Detecting fraudulent behavior on crowdfunding platforms: the role of linguistic and content-based cues in static and dynamic contexts. J. Manag. Inf. Syst. **33**(2), 421–455 (2016)
33. Huang, J., Boh, W.F., Goh, K.H.: A temporal study of the effects of online opinions: information sources matter. J. Manag. Inf. Syst. **34**(4), 1169–1202 (2017)
34. Ho, S.M., Hancock, J.T., Booth, C., Liu, X.: Computer-mediated deception: strategies revealed by language-action cues in spontaneous communication. J. Manag. Inf. Syst. **33**(2), 393–420 (2016)
35. Mai, F., Shan, Z., Bai, Q., Wang, X., Chiang, R.H.: How does social media impact bitcoin value? A test of the silent majority hypothesis. J. Manag. Inf. Syst. **35**(1), 19–52 (2018)

36. Dong, W., Liao, S., Zhang, Z.: Leveraging financial social media data for corporate fraud detection. J. Manag. Inf. Syst. **35**(2), 461–487 (2018)
37. Ludwig, S., Van Laer, T., De Ruyter, K., Friedman, M.: Untangling a web of lies: exploring automated detection of deception in computer-mediated communication. J. Manag. Inf. Syst. **33**(2), 511–541 (2016)
38. Liang, N., Biros, D.P., Luse, A.: An empirical validation of malicious insider characteristics. J. Manag. Inf. Syst. **33**(2), 361–392 (2016)
39. Huang, N., Hong, Y., Burtch, G.: Social network integration and user content generation: evidence from natural experiments. MIS Q. 17–001 (2016)
40. Shin, D., He, S., Lee, G.M., Whinston, A.B., Cetintas, S., Lee, K.C.: Enhancing social media analysis with visual data analytics: a deep learning approach. MIS Q. **44**(4), 1459–1492 (2020)
41. Ludwig, S., de Ruyter, K., Mahr, D., Wetzels, M., Brüggen, E., de Ruyck, T.: Take their word for it: the symbolic role of linguistic style matches in user communities. MIS Q. **38**(4), 1201–1217 (2014). https://doi.org/10.25300/MISQ/2014/38.4.12
42. Mudambi, S.: Research note: what makes a helpful online review? A study of customer reviews on Amazon.com. MIS Q. **34**(1), 185 (2010). https://doi.org/10.2307/20721420
43. Deng, S., Huang, Z.J., Sinha, A.P., Zhao, H.: The interaction between microblog sentiment and stock return: an empirical examination. MIS Q. **42**(3), 895–918 (2018)
44. Yin, D., Bond, S.D., Zhang, H.: Anxious or angry? Effects of discrete emotions on the perceived helpfulness of online reviews. MIS Q. **38**(2), 539–560 (2014)
45. Kuan, K.K., Hui, K.L.: What makes a review voted? An empirical investigation of review voting in online review systems. J. Assoc. Inf. Syst. **16**(1), 48–71 (2015)
46. Ho, S.Y., Choi, K.W.S., Yang, F.F.: Harnessing aspect-based sentiment analysis: how are tweets associated with forecast accuracy? J. Assoc. Inf. Syst. **20**(8), 2 (2019)
47. Lausen, J., Clapham, B., Siering, M., Gomber, P.: Who is the next "wolf of wall street"? Detection of financial intermediary misconduct. J. Assoc. Inf. Syst. **21**(5), 7 (2020)
48. Chou, C.H., Sinha, A.P., Zhao, H.: A hybrid attribute selection approach for text classification. J. Assoc. Inf. Syst. **11**(9), 1 (2010)
49. Chen, Y., Deng, S., Kwak, D.H., Elnoshokaty, A., Wu, J.: A multi-appeal model of persuasion for online petition success: a linguistic cue-based approach (2019)
50. Hong, Y., Huang, N., Burtch, G., Li, C.: Culture, conformity and emotional suppression in online reviews. J. Assoc. Inf. Syst. 16–020 (2016)
51. Bai, X., Marsden, J.R., Ross, W.T., Jr., Wang, G.: A note on the impact of daily deals on local retailers' online reputation: mediation effects of the consumer experience. Inf. Syst. Res. **31**(4), 1132–1143 (2020)
52. Pan, Y., Huang, P., Gopal, A.: Storm clouds on the horizon? New entry threats and R&D investments in the US IT industry. Inf. Syst. Res. **30**(2), 540–562 (2019)
53. Adamopoulos, P., Ghose, A., Todri, V.: The impact of user personality traits on word of mouth: Text-mining social media platforms. Inf. Syst. Res. **29**(3), 612–640 (2018)
54. Li, T., van Dalen, J., van Rees, P.J.: More than just noise? Examining the information content of stock microblogs on financial markets. J. Inf. Technol. **33**(1), 50–69 (2018)
55. Garcia-Crespo, A., Colomo-Palacios, R., Gomez-Berbis, J.M., Ruiz-Mezcua, B.: Semo: a framework for customer social networks analysis based on semantics. J. Inf. Technol. **25**(2), 178–188 (2010)
56. Benthaus, J., Risius, M., Beck, R.: Social media management strategies for organizational impression management and their effect on public perception. J. Strateg. Inf. Syst. **25**(2), 127–139 (2016)

Towards a Psychophysiological Investigation of Perceived Trustworthiness and Risk in Online Pharmacies: Results of a Pre-study

Anika Nissen[(✉)] and Semra Ersöz

University Duisburg-Essen, Essen, Germany
{anika.nissen,semra.ersoez}@uni-due.de

Abstract. Perceived trustworthiness and risk are crucial impact factors for a website's success. While they have been frequently applied to diverse e-commerce contexts, an investigation of these constructs for the special case of online pharmacies is still scarce. In an attempt to measure these constructs in a neural experiment, this paper offers a pre-study with the aim to gain first insights and select appropriate stimuli for the upcoming study. Therefore, five operating online pharmacies are tested in an online survey with 121 participants which rated scales of perceived trustworthiness, perceived risk, attitude towards the website, and use intention for each of the included pharmacies. Results show that online pharmacies with high reputation are rated higher in the included constructs. Consequently, reputation, perceived risk, and trustworthiness are crucial impact factors on attitude and use intention. Thus, two promising online pharmacies could be selected for the follow-up study.

Keywords: Online pharmacy · Perceived trustworthiness · Perceived risk · Attitude · Use intention

1 Introduction

Online pharmacies provide quick and convenient solutions for both the elderly and younger people because they offer a convenient solution to purchase medicines through home delivery [1]. Especially older people benefit from this, as they need to rely less on third parties. Unsurprisingly, the online pharmacy market is expected to grow, with the main drivers of growth being the increasing proportion of the elderly population, the growth in chronic diseases [2], the general increased availability of high-speed internet, and the improving healthcare infrastructure worldwide [3]. Next to North America, Europe has the largest online pharmacy market shares, with particularly the German market experiencing tremendous growth due to the increasing awareness of cheap over the counter (OTC) products which can be purchased online [2]. This is due to the relaxation of legal regulations for the sale of OTC products, but also the increased switch of prescription medicines to OTC status [4].

However, this development does not only come with opportunities. As illegal online pharmacies continue to be found worldwide despite increased regulatory controls, the

F. D. Davis et al. (Eds.): NeuroIS 2021, LNISO 52, pp. 9–19, 2021.
https://doi.org/10.1007/978-3-030-88900-5_2

online purchase of medicines poses an increased risk compared to other online shopping contexts. Consequently, the risk of purchasing from online pharmacies is twofold: on the one hand, it may be an illegal online pharmacy selling counterfeit medicines and, on the other hand, laypeople may purchase and use a medicine that is unsuitable or even harmful to their health due to a lack of advice and expertise [3]. Therefore, the consideration of perceived risk and trust in online pharmacies is of special interest as a misuse can harm the consumers' physical health.

The constructs of perceived risk and trust have been investigated in several e-commerce contexts and have shown to generally impact a website's acceptance [5–8]. Furthermore, prior NeuroIS research has also covered perceived trust and trustworthiness and has shown to reveal distinct neural activations associated with high perceived trustworthiness in e-commerce contexts [7, 9]. However, in the area of health services and medicines, studies considering website design and its impact on perceived risk and trust are still scarce. Since perceived risk might differ severely between an online shop selling clothes and an online pharmacy [10], this paper tackles this research gap by investigating the rated perceived risk and trustworthiness for five different operating online pharmacies. Thus, the goal of this work-in-progress is to introduce the constructs of perceived risk and trustworthiness in the context of online pharmacies and to provide a first data basis that is used to select stimuli for a follow-up FaceReader and neuroimaging experiment. To embed this goal in a research context, this working paper first presents the state of research on perceived trustworthiness and risk for purchasing medication primarily online and presents the context of NeuroIS research. Within this study, hypotheses were derived and tested for the non-neuronal part of the research. Five selected pharmacies were the subject of the empirical online survey. The results of the study are then discussed, and the study design for the follow-up experiment is presented.

2   Background: Perceived Trustworthiness and Risk

2.1   Perceived Trustworthiness and Risk in Online Pharmacies

Trust is an important component in the interaction between patients and healthcare providers such as pharmacies [11]. Patients trust that they will receive the right advice in pharmacies and that their condition will be cured with the recommended medicines. In turn, pharmacists trust that they will receive all the information truthfully to be able to recommend medicines to patients in the best possible way. With trust being a domain-specific construct in terms of operationalisation [10], even within the healthcare domain, several different definitions of trust can be found. Some studies define it as a stand-alone construct, while others define it in terms of sub-constructs like credibility and benevolence. The disagreement among researchers on the definition leads to a variation in the understanding and language use of the construct in literature [12]. In this working paper, trust is defined as the willingness to expose oneself to exploitation by another agent due to the prospect of potential benefits [13, 14], with the agent being the presented online pharmacy. This is based on the optimistic expectation that the other agent(s) will protect the rights of all involved [15]. In other words, it is based on the expectation that the commitments made by another person or organisation will be fulfilled [16], especially in relationships where the trusting party has no control over the other party but still needs to depend on them [15, 17–19]. This expectation is built on the other party's different characteristics, which result in its' perceived trustworthiness.

When considering online pharmacies, it can be seen that the interaction between patients and pharmacists has changed compared to offline pharmacies. While in offline pharmacies, trust is usually developed through the salesperson's expertise and the shop atmosphere [11, 20], these aspects are absent in online pharmacies. Consequently, in online environments, the website's reputation and visual design might ultimately impact the online pharmacy's perceived trustworthiness and, thus, the trust in it [11, 21]. Consequently, we hypothesise that there will be a positive correlation between the online pharmacy's rated reputation and the perceived trustworthiness:

H1. There is a positive correlation between reputation and perceived trustworthiness.

A meta-analysis of studies on the use of health websites, which include all websites with information on diseases and medicines, thus also online pharmacies, found backgrounds that are important for building trust [12]. Usability, beliefs, design elements, brand name and ownership, persuasion, and social influences [22–26] have emerged as major antecedents. All of the named constructs further add up to the attitude consumers build towards a website in general and an online pharmacy in particular [27, 28]. While they also have shown to impact the perceived trustworthiness of the website significantly, it is further hypothesised that the user's attitude towards an online pharmacy will have a positive correlation with the perceived trustworthiness and that both attitude and perceived trustworthiness have a positive impact on use intentions [28]:

H2. There is a positive correlation between perceived trustworthiness and attitude towards the website.
H3. There is a positive correlation between perceived trustworthiness and use intention.
H4. There is a positive correlation between attitude towards the website and use intention.

A closer look at the concept of trust and trustworthiness reveals an important component: uncertainty. This is made clear by the fact that the trusting party has no control over the other party, and thus the mutual relationship is characterised by uncertainty. Uncertainty about the occurrence of consequences is a dimension of perceived risk [29], which therefore poses an important counter-position to trust [30]. While being introduced as a concept by Bauer [31] in the context of online pharmacies, the perceived risk might be distinguished between product-related risk (the medicine to be purchased) and retailer risk (the online pharmacy) [31, 32]. With risk being a counter-position to perceived trustworthiness, we hypothesise the contradicting correlation of the first three hypotheses for perceived risk:

H5. There is a negative correlation between reputation and perceived risk.
H6. There is a negative correlation between perceived risk and attitude towards the website.
H7. There is a negative correlation between perceived risk and use intention.

2.2 Perceived Trustworthiness and Risk in NeuroIS Research

Based on two recent literature reviews about the state of the art in NeuroIS research [33, 34], related work for the constructs of perceived trustworthiness and risk is identified. Because our follow-up study focuses on decision making on online pharmacies, the following results are limited to brain regions in the prefrontal cortex (PFC). The main results are summarised in Table 1.

Table 1. Overview of NeuroIS Paper associated with perceived trustworthiness and risk (Abbreviations: functional magnetic resonance imaging (fMRI), electroencephalography (EEG), orbitofrontal cortex (OFC), dorsolateral prefrontal cortex (dlPFC), ventromedial prefrontal cortex (vmPFC))

Authors	Context	Method	Results (in areas of the frontal cortex)
Dimoka [9]	Trust in online shopping offers	fMRI	• Trust associated with lower activations in the (right) OFC
Riedl et al. [7]	Trust in Online shopping offers	fMRI	• Trustworthy offers activated right dlPFC in male, but not in female participants • Untrustworthy offers activated the left dlPFC in exclusively in women, and the left vmPFC exclusively in men
Riedl et al. [35]	Trust in humans vs. avatars	fMRI	• MPFC activated more for trustworthiness evaluation of humans compared to avatars
Vance et al. [36]	Risk in information security behavior	EEG	• Activations in approx. frontal (right) & motor cortex regions associated with risk perceptions (in P300)
Wang et al. [37]	Risk in social cues in ecommerce	EEG	• Frontal cortex (F3, FZ, F4) activated in purchase decision/associated with risk (in N2)

The provided overview shows that trust and risk evaluations in different research contexts trigger several areas in the PFC. However, since the context of the studies ranged from ecommerce to human-human interactions, a more thorough review of neuroscientific studies investigating perceived trust and risk is required to derive hypotheses for our main study. Nevertheless, this overview reveals two promising aspects relevant for the follow-up study: firstly, none of the reviewed studies used functional near-infrared spectroscopy (fNIRS) to investigate the selected constructs, which offer room to validate these prior findings by means of fNIRS. Secondly, even in the cited fMRI studies, several activations could be identified on cortical surfaces in the PFC, which are also assessable through EEG and fNIRS. This further calls for adding validation through research in different contexts that considers the constructs of perceived trust and/or risk and assess these evaluations through neural activations in PFC areas. This paper offers a first approach in this direction by evaluating potential stimulus material for a follow-up fNIRS study.

3 Method

Sample. We recruited 144 participants to fill out the online survey through the online platform clickworker. Out of the 144 participants, 23 data sets needed to be filtered out because participants either did not fill out the questionnaire completely, or they did not answer the attention check correctly (which was a statement hidden somewhere between the scales in which they were asked to click "4"). As a result, an included sample of $N = 121$ data sets is considered for further analyses. Average age is $M = 41.6$ years ($SD = 12.8$), 63.6% are male, 36.4% are female. 50.4% of the sample are currently employed, 23.1% are freelancers, 12.4% are students, 9.1% are searching for employment, and 3.3% are retired. We further included questions regarding the participants' disposition to trust and familiarity with online pharmacies with 5-point Likert scales (adapted from [33, 34]). Mean disposition to trust was $M = 3.52$ ($SD = 1.02$) and familiarity was $M = 3.38$ ($SD = 1.12$).

Stimuli. To test the here stated hypotheses and select appropriate, real-life stimuli that will be extreme opposites regarding their associated perceived risk and trustworthiness, we took a look at the top ten online pharmacies operating in Germany [9]. From this, we selected the top 2, namely *DocMorris* and *Shop-Apotheke*, the bottom 2, namely *Aponeo* and *Eurapon*, and one from the middle being *Apotal*. For each online pharmacy, we used the product page of *IBU-ratiopharm* 400 mg pain killers with 10 pills per package and manipulated the prices on the screenshots to avoid biases due to different pricing (two examples of stimuli can be seen in Fig. 1).

Fig. 1. Two examples for used stimuli

In the survey, participants were shown the online pharmacy screenshot and below, they had to rate the scales explained in the next paragraph. The order in which the questions and the websites were shown was completely randomised – only the scale for reputation was always asked first before the other scales came in randomised order. After having rated all included websites, participants had to give their demographics and answer the control questions for disposition to trust and familiarity, after which they were thanked and debriefed.

Table 2. Questionnaire Items and Cronbach's Alpha

Questionnaire items	CA (*if dropped*)
Perceived Risk [38]	**.837**
Do not trust that my credit card number will be secure on this online pharmacy	*.794*
It is difficult to judge quality of a product/service on this online pharmacy	*.831*
Do not trust that my personal information will be kept private by this online pharmacy	*.790*
I loose too much time when buying products on this online pharmacy	*.823*
Overall, I feel making a purchase on this online pharmacy is risky	*.781*
Perceived Trustworthiness [39]	**.895**
I can trust this online pharmacy	*.815*
I trust the information presented on the online pharmacy	*.859*
I trust the transaction process on this online pharmacy	*.875*
Attitude Towards the Website [27]	**.930**
I liked the online pharmacy	*.908*
I enjoyed being on the online pharmacy	*.918*
I would like to return to the online pharmacy	*.901*
I will recommend to others to browse the online pharmacy	*.909*
The online pharmacy made me feel like buying	*.933*
Use Intention [40]	**.892**
I would use this online pharmacy to inquire what others think of a medicine	*.856*
I would use this online pharmacy to inform myself about the effects of a medicine	*.853*
I would use this online pharmacy to read the reviews of a medicine	*.856*
I would use my use my credit card to purchase from this online pharmacy	*.896*
I am very likely to purchase medicine from this online pharmacy	*.879*

Scales. We asked participants how they would rate that "the online pharmacy has a very good reputation" on a 5-point Likert scale (strongly agree to strongly disagree) for each online pharmacy. After that followed scales for perceived risk adapted from [38], perceived trustworthiness adapted from [39], attitude towards the website adapted from [27], and finally, use intention adapted to online pharmacies from [40, 41]. All included scales were measured with a 5-point Likert scale. The used constructs and items provided sufficient reliability through Cronbach's Alpha (CA) and were all included in the following analysis (Table 2).

4 Results

In order to test our hypotheses and draw conclusions regarding the prior stated hypotheses, we ran one-way ANOVAs with post-hoc Tukey tests to identify significant differences between the five included online pharmacies in the included constructs. Further, since we were interested in correlations between the included constructs, we ran Pearson's correlation analyses across all five included pharmacies for all constructs. One-way ANOVAs resulted in significant differences between the included online pharmacies for all selected constructs ($F_{reputation}$ (4, 600) = 18.5; F_{risk} (4, 600) = 3.73, $F_{trustworthiness}$ (4, 600) = 10.45; $F_{attitude}$ (4, 600) = 12.65; F_{use} (4, 600) = 9.27, all p's < .005). Since perceived risk and trust may be rated differently between men and women, we ran a MANOVA including our dependent variables to check for effects due to gender. Results show that, although there are significant differences between male and female participants in the constructs of risk, trust, and use intention (F(4, 592) = 6.916, p < .001), no significant interaction effect between online pharmacy and gender was identified (F(16, 1809) = 0.417, p > .05).

As a result, *DocMorris* was consistently rated highest for all included constructs, while *Apotal* was consistently rated lowest. Simultaneously, significant difference between *Shop-Apotheke* and *DocMorris* could only be identified for reputation. Differences between *Aponeo* and *Eurapon* compared to *DocMorris* were found for reputation, perceived trustworthiness, attitude, use intention, but not for perceived risk.

To test our hypotheses, we further ran Pearson's correlation analysis (Table 2). As a result of this analysis, all of our hypotheses are supported by showing significant positive correlations between reputation, perceived trustworthiness, attitude, and use intention, while the correlations with perceived risk are continuously negative. The next steps are briefly discussed in the following section.

5 Discussions of the NeuroIS Follow-Up Study and Conclusion

This study has shown that an online pharmacy's perceived risk and trustworthiness are crucial predictors for use intention. Furthermore, as pointed to in the introduction, online pharmacies pose the risk to sell illegal or counterfeit drugs which may have a harming impact on the buyer's health. This might explain the significant correlation of reputation with the other included constructs. As a result, and along with the included constructs, this study allows us to select two operating pharmacies as stimuli for the upcoming study. Since *DocMorris* was rated highest in the included constructs and *Apotal* was

Table 3. Pearson's r correlation matrix and related hypothesis support (grey numbers not used for hypothesis' support, ***p < .001)

	Risk	Trustworthiness	Attitude	Use	Results
Reputation	- .333***	.637***	.666***	.588***	H5 ✓, H1 ✓
Risk	-	- .584***	- .378***	- .324***	H6 ✓, H7 ✓
Trust-worthiness		-	.695***	.668***	H2 ✓, H3 ✓
Attitude			-	.769***	H4 ✓

rated significantly lowest, we will use these two websites as opposing stimuli regarding their associated perceived risk and trustworthiness.

Needless to say, this pre-study also comes with limitations. One major limitation is that the included constructs were evaluated using only one screenshot of each pharmacy, and thus, actual use of the websites, as well as risks due to different medication types are not considered. Consequently, we wonder in our follow-up study how the risk and trustworthiness of the selected websites is evaluated during and after actual use. As means to assess this, emotional reactions will be measured through facial expression capturing during actual use of the selected websites in two different buying situations: a prescription and an OTC drug. We thereby control the effect of a low- and high-risk drug and thus, an additional risk dimension which is exclusive for online pharmacies.

The participants' faces will be recorded during the purchases and analysed via automated facial expression recognition software Facereader™ Version 8.0. The software recognises the emotions happy, sad, angry, disgusted, surprised, scared and contempt. It also measures a neutral state. We want to examine the emotions' effects on perceived trust and risk by analysing the emotions durations and frequencies. Additional information about emotions will be provided by surveying the participants with PANAS scale for both conditions. Correlations will be figured out between experienced and rated emotions, experienced emotions and perceived risk and trust, and rated emotions and perceived risk and trust. The conditions of low- and high-risk drug purchases will be compared to ascertain differences.

After that, the pharmacies' screenshots are to be shown in an event-related experimental paradigm together with the questionnaire scales used in this study. During the latter, neural activity is to be assessed through fNIRS to get additional insights into neural correlates of perceived risk, perceived trustworthiness, attitude, and use intention in decision making related brain areas – namely the PFC. As pointed out in prior related literature, several different activation patterns can be observed in frontal areas related to risk and trust evaluations [7, 9, 35–37]. In order to derive hypotheses for the fNIRS experiment, a more thorough literature review is planned that is not reduced to NeuroIS research, but also considers neuroeconomics literature as well. From the latter, prior related studies have found i.e. that activations in the OFC are strongly related to increased purchase intentions [42, 43]). Furthermore, within neuroeconomic literature, fNIRS has proven to be a reliable method for assessing neural activations on cortical surfaces in the frame of (economic) decision making [44, 45], which is also transferable to online shopping contexts [46, 47]. Yet, although fNIRS has proven to be more robust against artefacts in field studies, EEG is still the most commonly applied method to investigate neural correlates in similar contexts [33, 48].

Therewith, although the included constructs have been already investigated with neuroimaging methods in other ecommerce contexts, both the assessment of facial expressions and subsequent measurement of neural activity with fNIRS during and after use, are two rather novel approaches. Both of these methods offer rich data and might provide further insights not only into risk perceptions of online pharmacies, but also into the relation between facial expressions, neural activity, and self-reported scales in a timely research context.

References

1. Fortune Business Insights: ePharmacy Market Size, Share and Industry Analysis By Product (Over-The-Counter Products, Prescription Medicine) and Regional Forecast, 2019–2026 (2019)
2. Research and Markets: Europe's Online Pharmacy Industry, 2020 Analysis by Platform, Type and Geography, Dublin (2020)
3. Grand View Research: Global ePharmacy Market Size, Share_Industry Analysis Report, 2025 (2017)
4. Wieringa, J.E., Reber, K.C., Leeflang, P.: Improving pharmacy store performance: the merits of over-the-counter drugs. Eur. J. Mark. 49, 1276–1299 (2015). https://doi.org/10.1108/EJM-06-2013-0331
5. Erdil, T.S.: Effects of customer brand perceptions on store image and purchase intention: an application in apparel clothing. Procedia Soc. Behav. Sci. 207, 196–205 (2015). https://doi.org/10.1016/j.sbspro.2015.10.088
6. Gefen, D.: Customer loyalty in e-commerce. JAIS 3, 27–53 (2002). https://doi.org/10.17705/1jais.00022
7. Riedl, R., Hubert, M., Kenning, P.: Are there neural gender differences in online trust? An fMRI study on the perceived trustworthiness of eBay offers. MIS Q. 34, 397–428 (2010). https://doi.org/10.2307/20721434
8. McKnight, D.H., Choudhury, V., Kacmar, C.: Developing and validating trust measures for e-commerce: an integrative typology. Inf. Syst. Res. 13, 334–359 (2002). https://doi.org/10.1287/isre.13.3.334.81
9. Dimoka, A.: What does the brain tell us about trust and distrust? Evidence from a functional neuroimaging study. MIS Q. 34, 373–396 (2010). https://doi.org/10.2307/20721433
10. Montague, E.N.H., Winchester, W.W., Kleiner, B.M.: Trust in medical technology by patients and health care providers in obstetric work systems. Behav. Inf. Technol. 29, 541–554 (2010). https://doi.org/10.1080/01449291003752914
11. Castaldo, S., Grosso, M., Mallarini, E., et al.: The missing path to gain customers loyalty in pharmacy retail: the role of the store in developing satisfaction and trust. Res. Soc. Adm. Pharm. 12, 699–712 (2016). https://doi.org/10.1016/j.sapharm.2015.10.001
12. Vega, L.C., Montague, E., Dehart, T.: Trust between patients and health websites: a review of the literature and derived outcomes from empirical studies. Health Technol. (Berl) 1, 71–80 (2011). https://doi.org/10.1007/s12553-011-0010-3

13. Rousseau, D.M., Sitkin, S.B., Burt, R.S., et al.: Not so different after all: a cross-discipline view of trust. AMR **23**, 393–404 (1998). https://doi.org/10.5465/amr.1998.926617
14. Schlösser, T., Fetchenhauer, D., Dunning, D.: Trust against all odds? Emotional dynamics in trust behavior. Decision **3**, 216–230 (2016). https://doi.org/10.1037/dec0000048
15. Hosmer, L.T.: Trust: the connecting link between organizational theory and philosophical ethics. Acad. Manag. Rev. 379–403 (1995)
16. Rotter, J.B.: Generalised expectancies for interpersonal trust. Am. Psychol. **26**, 443–452 (1971). https://doi.org/10.1037/h0031464
17. Deutsch, M.: Trust and suspicion. J. Conflict Resolut. **2**, 265–279 (1958). https://doi.org/10.1177/002200275800200401
18. Fukuyama, F.: Trust: The Social Vitues and the Creation of Prosperity. Free Press, New York (1995)
19. Hart, K.M., Capps, H.R., Cangemi, J.P., Caillouet, L.M.: Exploring organisational trust and its multiple dimensions: a case study of General Motors. Organ. Dev. J. 31–39 (1986)
20. Doney, P.M., Cannon, J.P.: An examination of the nature of trust in buyer-seller relationships. J. Mark. 35–51 (1997)
21. Hsu, M.-H., Chang, C.-M., Chu, K.-K., et al.: Determinants of repurchase intention in online group-buying: the perspectives of DeLone & McLean IS success model and trust. Comput. Hum. Behav. **36**, 234–245 (2014). https://doi.org/10.1016/j.chb.2014.03.065
22. Rains, S.A., Karmikel, C.D.: Health information-seeking and perceptions of website credibility: examining Web-use orientation, message characteristics, and structural features of websites. Comput. Hum. Behav. **25**, 544–553 (2009). https://doi.org/10.1016/j.chb.2008.11.005
23. Fruhling, A.L., Lee, S.M.: The influence of user interface usability on rural consumers' trust of e-health services. Int. J. Electron. Healthc. **2**, 305–321 (2006). https://doi.org/10.1504/IJEH.2006.010424
24. Song, J., Zahedi, F."M.": Trust in health infomediaries. Decis. Support Syst. **43**, 390–407(2007). https://doi.org/10.1016/j.dss.2006.11.011
25. Fisher, J., Burstein, F., Lynch, K., et al.: "Usability + usefulness = trust": an exploratory study of Australian health web sites. Internet Res. **18**, 477–498 (2008). https://doi.org/10.1108/10662240810912747
26. Rosenbaum, S.E., Glenton, C., Cracknell, J.: User experiences of evidence-based online resources for health professionals: user testing of The Cochrane Library. BMC Med. Inform. Decis. Making 1–11 (2008)
27. Porat, T., Tractinsky, N.: It's a pleasure buying here: the effects of web-store design on consumers' emotions and attitudes. Hum.-Comput. Interact. 235–276 (2012)
28. Kumar, A., Sikdar, P., Alam, M.M.: E-retail adoption in emerging markets. Int. J. E-Bus. Res. **12**, 44–67 (2016). https://doi.org/10.4018/ijebr.2016070104
29. Cunningham, S.M.: The major dimensions of perceived risk. In: Cox, D.F. (ed.) Risk-Taking and Information Handling in Consumer Behavior. Harvard University Press, Boston (1967)
30. Das, T.K., Teng, B.-S.: The risk-based view of trust: a conceptual framework. J. Bus. Psychol. **19**, 85–116 (2004). https://doi.org/10.1023/B:JOBU.0000040274.23551.1b
31. Bauer, R.A.: Consumer behaviour as risk taking. Dynamic Marketing for a Changing World, pp. 389–398 (1960)
32. Büttner, O.B., Schulz, S., Silberer, G.: Vertrauen, Risiko und Usability bei der Nutzung von Internetapotheken. In: Bauer, H., Neumann, M.M., Schüle, A. (eds.) Konsumentenverhalten. Konzepte und Anwendungen für ein nachhaltiges Kundenbindungsmanagement, Vahlen, München, pp. 355–366 (2006)
33. Riedl, R., Fischer, T., Léger, P.-M., Davis, F.D.: A decade of NeuroIS research: progress, challenges, and future directions. Data Base Adv. Inf. Syst. **51**(3), 13–54 (2020)

34. Xiong, J., Zuo, M.: What does existing NeuroIS research focus on? Inf. Syst. **89**(March 2020) (2020). https://doi.org/10.1016/j.is.2019.101462
35. Riedl, R., Mohr, P., Kenning, P., et al.: Trusting humans and avatars: a brain imaging study based on evolution theory. J. Manag. Inf. Syst. **30**, 83–114 (2014). https://doi.org/10.2753/MIS0742-1222300404
36. Vance, A., Anderson, B.B., Kirwan, C.B., Eargle, D.: Using measures of risk perception to predict information security behavior: insights from electroencephalography (EEG). J. Assoc. Inf. Syst. **15**, 679–722 (2014)
37. Wang, Q., Meng, L., Liu, M., Wang, Q., Ma, Q.: How do social-based cues influence consumers' online purchase decisions? An event-related potential study. Electron. Commer. Res. **16**(1), 1–26 (2015). https://doi.org/10.1007/s10660-015-9209-0
38. Forsythe, S.M., Shi, B.: Consumer patronage and risk perceptions in Internet shopping. J. Bus. Res. **56**, 867–875 (2003). https://doi.org/10.1016/S0148-2963(01)00273-9
39. Cyr, D., Head, M., Larios, H.: Colour appeal in website design within and across cultures: a multi-method evaluation. Int. J. Hum. Comput. Stud. **68**, 1–21 (2010). https://doi.org/10.1016/j.ijhcs.2009.08.005
40. Gefen, D., Karahanna, E., Straub, D.W.: Trust and TAM in online shopping: an integrated model. MIS Q. 51–90 (2003)
41. Gefen, D.: E-commerce: the role of familiarity and trust. Omega **28**, 725–737 (2000). https://doi.org/10.1016/S0305-0483(00)00021-9
42. Plassmann, H., O'Doherty, J., Rangel, A.: Orbitofrontal cortex encodes willingness to pay in everyday economic transactions. J. Neurosci. **27**, 9984–9988 (2007). https://doi.org/10.1523/jneurosci.2131-07.2007
43. Plassmann, H., O'Doherty, J.P., Rangel, A.: Appetitive and aversive goal values are encoded in the medial orbitofrontal cortex at the time of decision making. J. Neurosci. **30**, 10799–10808 (2010). https://doi.org/10.1523/JNEUROSCI.0788-10.2010
44. Krampe, C., Strelow, E., Haas, A., Kenning, P.: The application of mobile fNIRS to "shopper neuroscience" – first insights from a merchandising communication study. Eur. J. Mark. **52**, 244–259 (2018). https://doi.org/10.1108/EJM-12-2016-0727
45. Gier, N., Kurz, J., Kenning, P.: Online reviews as marketing placebo? First insights from Neuro-IS utilizing fNIRS. In: Proceedings of the Twenty-Eighth European Conference on Information Systems (ECIS) (2020)
46. Nissen, A., Kenning, P., Krampe, C., Schütte, R.: Utilizing mobile fNIRS to investigate neural correlates of the TAM in ecommerce. In: 40th International Conference on Information Systems, ICIS 2019 (2019)
47. Nissen, A.: Exploring the neural correlates of visual aesthetics on websites. In: Davis, F.D., Riedl, R., vom Brocke, J., Léger, P.-M., Randolph, A., Fischer, T. (eds.) Information Systems and Neuroscience. LNISO, vol. 32, pp. 211–220. Springer, Cham (2020). https://doi.org/10.1007/978-3-030-28144-1_23
48. Hirshfield, L.M., Bobko, P., Barelka, A., et al.: Using noninvasive brain measurement to explore the psychological effects of computer malfunctions on users during human-computer interactions. Adv. Hum.-Comput. Interact. **2014** (2014). https://doi.org/10.1155/2014/101038

Exploring the Influence of Personality Traits on Affective Customer Experiences in Retailing: Combination of Heart Rate Variability (HRV) and Self-report Measures

Anna Hermes[1](✉) and René Riedl[1,2]

[1] Department of Business Informatics–Information Engineering, Johannes Kepler University Linz, Linz, Austria
anna.hermes@jku.at
[2] University of Applied Sciences Upper Austria, Steyr, Austria
rene.riedl@fh-steyr.at

Abstract. As a result of changes in customers' shopping behaviors and a corresponding increase in omnichannel behavior (i.e., a blend of online and offline channels), a good customer experience (CX) is crucial for retailers' success. Affective CX responses are especially crucial in impacting a company's marketing outcomes, such as a high level of future purchase intentions. Here, we hypothesize that a customer's affective CX is significantly influenced by his or her personality traits. Based on this hypothesis, we plan to collect physiological data (heart rate variability) and self-report data to study affective CX. Specifically, we will examine the relationship between personality traits, affective CX, and future purchase channel choice intentions. Based on the findings, we will then formulate academic and managerial implications.

Keywords: Customer experience · NeuroIS · HRV · Retail · Emotions

1 Introduction

Due to newly available shopping channels (e.g., the internet, mobile apps, or augmented reality), Customer Experience (CX) research including various channels (so-called omnichannel experiences) is of growing importance for academics and retailers alike [1–5]. In particular, affective CX is highly researched and can significantly influence customers' purchase intention (e.g., [4, 6–8]) and satisfaction [9, 10]. Yet, the field is dominated by traditional methods such as self-reports [2, 11, 12]. Against this background, we follow explicit calls (e.g., [2, 12–14]) to advance the CX research field in the Neuro-Information-Systems (NeuroIS) direction by combining heart rate variability (HRV) and self-report measurements. HRV indicates the variation of the time between consecutive heartbeats and reflects the heart's ability to quickly respond to changing circumstances [15]. As such, it can help to understand the current status of the autonomic nervous system (ANS) [15] as well as affective responses (i.e., stress [16] or

© The Author(s), under exclusive license to Springer Nature Switzerland AG 2021
F. D. Davis et al. (Eds.): NeuroIS 2021, LNISO 52, pp. 20–29, 2021.
https://doi.org/10.1007/978-3-030-88900-5_3

other affective states such as amusement [17]). Importantly, personality traits such as the Big Five can also shape emotions [18], and, at least partially, predict HRV responses [19–21] (e.g., *Neuroticism* is linked to reduced HRV [22]).

A wide range of literature investigates the influence of personality on stress (i.e., the impact of *Neuroticism* on physiological stress responses; for a meta-analysis, see [16]). However, literature examining the impacts of the Big Five on affective CX in general, and positive affective CX in particular, is still limited [11, 23], and further calls have been made to systematically study the influences of individual factors such as personality on CX (e.g., [2, 24–26]). Within this context, the proposed study aims to extend the literature by (i) measuring CX through questionnaires *and* physiological (HRV) measurement, (ii) examining the effects of the Big Five on affective CX and future purchase channel choice intentions, and (iii) including online as well as in-store customers. Hence, the study addresses two research questions:

RQ1: How can HRV and self-report methods be combined to measure affective CX?
RQ2: How do the Big Five influence affective CX and future purchase channel choice intentions in online and in-store shopping settings?

2   Theoretical Background

Personality Traits and Affective CX. In 1963, Eysenck [27] formulated the theory that personality has a biological basis, and the extended literature has provided evidence that differences in HRV can provide valuable insights for personality and emotions research (e.g., [28, 29]). In a meta-analysis, Chida and Hamer [16] analyzed 27 studies examining determinants of HRV variations and found that anxiety, *Neuroticism*, and negative affect were associated with decreased HRV in stress situations. This finding is in line with a study by Čukić and Bates [22], who concluded that a higher level of *Neuroticism* was associated with a decreased HRV in both a resting- and stress-condition. However, other studies reported nonsignificant findings for a relationship between personality and HRV (e.g., in a stress context [21, 30, 31], or in daily life [29]). Evans et al. [30] did not find that *Neuroticism* and *Extraversion* impacted HRV, yet extroverts showed lower levels of cortisol reactivity.

Personality may not only exert influence on biological processes such as HRV, but may also shape a person's affective tendencies [18] as well as purchase channel choices (e.g., [32]). Extroverts are characterized by sociability, activity, and seeking pleasurable and enjoyable experiences [18]. Initial evidence suggests that extroverts share hedonic shopping values; namely, they enjoy shopping and feel excited about it [33]. People high in *Extraversion* were found to spend more time and money at shopping malls [34] and did not only shop for necessities (e.g., groceries) but were prone to shop for other goods like shoes and clothing [35]. Hence, we hypothesize:

H1: *Extraversion* positively affects a customer's affective CX as measured through HRV and self-report for (a) online and (b) in-store customers.

Agreeableness was also linked to hedonic shopping values; people high in *Agreeableness* showed a preference for ludic and aesthetic shopping environments [33]. Additionally, Goldsmith [35] found that *Agreeableness* indirectly (via happiness) influenced the preference to go shopping for shoes and clothing. Hence, we propose:

H2: *Agreeableness* positively affects the customer's affective CX as measured through HRV and self-report for (a) online and (b) in-store customers.

Openness to Experience is defined as being imaginative, sensitive, and intellectually curious, as well as loving the arts and beauty [18]. This characterization is in line with preliminary findings by Guido [33], who linked this personality trait with hedonic shopping values. The researcher concluded that people high in *Openness to Experience* valued aesthetically appealing shopping environments. Moreover, findings by Goldsmith [35] indicated that people high in *Openness to Experience* did not only shop for groceries, but enjoyed shopping for other goods such as clothing and shoes, indicating that shopping was not only seen as a necessity. Hence, we argue that people high in *Openness to Experience* show a higher tendency to experience a positive CX and propose:

H3: *Openness to Experience* positively affects the customer's affective CX as measured through HRV and self-report for (a) online and (b) in-store customers.

Conscientiousness is characterized by being well-organized and structured [18]. In line with these tendencies, Guido [33] and Gohary and Hanzaee [36] suggest that people high in *Conscientiousness* are functional, task-, and goal-oriented (utilitarian) shoppers. Hence, in contrast to, for example, extraverts, we suggest that people high in *Conscientiousness* experience a less positive CX:

H4: *Conscientiousness* negatively affects the customer's affective CX as measured through HRV and self-report for (a) online and (b) in-store customers.

Neuroticism is especially linked to strong emotional responses [27] such as anxiety, negative emotions, and psychological distress [18]. Guido [33] concluded that people high in *Neuroticism* were utilitarian shopping motivated; hence, their reason to shop was often goal- and not enjoyment-driven. Hence, we propose:

H5: *Neuroticism* negatively affects the customer's affective CX as measured through HRV and self-report for (a) online and (b) in-store customers.

Affective CX and Neurophysiological Measurements in Consumer Emotions Research. A few researchers have used physiological tools to measure customer emotions (see [11, 37], and [14] for a review). Ahn and Picard [38] as well as Popa et al. [39], for example, observed facial valence to study product preferences (i.e., a customer's liking of a product). Further, Kindermann and Schreiner [13] used the implicit association test (IAT) to measure customers' emotions towards brand stimuli (e.g., logos). Additionally, Guerreiro et al. [40] used skin conductance and eye-tracking to determine customers' visual attention and emotional arousal in product choices. However, using physiological measurements in the context of affective CX is still limited, and researchers call for future research in this domain (e.g., [2, 12–14]).

HRV is a standard indicator when examining parasympathetic influences on the heart [41]. Researchers suggest that HRV can be used to assess emotional responses [42]. A meta-analysis of HRV studies suggests that most positive emotions such as amusement and joy show increased HRV, while most negative emotions such as anger, anxiety, and fear show decreased HRV [43]. Recently, Wu et al. [17] found that participants' root mean square of successive differences (RMSSD), a major HRV measure, was significantly larger for amused participants, and amusement led to a decrease in a participant's heart rate. Yet, the researchers did not find differences in HRV for fearful, neutral, or angry participants. This finding is in line with Steenhaut et al. [31], who asked participants to watch sad and happy film clips. Results showed that the higher a participant's reported negative emotionality, the higher their HRV. This finding, however, is not in line with most existing studies. Thus, we propose:

H6: People with higher (lower) positive affective CX (as measured through a questionnaire) show an increased (decreased) HRV for (a) online and (b) in-store customers.

Affective CX and Future Purchase Intentions. The extant literature shows that affective CX can impact a customer's level of satisfaction with online shopping as well as with retailers' apps [9, 10, 44–47]. Moreover, customers with positive emotions such as pleasure and enjoyment were more likely to shop at a grocery store and also perceived higher hedonic value (e.g., a more enjoyable interaction with the retailer) [48, 49]. Hence, mounting evidence suggests that a positive affective CX positively impacts future customer shopping intentions (see [50] for a review). Hence, for online and in-store shoppers we hypothesize:

H7: A customer's level of positive affective CX as measured through HRV and self-report positively impacts future purchase intentions for (a) online and (b) in-store customers.

3   Proposed Method

Research Model. The goal of this study is three-fold. First, we investigate the role of the Big Five on affective CX and purchase channel choice intentions. Second, our research design examines online as well as in-store behavioral intentions. Third, we will measure affective CX through self-report and HRV. Figure 1 visualizes the overall research framework.

Participants and Procedure. This study will be conducted together with a grocery retailer and participants will be asked to shop at the retailer's online and physical shop. In cooperation with this company, we will recruit 40 participants, of which 20 participants will shop in-store and 20 online (for further information on sample size in HRV research, see [51]). Participants must be healthy without neurological and psychiatric disorders, not using any medications, and show a normal level of physical activity. As demonstrated by previous studies (e.g., [22]), we will control for age, sex, height, weight, and BMI. On the day of the experiment, we will first inform the participants that this study focuses

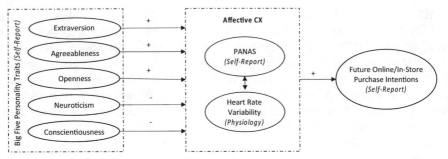

Fig. 1. Envisioned overall research framework. CX = Customer Experience, PANAS = Positive and Negative Affect Schedule

on CX and shopping behavior and then participants will be asked to sign the informed consent if they agree to take part. Next, we will measure the baseline heart rate (the participant not performing any task) for each participant. Because a participant's posture (e.g., sitting vs. lying [52]), as well as walking (e.g., [53–55]) affects HRV, the baseline will be taken for the online group in a sitting condition and for the in-store group in a normal walking condition. In both conditions (online as well as in-store), participants will then be given a shopping task. All in-store shoppers will be given the same shopping list (the list will include a fixed number of light-weight grocery products, e.g., an apple or a yogurt). Next, participants will be asked to enter the store, collect the products, and place them directly into the shopping cart. After collecting all products, participants will check out at the cash register. After the check-out, the in-store shoppers will be asked to complete a questionnaire. Online shoppers will be provided with a computer (with the grocers' website already opened) and they will receive the same shopping list as the in-store shoppers. Next, the online shoppers will be asked to find and put all the products from the list into the online shopping basket. Finally, participants will be asked to check out. After the check-out, the online shoppers will also be asked to complete a questionnaire.

Measures. As demonstrated by other researchers (e.g., [56, 57]), we will collect HRV through a chest belt (Polar H7, linked to a smartphone app) to measure affective CX physiologically. We choose HRV because it is less obtrusive than other physiological measures, easier to use, widely accessible, and inexpensive [51]. Thus, the introduced methods enable scientific researchers, as well as retail marketing and CX practitioners, to employ these methods in future research and evaluation studies. At the end of the experiment, participants will be asked to complete a questionnaire to assess the constructs via self-report measurements. We will use established and validated scales, including the German Big Five Scale by Rammstedt and Danner [58] as well as Breyer and Bluemke's [59] German translation of the Positive and Negative Affect Schedule (PANAS, [60]) to measure affective CX. Lastly, purchase intentions will be assessed (scale adapted from [61, 62]).

Data Analysis. Data will be separately analyzed for in-store and online participants. Various indicators from electrocardiography (ECG) recordings can help to analyze HRV data [41, 63]. Most of these indicators can be divided into a time-domain or a frequency-domain of the signal [51]. Indicators within the time-domain mainly focus on the time intervals between subsequent normal QRS complexes such as the duration of a heart-beat (also known as normal-to-normal R-R interval, NN) [63]. Following Baumgartner et al. [57], we will focus on five indicators, three from the time-domain and two from the frequency-domain. As time-domain indicators, we will calculate the SDNN as the standard deviation of the signal's NN intervals as well as the SDANN as the standard deviation of the averages of the signals' NN intervals (calculated over specific periods). As the last indicator from the time-domain, we will calculate the RMSSD, which considers the differences of subsequent NN intervals. These differences are first squared and, in a second step, these squares are used to calculate the square root of the arithmetic mean [57]. Moreover, as suggested by Baumgartner et al. [57], we will consider two indicators from the frequency-domain, namely the LF as the signal's power in the low-frequency spectrum (0.04 to 0.15 Hz) and the HF as the signal's power in the high-frequency spectrum (0.15 to 0.5 Hz). As demonstrated by Baumgartner et al. [57], we plan to clean the data through the Kubios HRV software, which allows for artifact removal/correction based on certain thresholds (e.g., missed beats), and further provides the ability to analyze time and frequency domains. It should be noted that data cleaning is especially important when employing consumer-grade devices to measure HRV. Even though automatic cleaning is possible, a human investigator is necessary to identify appropriate threshold values for the particular data set to avoid over-correction (see also [57]).

4 Expected Outcomes and Conclusion

This study aims to answer the research questions "RQ1: How can HRV and self-report methods be combined to measure affective CX?" and "RQ2:

How do the Big Five influence affective CX and future purchase channel choice intentions in online and in-store shopping settings?". To do so, we invite participants to perform a shopping task in either a physical retailing location (in-store shopper condition) or an online shop (online shopper condition). We plan to measure participants' affective CX through HRV and self-report measures. Additionally, we will examine participants' Big Five personalities as well as their future purchase intentions.

The findings from the study will provide implications for other researchers as well as retailers. We hope to provide theoretical contributions for future CX researchers, and we hope this study will spark an interest in more research within the domain of how personality impacts CX and how neurophysiological methods such as HRV can be used to measure affective CX. Further, if retailers know how personality influences affective CX and purchase channel choice, they can optimize their channels accordingly.

Acknowledgments. This study has been conducted within the training network project PER-FORM funded by the European Union's Horizon 2020 research and innovation program under the Marie Skłodowska-Curie grant agreement No. 765395. Note: This research reflects only the authors' views. The Agency is not responsible for any use that may be made of the information it contains.

References

1. von Briel, F.: The future of omnichannel retail: a four-stage Delphi study. Technol. Forecast. Soc. Chang. **132**, 217–229 (2018)
2. Lemon, K.N., Verhoef, P.C.: Understanding customer experience throughout the customer journey. J. Mark. **80**, 69–96 (2016)
3. Puccinelli, N.M., Goodstein, R.C., Grewal, D., Price, R., Raghubir, P., Stewart, D.: Customer experience management in retailing: understanding the buying process. J. Retail. **85**, 15–30 (2009)
4. Gentile, C., Spiller, N., Noci, G.: How to sustain the customer experience: an overview of experience components that co-create value with the customer. Eur. Manag. J. **25**, 395–410 (2007)
5. Hilken, T., Heller, J., Chylinski, M., Keeling, D.I., Mahr, D., de Ruyter, K.: Making omnichannel an augmented reality: the current and future state of the art. J. Res. Interact. Mark. **12**, 509–523 (2018)
6. Holbrook, M.B., Hirschman, E.C.: The experiential aspects of consumption: consumer fantasies, feelings, and fun. J. Consum. Res. **9**, 132–140 (1982)
7. Laros, F.J.M., Steenkamp, J.B.E.M.: Emotions in consumer behavior: a hierarchical approach. J. Bus. Res. **58**, 1437–1445 (2005)
8. Achar, C., So, J., Agrawal, N., Duhachek, A.: What we feel and why we buy: the influence of emotions on consumer decision-making. Curr. Opin. Psychol. **10**, 166–170 (2016)
9. Rose, S., Clark, M., Samouel, P., Hair, N.: Online customer experience in e-retailing: an empirical model of antecedents and outcomes. J. Retail. **88**, 308–322 (2012)
10. Martin, J., Mortimer, G., Andrews, L.: Re-examining online customer experience to include purchase frequency and perceived risk. J. Retail. Consum. Serv. **25**, 81–95 (2015)
11. Hermes, A., Riedl, R.: How to measure customers' emotional experience? A short review of current methods and a call for neurophysiological approaches. In: Davis, F.D., Riedl, R., vom Brocke, J., Léger, P.-M., Randolph, A.B., Fischer, T. (eds.) NeuroIS 2020. LNISO, vol. 43, pp. 211–219. Springer, Cham (2020). https://doi.org/10.1007/978-3-030-60073-0_25
12. De Keyser, A., Verleye, K., Lemon, K.N., Keiningham, T.L., Klaus, P.: Moving the customer experience field forward: introducing the touchpoints, context, qualities (TCQ) nomenclature. J. Serv. Res. **23**, 433–455 (2020)
13. Kindermann, H., Schreiner, M.: IAT measurement method to evaluate emotional aspects of brand perception—a pilot study. In: Davis, F.D., Riedl, R., vom Brocke, J., Léger, P.-M., Randolph, A.B. (eds.) Information Systems and Neuroscience. LNISO, vol. 25, pp. 167–173. Springer, Cham (2018). https://doi.org/10.1007/978-3-319-67431-5_19
14. Verhulst, N., De Keyser, A., Gustafsson, A., Shams, P., Van Vaerenbergh, Y.: Neuroscience in service research: an overview and discussion of its possibilities. J. Serv. Manag. **30**, 621–649 (2019)
15. Acharya, U.R., Joseph, K.P., Kannathal, N., Lim, C.M., Suri, J.S.: Heart rate variability: a review. Med. Biol. Eng. Comput. **44**, 1031–1051 (2006)
16. Chida, Y., Hamer, M.: Chronic psychosocial factors and acute physiological responses to laboratory-induced stress in healthy populations: a quantitative review of 30 years of investigations. Psychol. Bull. **134**, 829–885 (2008)
17. Wu, Y., Gu, R., Yang, Q., Luo, Y.J.: How do amusement, anger and fear influence heart rate and heart rate variability? Front. Neurosci. **13**, 1131 (2019)
18. Costa, P.T., McCrae, R.R.: Normal personality assessment in clinical practice: the NEO personality inventory. Psychol. Assess. **4**, 5–13 (1992)
19. Gallagher, S., O'Riordan, A., McMahon, G., Creaven, A.-M.: Evaluating personality as a moderator of the association between life events stress and cardiovascular reactivity to acute stress. Int. J. Psychophysiol. **126**, 52–59 (2018)

20. Bibbey, A., Carroll, D., Roseboom, T.J., Phillips, A.C., de Rooij, S.R.: Personality and physiological reactions to acute psychological stress. Int. J. Psychophysiol. **90**, 28–36 (2013)
21. Coyle, D.K.T., Howard, S., Bibbey, A., Gallagher, S., Whittaker, A.C., Creaven, A.M.: Personality, cardiovascular, and cortisol reactions to acute psychological stress in the Midlife in the United States (MIDUS) study. Int. J. Psychophysiol. **148**, 67–74 (2020)
22. Čukić, I., Bates, T.C.: The association between neuroticism and heart rate variability is not fully explained by cardiovascular disease and depression. PLoS One **10**, e0125882 (2015)
23. Hermes, A., Riedl, R.: The nature of customer experience and its determinants in the retail context: literature review. In: Gronau, N., Heine, M., Krasnova, Pousttchi, K. (eds.) WI2020 Zentrale Tracks, pp. 1738–1749. GITO Verlag, Potsdam (2020)
24. Handarkho, Y.D.: The intentions to use social commerce from social, technology, and personal trait perspectives: analysis of direct, indirect, and moderating effects. J. Res. Interact. Mark. **14**, 305–336 (2020)
25. Piroth, P., Ritter, M.S., Rueger-Muck, E.: Online grocery shopping adoption: do personality traits matter? Br. Food J. **122**, 957–975 (2020)
26. Grewal, D., Roggeveen, A.L.: Understanding retail experiences and customer journey management. J. Retail. **96**, 3–8 (2020)
27. Eysenck, H.J.: Biological basis of personality. Nature **199**, 1031–1034 (1963)
28. Appelhans, B.M., Luecken, L.J.: Heart rate variability as an index of regulated emotional responding. Rev. Gen. Psychol. **10**, 229–240 (2006)
29. Ode, S., Hilmert, C.J., Zielke, D.J., Robinson, M.D.: Neuroticism's importance in understanding the daily life correlates of heart rate variability. Emotion **10**, 536–543 (2010)
30. Evans, B.E., et al.: Neuroticism and extraversion in relation to physiological stress reactivity during adolescence. Biol. Psychol. **117**, 67–79 (2016)
31. Steenhaut, P., Demeyer, I., De Raedt, R., Rossi, G.: The role of personality in the assessment of subjective and physiological emotional reactivity: a comparison between younger and older adults. Assessment **25**, 285–301 (2018)
32. McElroy, J.C., Hendrickson, A.R., Townsend, A.M., DeMarie, S.M.: Dispositional factors in internet use: personality versus cognitive style. MIS Q. **31**, 809–820 (2007)
33. Guido, G.: Shopping motives, big five factors, and the hedonic/utilitarian shopping value: an integration and factorial study. Innov. Mark. **2**, 57–67 (2006)
34. Breazeale, M., Lueg, J.E.: Retail shopping typology of American teens. J. Bus. Res. **64**, 565–571 (2011)
35. Goldsmith, R.: The Big Five, happiness, and shopping. J. Retail. Consum. Serv. **31**, 52–61 (2016)
36. Gohary, A., Hanzaee, K.H.: Personality traits as predictors of shopping motivations and behaviors: a canonical correlation analysis. Arab Econ. Bus. J. **9**, 166–174 (2014)
37. Caruelle, D., Gustafsson, A., Shams, P., Lervik-Olsen, L.: The use of electrodermal activity (EDA) measurement to understand consumer emotions – a literature review and a call for action. J. Bus. Res. **104**, 146–160 (2019)
38. Ahn, H., Picard, R.W.: Measuring affective-cognitive experience and predicting market success. IEEE Trans. Affect. Comput. **5**, 173–186 (2014)
39. Popa, M.C., Rothkrantz, L.J.M., Wiggers, P., Shan, C.: Assessment of facial expressions in product appreciation. Neural Netw. World **27**, 197–213 (2017)
40. Guerreiro, J., Rita, P., Trigueiros, D.: Attention, emotions and cause-related marketing effectiveness. Eur. J. Mark. **49**, 1728–1750 (2015)
41. Task Force of the European Society of Cardiology and the North American Society of Pacing and Electrophysiology: Heart Rate Variability: Standards of Measurement, Physiological Interpretation, and Clinical Use. Circulation **93**, 1043–1065 (1996)

42. Choi, K.H., Kim, J., Kwon, O.S., Kim, M.J., Ryu, Y.H., Park, J.E.: Is heart rate variability (HRV) an adequate tool for evaluating human emotions? – a focus on the use of the International Affective Picture System (IAPS). Psychiatry Res. **251**, 192–196 (2017)
43. Kreibig, S.D.: Autonomic nervous system activity in emotion: a review. Biol. Psychol. **84**, 394–421 (2010)
44. Molinillo, S., Navarro-García, A., Anaya-Sánchez, R., Japutra, A.: The impact of affective and cognitive app experiences on loyalty towards retailers. J. Retail. Consum. Serv. **54**, 101948 (2020)
45. Micu, A.E., Bouzaabia, O., Bouzaabia, R., Micu, A., Capatina, A.: Online customer experience in e-retailing: implications for web entrepreneurship. Int. Entrepreneurship Manag. J. **15**(2), 651–675 (2019). https://doi.org/10.1007/s11365-019-00564-x
46. Terblanche, N.S.: Revisiting the supermarket in-store customer shopping experience. J. Retail. Consum. Serv. **40**, 48–59 (2018)
47. Pandey, S., Chawla, D.: Online customer experience (OCE) in clothing e-retail. Int. J. Retail Distrib. Manag. **46**, 323–346 (2018)
48. Anninou, I., Foxall, G.R.: The reinforcing and aversive consequences of customer experience. The role of consumer confusion. J. Retail. Consum. Serv. **51**, 139–151 (2019)
49. Högberg, J., Ramberg, M.O., Gustafsson, A., Wästlund, E.: Creating brand engagement through in-store gamified customer experiences. J. Retail. Consum. Serv. **50**, 122–130 (2019)
50. Hermes, A., Riedl, R.: Dimensions of retail customer experience and its outcomes: a literature review and directions for future research. In: Nah, F.-H., Siau, K. (eds.) HCII 2021. LNCS, vol. 12783, pp. 71–89. Springer, Cham (2021). https://doi.org/10.1007/978-3-030-77750-0_5
51. Massaro, S., Pecchia, L.: Heart rate variability (HRV) analysis: a methodology for organizational neuroscience. Organ. Res. Methods. **22**, 354–393 (2019)
52. Acharya, U.R., Kannathal, N., Hua, L.M., Yi, L.M.: Study of heart rate variability signals at sitting and lying postures. J. Bodyw. Mov. Ther. **9**, 134–141 (2005)
53. Mujib Kamal, S., Babini, M.H., Krejcar, O., Namazi, H.: Complexity-based decoding of the coupling among heart rate variability (HRV) and walking path. Front. Physiol. **11**, 602027 (2020)
54. Hooper, T.L., Dunn, D.M., Props, J.E., Bruce, B.A., Sawyer, S.F., Daniel, J.A.: The effects of graded forward and backward walking on heart rate and oxygen consumption. J. Orthop. Sport. Phys. Ther. **34**, 65–71 (2004)
55. de Brito, J.N., et al.: The effect of green walking on heart rate variability: a pilot crossover study. Environ. Res. 185, 109408 (2020)
56. Kalischko, T., Fischer, T., Riedl, R.: Techno-unreliability: a pilot study in the field. In: Davis, F.D., Riedl, R., vom Brocke, J., Léger, P.-M., Randolph, A., Fischer, T. (eds.) Information Systems and Neuroscience. LNISO, vol. 32, pp. 137–145. Springer, Cham (2020). https://doi.org/10.1007/978-3-030-28144-1_15
57. Baumgartner, D., Fischer, T., Riedl, R., Dreiseitl, S.: Analysis of heart rate variability (HRV) feature robustness for measuring technostress. In: Davis, F.D., Riedl, R., vom Brocke, J., Léger, P.-M., Randolph, A.B. (eds.) Information Systems and Neuroscience. LNISO, vol. 29, pp. 221–228. Springer, Cham (2019). https://doi.org/10.1007/978-3-030-01087-4_27
58. Rammstedt, B., Danner, D.: Die Facettenstruktur des Big Five Inventory (BFI). Diagnostica **63**, 70–84 (2017)
59. Breyer, B., Bluemke, M.: Deutsche Version der Positive and Negative Affect Schedule PANAS (GESIS Panel). Zusammenstellung sozialwissenschaftlicher Items und Skalen (2016)
60. Watson, D., Clark, L.A., Tellegen, A.: Development and validation of brief measures of positive and negative affect: the PANAS scales. J. Pers. Soc. Psychol. **54**, 1063–1070 (1988)
61. Bleier, A., Harmeling, C.M., Palmatier, R.W.: Creating effective online customer experiences. J. Mark. **83**, 98–119 (2019)

62. Ajzen, I., Fishbein, M.: Understanding Attitudes and Predicting Social Behaviour. Prentice Hall, Englewood Cliffs (1980)
63. Xhyheri, B., Manfrini, O., Mazzolini, M., Pizzi, C., Bugiardini, R.: Heart rate variability today. Prog. Cardiovasc. Dis. **55**, 321–331 (2012)

Motor Dysfunction Simulation in Able-Bodied Participants for Usability Evaluation of Assistive Technology: A Research Proposal

Felix Giroux[1](✉), Jared Boasen[1,4], Charlotte J. Stagg[5], Sylvain Sénécal[1,3], Constantinos Coursaris[1,2], and Pierre-Majorique Léger[1,2]

[1] Tech3Lab, HEC Montréal, Montréal, QC, Canada
{felix.giroux,jared.boasen,sylvain.senecal,
constantinos.coursaris,pierre-majorique.leger}@hec.ca
[2] Department of Information Technologies, HEC Montréal, Montréal, QC, Canada
[3] Department of Marketing, HEC Montréal, Montréal, QC, Canada
[4] Faculty of Health Sciences, Hokkaido University, Sapporo, Japan
[5] Wellcome Centre for Integrative Neuroimaging, Nuffield Department of Clinical Neuroscience, University of Oxford, Oxford, UK

Abstract. The development of assistive technologies and guidelines for their accessibility is impeded by the limited access to disabled participants. As a consequence, performing disability simulations on able bodied participants is a common practice in usability evaluation of assistive technologies. However, still little is known about how disability simulation can influence the usability evaluation of assistive technologies by able-bodied participants. This research proposal explores the effect of a motor dysfunction simulation in able-bodied participants that impedes use of a mouse input, but supports gesture-based assistive technology. Results of this study may provide insights on how to improve, via the experimental design, the meaning and the validity of usability evaluation of assistive technologies by able-bodied participants.

Keywords: Accessibility · Usability evaluation · Disability simulation · Gesture-based interface · Assistive technology

1 Introduction

Over a billion people worldwide have some form of permanent or temporary disability that restricts their access to many Information Systems' (IS) inputs such as mice and keyboards [1]. In recent years, growing research in and development of assistive technologies has strived at maximizing the accessibility of IS via special interfaces. However, assessing the usability of assistive technologies with disabled individuals poses challenges for labs to identify and recruit disabled subjects, and for the latter to participate in research experiments living with a disability. Therefore, usability evaluations are often performed on able bodied participants, which has permitted research across a wide range of technologies using various methodological approaches [2–5]. For instance, the effectiveness

F. D. Davis et al. (Eds.): NeuroIS 2021, LNISO 52, pp. 30–37, 2021.
https://doi.org/10.1007/978-3-030-88900-5_4

of assistive technologies, the efficiency and the satisfaction with which the task is performed, have been respectively assessed via objective measures of task performance, subjective reports of mental and physical demand such as the NASA TLX, and subjective reports of user's perceptions and attitudes towards the technology [2–5]. Methods targeting neurophysiological measures such as electroencephalographic (EEG) coherence have also been used in Human-Computer Interaction (HCI) research to index neural processing effort related to psychomotor efficiency during video games [6] and interactive media [7] use. However, similar neurophysiological approaches have not been reported in usability research on assistive technology, thus constituting an important research gap.

Another very important shortcoming of assistive technology research assessing usability with able-bodied participants is the very fact that able-bodied participants arguably bias the results of these studies and may thereby negatively affect the development of appropriate accessibility guidelines [2, 8–10]. Indeed, one Brain-Computer Interface (BCI) application study reported that a usability evaluation of an assistive technology by able-bodied participants was deemed to be generally low in terms of efficiency [8]. This was attributable to the participants mentally comparing the BCI's efficiency with traditional input modes (i.e., a mouse) using their functioning limb. In another BCI study, a negative correlation was found between able bodied participants' performance and satisfaction with a BCI [9]. In contrast, prior reports have also shown high subjective ratings of satisfaction by users with disabilities despite their reported low performance in using a BCI [10]. The previous results can be seen as an extension of the disability paradox that may explain the differences between the positive self-perception of well-being and satisfaction by disabled individuals and the negative perception by able-bodied participants [2, 11]. To improve the meaning and the validity of usability evaluations of assistive technologies with able-bodied participants, HCI and accessibility research has often used disability simulation [12–18]. However, little is known about the effect of disability simulations on usability evaluation with able-bodied participants. To our knowledge, research that uses disability simulations to impede the normal use of a computer input (i.e., mouse and keyboard) by able-bodied participants, and has them assessing the usability of an assistive technology that supports the simulated disability has not been attempted.

The present research proposal addresses the previous gaps by investigating the effect of a hand disability simulation on able bodied participants' usability evaluation of a traditional input (i.e., mouse) and a gesture-based input device. Drawing on the task-technology fit (TTF) literature [19, 20] and a framework on individual-technology fit (ITF) applied to BCI [21], we will design a disability simulation that will restrict able bodied participants' ability to naturally use a mouse, but this without impacting the use of the gesture-based device investigated in this study. Therefore, our first objective is to compare the usability evaluation between the mouse-based task (i.e., low ITF) and the gesture-based task (i.e., high ITF) to validate our ITF design. Considering extant approaches and the need for neurophysiological insight, we will use task performance, EEG theta band inter-cortical coherence, and self-reported measures to compare the two input devices' usability respectively according to the following dimensions: 1) input device effectiveness, 2) input device efficiency and 3) user satisfaction with the input device. Correspondingly, we predict that, as compared with the mouse control, the gesture-based inputs will allow participants to achieve superior task performance (i.e., higher effectiveness), with better motor efficiency, in association with increased inter-cortical coherence between motor and sensory processing regions, and increased self-reported satisfaction with the technology. Our second and main objective is to show that when able bodied participants can first experience the limit of their normal input controls (i.e., mouse and keyboards) with a disability simulation prior to testing an assistive technology well suited for their simulated disability, their perception and thus evaluation of this technology may be positively influenced as they can fully grasp its usability. Therefore, we further predict that the order in which the experimental tasks are performed will affect the extent to which the assistive technology has an enhancing effect on the satisfaction dimension of usability. More specifically, we expect that the ratings of perceptions and attitudes towards the gesture-based input device will be more positive when the gesture-based task follows the mouse-based task than when it precedes it.

2 Methods

2.1 Input Technologies

The two input technologies that will be compared in this study are a traditional mouse, and the leap motion controller (LMC). The LMC is a popular gesture-based system targeting hand and finger movements. This device uses three infrared LED lights and two infrared cameras to track the position and movements of defined objects like hands, fingers, or tools that are within a range of approximately 25 to 600 mm above it [22, 23] (See Fig. 1). HCI research has validated the LMC as a contact-free pointing device using standard tests such as the Fitts paradigm or the ISO 9241-9 serial point-selection task [22–26]. This test provides a widely recognized metric in research and in the industry, the Fitts' throughput, which combines measures of movement time, distance and accuracy with a pointing device [26]. The Fitts paradigm is applicable to the pointing and dragging interactions of various standard devices, including a mouse, trackball, stylus, or contact-free devices like the LMC [23, 24]. Research comparing the usability of the LMC to traditional mouse input by able-bodied participants have generally shown superior task performance, with higher efficiency and satisfaction [24, 27].

2.2 Disability Simulation

Disability simulations in HCI include reduced dexterity and tactile loss with a splint or special glove, or reduced vision and audition with eye patches or earplugs [12, 14, 15]. It may appear ironic to use a gesture-based control as an assistive technology for a user with motor disability. However, it is important to distinguish severely impaired users, who could be supported by systems like BCI, and moderately impaired users, who could be using part of their motor function. As an example, one disability simulation that could impede the use of a traditional mouse but would be supported by the LMC is the Cambridge simulation gloves [28]. This glove could allow restricting the ability to grip and use a mouse without affecting LMC control with open hand movements (See Fig. 2).

2.3 Task

As mentioned earlier, the Fitts paradigm or the ISO 9241-9 serial point-selection task has been traditionally used to compare and assess novel pointing devices in HCI [23, 24] (See Fig. 1). In this task, users need to horizontally move a cursor to click on instrumental targets while the distance between the two targets and their horizontal size vary from one click to another. Using the LMC, the serial point-selection task has typically been performed with a pointing finger (i.e., the index). The LMC also allows using our open hand to control the direction of a cursor in a 2D plane with wrist flexion and rotation. Therefore, the ISO 9241-9 serial point-selection task is well suited to address our research question. Other tasks that could be of interest to our research are a game or a more naturalistic interface navigation or communication task. Indeed, while the LMC was not specifically designed for disabled users, this tool has been widely used as a therapy-driven serious game for hand rehabilitation [29–34]. These games involve open hand gestures that are based on specific hand motor rehabilitation therapy movement [32] to control objects (e.g., plane or ball) in a 3D environment [29] or to perform a drawing task. Some games also allow recording performance metrics (e.g., score, time to complete) that can be gathered via Application Programming Interfaces (APIs) [22].

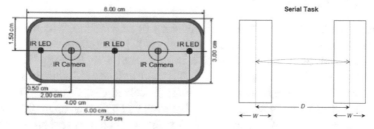

Fig. 1. The left image shows the LMC features and dimensions in centimeters [23]. The right image shows the Fitts paradigm or the ISO 9241-9 serial point-selection task with a given distance (D) between the targets with given width (W) [24].

2.4 Experimental Design

We propose a two-factor within-subject experiment with repeated measures using a sample size of 20 able bodied participants. The first experimental factor will be the input technology, and more specifically the task-technology fit determined by the extent to which either the mouse or the LMC is adapted for the task performed with a simulated hand disability. The second factor will be the order in which the technologies are experienced.

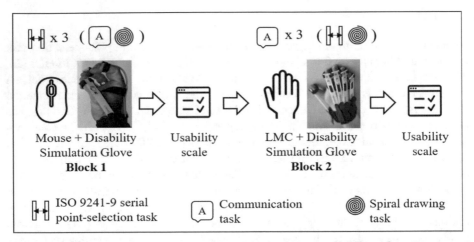

Fig. 2. Experimental design that consists of two blocks (mouse and LMC) of three iterations of the same task. A usability scale is administered after each block. The order of the blocks and task assignment within them are randomized. Both blocks are performed with the same disability simulation glove.

2.5 Subjective Measurement

Our usability scale focuses on the broad category of satisfaction, which is determined by perceptions and attitudes towards the technology [2]. Therefore, we will select questionnaire items borrowed from IS literature on Information Technology (IT) perception and acceptance [35, 36], self-efficacy [37, 38], and task- or individual-technology fit [19–21]. We will assess the following variables: Perceived Usefulness (PU) and Ease of Use (PEOU) as the extent to which the user believes that using the technology enhances his or her task performance (4 items), and the extent to which the user believes that using the technology is free from effort (5 items) [35]; Self-Efficacy (SE) as the extent to which the user believes that he or she has the ability to accomplish the task (3 items) [36–38]; Perceived Individual-Technology Fit (PITF) the extent to which the user believes that he or she has characteristics that match with the technology to enable its use (1 item) [21]; Satisfaction (S) as the extent to which the user is satisfied with the technology (1 item); and Intention to Recommend (IR) the technology for an individual that has similar motor disability as the one simulated by the splint (1 item).

2.6 Objective Measurement

We will assess task performance via Fitts' throughput for the ISO 9241-9 serial point-selection task, and other performance metrics (i.e., score) for the navigation game or task. We will also record EEG signals at a 500 Hz sampling frequency using a 32-channel BrainVision Recorder system (Brain Products GmbH). We will measure theta band inter-cortical coherence as an index of neural effort related to psychomotor efficiency. This will allow us to infer the efficiency of the technology in achieving the task.

2.7 Anticipated Analysis

Our two hypotheses will be investigated using a single repeated measures (RM) ANOVA separately for each dependant variable: Fitts' throughput (task performance), inter-cortical coherence (psychomotor efficiency), and the six satisfaction subscales (PU, PEOU, SE, PTTF, S, IR).

3 Expected Contributions

In line with the call for societal contributions in NeuroIS research [39], this proposal aims at testing a novel disability simulation approach to impair traditional input (i.e., mouse) in order to enhance the perception of assistive technology by able-bodied participants in usability evaluation. By addressing an extension of the disability paradox [2], the results of this study may show that well-designed disability simulations and scenarios can positively influence usability evaluation by able-bodied participants of assistive technologies. Consequently, this may lead to the emergence of richer and more valid insights from able bodied participants that will contribute to the development of contextually relevant accessibility guidelines.

Finally, we hope that the proposed design will allow further investigation with non-invasive brain stimulation methods. Transcranial direct current stimulation (tDCS) is a type of neuromodulation that delivers a constant and low direct current on the head via electrodes and has been used in few NeuroIS studies [40–42]. Related to this research proposal, a recent systematic review and meta-analysis recently revealed the potential for Anodal tDCS in increasing measures of cognitive empathy experienced by healthy adults in lab experiments [43]. Therefore, it may be interesting to explore the potential of tDCS in positively influencing able bodied participants' empathy towards disabled people, and thus their perception of an assistive technology as they perform a computer-based task with a disability simulation. tDCS was also shown to enhance motor learning, and has important applications in rehabilitation purposes, including training for recovery of motor functions after a stroke [44–46]. Although the LMC has been widely used as therapy for motor function rehabilitation [29–34], no research has explored the potential of combining both tDCS and LMC. Therefore, this study may pave the way to the development of novel therapy for motor dysfunction recovery.

References

1. World Health Organization: Disability and Health. https://www.who.int/en/news-room/fact-sheets/detail/disability-and-health
2. Bajcar, B., Borkowska, A., Jach, K.: Asymmetry in usability evaluation of the assistive technology among users with and without disabilities. Int. J. Hum.-Comput. Interact. 36(19), 1849–1866 (2020)
3. Kübler, A., et al.: The user-centered design as novel perspective for evaluating the usability of BCI-controlled applications. PLoS One 9(12), e112392 (2014)
4. Choi, I., Rhiu, I., Lee, Y., Yun, M.H., Nam, C.S.: A systematic review of hybrid brain-computer interfaces: taxonomy and usability perspectives. PloS One 12(4), e0176674 (2017)

5. Lorenz, R., Pascual, J., Blankertz, B., Vidaurre, C.: Towards a holistic assessment of the user experience with hybrid BCIs. J. Neural Eng. **11**(3), 035007 (2014)
6. Rietschel, J.C., Miller, M.W., Gentili, R.J., Goodman, R.N., McDonald, C.G., Hatfield, B.D.: Cerebral-cortical networking and activation increase as a function of cognitive-motor task difficulty. Biol. Psychol. **90**(2), 127–133 (2012)
7. Boasen, J., Giroux, F., Duchesneau, M.O., Sénécal, S., Léger, P.M., Ménard, J.F.: High-fidelity vibrokinetic stimulation induces sustained changes in intercortical coherence during a cinematic experience. J. Neural Eng. **17**(4), 046046 (2020)
8. Pinegger, A., Hiebel, H., Wriessnegger, S.C., Müller-Putz, G.R.: Composing only by thought: novel application of the P300 brain-computer interface. PloS One **12**(9), e0181584 (2017)
9. Eidel, M., Kübler, A.: Wheelchair control in a virtual environment by healthy participants using a P300-BCI based on tactile stimulation: training effects and usability. Front. Hum. Neurosci. **14** (2020)
10. Holz, E.M., Botrel, L., Kübler, A.: Independent home use of Brain Painting improves quality of life of two artists in the locked-in state diagnosed with amyotrophic lateral sclerosis. Brain-Comput. Interfaces **2**(2–3), 117–134 (2015)
11. Ubel, P.A., Loewenstein, G., Schwarz, N., Smith, D.: Misimagining the unimaginable: the disability paradox and health care decision making. Health Psychol. **24**(4), 57–62 (2005)
12. Burgstahler, S., Doe, T.: Disability-related simulations: if when and how to use them in professional development. Rev. Disabil. Stud. **1**(2), 8–18 (2004)
13. Cardoso, C., Clarkson, P.J.: Simulation in user-centred design: helping designers to empathise with atypical users. J. Eng. Des. **23**(1), 1–22 (2012)
14. Dahl, Y., Alsos, O.A., Svanæs, D.: Fidelity considerations for simulation-based usability assessments of mobile ICT for hospitals. Int. J. Hum.-Comput. Interact. **26**(5), 445–476 (2010)
15. Giakoumis, D., Kaklanis, N., Votis, K., Tzovaras, D.: Enabling user interface developers to experience accessibility limitations through visual, hearing, physical and cognitive impairment simulation. Univ. Access Inf. Soc. **13**(2), 227–248 (2013). https://doi.org/10.1007/s10209-013-0309-0
16. Mankoff, J., Fait, H., Juang, R.: Evaluating accessibility by simulating the experiences of users with vision or motor impairments. IBM Syst. J. **44**(3), 505–517 (2005)
17. Petrie, H., Bevan, N.: The evaluation of accessibility, usability, and user experience. Univ. Access Handb. **1**, 1–16 (2009)
18. Barsalou, L.W.: Grounded cognition. Annu. Rev. Psychol. **59**, 617–645 (2008)
19. Goodhue, D.L., Thompson, R.L.: Task-technology fit and individual performance. MIS Q. 213–236 (1995)
20. Dishaw, M.T., Strong, D.M.: Extending the technology acceptance model with task–technology fit constructs. Inf. Manag. **36**(1), 9–21 (1999)
21. Randolph, A.B.: Not all created equal: individual-technology fit of brain-computer interfaces. In: 2012 45th Hawaii International Conference on System Sciences, pp. 572–578. IEEE (2012)
22. Guna, J., Jakus, G., Pogačnik, M., Tomažič, S., Sodnik, J.: An analysis of the precision and reliability of the leap motion sensor and its suitability for static and dynamic tracking. Sensors **14**(2), 3702–3720 (2014)
23. Bachmann, D., Weichert, F., Rinkenauer, G.: Review of three-dimensional human-computer interaction with focus on the leap motion controller. Sensors **18**(7), 2194 (2018)
24. Bachmann, D., Weichert, F., Rinkenauer, G.: Evaluation of the leap motion controller as a new contact-free pointing device. Sensors **15**(1), 214–233 (2015)
25. Smeragliuolo, A.H., Hill, N.J., Disla, L., Putrino, D.: Validation of the Leap Motion Controller using markered motion capture technology. J. Biomech. **49**(9), 1742–1750 (2016)
26. Weichert, F., Bachmann, D., Rudak, B., Fisseler, D.: Analysis of the accuracy and robustness of the leap motion controller. Sensors **13**(5), 6380–6393 (2013)

27. Jones, K.S., McIntyre, T.J., Harris, D.J.: Leap motion-and mouse-based target selection: productivity, perceived comfort and fatigue, user preference, and perceived usability. Int. J. Hum.-Comput. Interact. **36**(7), 621–630 (2020)
28. Cambridge Simulation Gloves: Inclusive Design Toolkit. http://www.inclusivedesigntoolkit. com/betterdesign2/gloves/gloves.html
29. Afyouni, I., et al.: A therapy-driven gamification framework for hand rehabilitation. User Model. User-Adap. Inter. **27**(2), 215–265 (2017). https://doi.org/10.1007/s11257-017-9191-4
30. Barrett, N., Swain, I., Gatzidis, C., Mecheraoui, C.: The use and effect of video game design theory in the creation of game-based systems for upper limb stroke rehabilitation. J. Rehabil. Assistive Technol. Eng. **3**, 2055668316643644 (2016)
31. Fernández-González, P., et al.: Leap motion-controlled video game-based therapy for upper limb rehabilitation in patients with Parkinson's disease: a feasibility study. J. Neuroeng. Rehabil. **16**(1), 1–10 (2019)
32. Grubišić, I., Skala Kavanagh, H.A.N.A., Grazio, S.: Novel approaches in hand rehabilitation. Period. Biol. **117**(1), 139–145 (2015)
33. Tarakci, E., Arman, N., Tarakci, D., Kasapcopur, O.: Leap Motion Controller–based training for upper extremity rehabilitation in children and adolescents with physical disabilities: a randomized controlled trial. J. Hand Ther. **33**(2), 220–228 (2020)
34. Vanbellingen, T., Filius, S.J., Nyffeler, T., van Wegen, E.E.: Usability of videogame-based dexterity training in the early rehabilitation phase of stroke patients: a pilot study. Front. Neurol. **8**, 654 (2017)
35. Davis, F.: Perceived usefulness, perceived ease of use, and user acceptance of information technology. MIS Q. **13**(3), 319–340 (1989)
36. Venkatesh, V., Morris, M.G., Davis, G.B., Davis, F.D.: User acceptance of information technology: toward a unified view. MIS Q. **27**(3), 425–478 (2003)
37. Maddux, J.E.: Self-efficacy Theory. Springer, New York (1995)
38. Compeau, D., Higgins, C.: Application of social cognitive theory to training for computer skills. Inf. Syst. Res. **6**(2), 118e143 (1995). Comput. Hum. Behav. **17**, 21–33
39. vom Brocke, J., Hevner, A., Léger, P.M., Walla, P., Riedl, R.: Advancing a NeuroIS research agenda with four areas of societal contributions. Eur. J. Inf. Syst. **29**(1), 9–24 (2020)
40. Dumont, L., Larochelle-Brunet, F., Théoret, H., Riedl, R., Sénécal, S., Léger, P.M.: Non-invasive brain stimulation in information systems research: a proof-of-concept study. PloS One. **13**(7), e0201128 (2018)
41. Dumont, L., Larochelle-Brunet, F., Théoret, H., Sénécal, S., Léger, P.M., Riedl, R.: Using transcranial direct current stimulation (tDCS) to assess the role of the dorsolateral prefrontal cortex in technology acceptance decisions: a pilot study. In: Proceedings of the Gmunden Retreat on NeuroIS (2014)
42. Dumont, L., El Mouderrib, S., Théoret, H., Sénécal, S., Léger, P.M.: Non-invasive brain stimulation as a set of research tools in NeuroIS: opportunities and methodological considerations. Commun. Assoc. Inf. Syst. **43**(1), 5 (2018)
43. Bahji, A., Forth, E., Yang, C.C., Khalifa, N.: Transcranial direct current stimulation for empathy: a systematic review and meta-analysis. Soc. Neurosci. 1–24 (2021)
44. Nitsche, M.A., Paulus, W.: Excitability changes induced in the human motor cortex by weak transcranial direct current stimulation. J. Physiol. **527**(3), 633–639 (2000)
45. Stagg, C.J., Nitsche, M.A.: Physiological basis of transcranial direct current stimulation. Neuroscientist **17**(1), 37–53 (2011)
46. Nitsche, M.A., et al.: Facilitation of implicit motor learning by weak transcranial direct current stimulation of the primary motor cortex in the human. J. Cogn. Neurosci. **15**(4), 619–626 (2003)

Exploring the Potential of NeuroIS in the Wild: Opportunities and Challenges of Home Environments

Anke Greif-Winzrieth[1]([⊠]), Christian Peukert[1], Peyman Toreini[1],
and Marc T. P. Adam[2]

[1] Karlsruhe Institute of Technology (KIT), Karlsruhe, Germany
{anke.greif-winzrieth,christian.peukert,peyman.toreini}@kit.edu
[2] University of Newcastle, Newcastle, Australia
marc.adam@newcastle.edu.au

Abstract. At this stage, empirical studies in the NeuroIS field have been conducted primarily in laboratory environments. However, the continuing advances in sensor technologies and software interfaces have created novel opportunities to explore the potential of NeuroIS not only in highly controlled lab environments but also in the wild. In this exploratory study, we focus particularly on the potential of conducting NeuroIS studies in remote home environments (NeuroIS@Home) by physically sending equipment (e.g., sensors) to the participant's location and/or utilizing existing equipment in the participants' environment (e.g., cameras, input devices). To explore the potential of NeuroIS@Home, we conducted an online expert survey with 16 respondents. We identify higher external/ecological validity of experimental results and the potential of scalability as the most promising opportunities, whereas the lack of control over environmental factors and data quality turned out to be the most severe challenges.

Keywords: NeuroIS · Expert Survey · NeuroIS@Home · Home Environments

1 Introduction

Over the past fifteen years, scholars in the field of NeuroIS have built on neuroscience theories, methods, and tools to advance our theoretical and practical understanding of the development, adoption, and impact of information systems in various application domains [1, 2]. Notwithstanding the insights that have been gained from this pioneering work, it is important to note that up to this point, NeuroIS studies have primarily relied on experiments in controlled lab environments [2, 3]. It is safe to state that this reliance on controlled lab environments is inherently linked to the technical and methodological requirements of neurophysiological measurements. However, complementing the controlled environment of lab experiments with experiments in less controlled environments in the wild (e.g., online, field) can allow researchers to consider different aspects of validity. Each form of experimentation manifests in different levels of internal, external, and ecological validity (known as the "three-horned dilemma") [4]. Against this

F. D. Davis et al. (Eds.): NeuroIS 2021, LNISO 52, pp. 38–46, 2021.
https://doi.org/10.1007/978-3-030-88900-5_5

backdrop, it is not surprising that there have been repeated calls in the NeuroIS field for conducting more experiments in the wild [2, 5].

In this paper, we explore opportunities and challenges of conducting NeuroIS studies in the wild, particularly in remote home environments, which we call "NeuroIS@Home." Instead of participants coming to a laboratory, all necessary equipment along with instructions would be sent out to the participants and/or participants use their own sensor equipment to conduct the experiment remotely without on-site research staff (e.g., in the participant's home). Thereby, we build on the recent developments in the proliferation of consumer-grade sensor technologies (e.g., heart rate measurements using smartwatches [3]) and analytics libraries (e.g., eye tracking and remote photoplethysmography from consumer-grade webcams [6]). Beyond that, shutdowns due to the COVID-19 pandemic impede or even prevent data collection in laboratories [7], further driving the exploration of complementary approaches outside conventional lab facilities. To investigate the potential of NeuroIS@Home in more detail, we have conducted an exploratory online expert survey. Based on the opportunities and challenges identified by 16 experts, we derive superordinate categories in an iterative process that shed light on the experts' current view on NeuroIS@Home.

2 Background

Most NeuroIS studies performed in the past decade rely on laboratory experiments employing high-end equipment [2], thereby profiting from a range of methodological advantages associated with controlled laboratory conditions – most prominently high internal validity. Complementary to these conventional studies, there have also been efforts to conduct NeuroIS studies in field settings [8, 9]. Thereby, the use of portable consumer-grade devices, e.g., to assess heart rate variability (HRV) [10], eye movements [11, 12], and even electroencephalography (EEG) [13] was proposed to facilitate NeuroIS experiments in the wild. Although technology is developing rapidly and such devices are already widely available, their adoption in NeuroIS research is still scarce, which raises the question of why researchers are still reluctant to conduct such studies at large scale. Therefore, it is essential to explore researchers' considerations with regard to conducting NeuroIS studies in the wild.

One setting for conducting such remote NeuroIS studies could be to send out equipment and instructions to the participants and let them conduct the experiments on their own (e.g., at home), or even let them use their own devices to collect data as suggested by [5]. Following the NeuroIS research frameworks introduced by [14] and [15], this will mainly affect the identification of research questions (e.g., allowing for investigation of situations that cannot be simulated in the lab), the experimental design (e.g., need to take participants' abilities into account and provision of appropriate assistance), and the data collection (e.g., regarding the choice of devices, troubleshooting strategies, and ensuring data quality) (see Fig. 1). Naturally, conducting NeuroIS studies in remote home environments will come with its own opportunities but also challenges within each of these phases that need to be considered in an early stage of developing appropriate methods and tools.

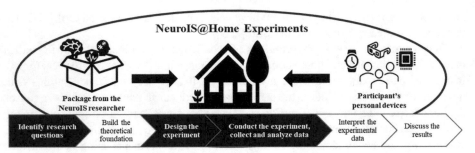

Fig. 1. Phases in the NeuroIS research framework [14, 15] potentially affected by NeuroIS@Home experiments

3 Expert Survey

To explore the potential of NeuroIS@Home, we conducted an online expert survey. We invited 24 colleagues from our scientific network who at least conducted one experiment using NeuroIS tools to participate, from whom 16 filled out the survey. On average, it took 13.3 min to complete the survey and among our respondents were 6 professors, 5 postdoctoral researchers, 3 Ph.D. students, and 2 master students.

Expertise: After a short introduction of the survey aims, respondents were asked to state their expertise in 8 commonly-used NeuroIS tools[1] taken from [1] (on a scale from 1 (very low) to 10 (very high), or "no expertise"). In our sample, expertise was highest in Eye Tracking (13 respondents, 9 with rating > 5), followed by SCR (12, 4), ECG (11, 7), EEG (11, 5), fEMG (8, 0), fMRI (7, 1), PET (6, 1), and MEG (6, 1). Two respondents stated expertise with further tools (Photoplethysmography, Functional Near-Infrared Spectroscopy, Functional Magnetic Resonance Spectroscopy). All experts reported expertise with at least one of the tools for which wearable sensors exist such that they can be considered relevant for the NeuroIS@Home scenario (i.e., Eye Tracking, SCR, fEMG, ECG, or EEG). Further, 37% of our sample had experience in conducting a NeuroIS experiment outside a controlled lab environment.

Scenario: The following scenario was presented to the respondents: "Please imagine that you have a promising research idea that involves using a neurophysiological tool to conduct an experiment. As an alternative to inviting participants to a laboratory, one of your colleagues pitches the idea to put all the necessary equipment along with instructions in a box and send it out to a participant. In this setup, you should let the participants conduct the experiment remotely on their own (e.g., at home)."

Challenges & Opportunities: Based on the scenario, respondents were asked to share at least three opportunities and challenges they see with respect to the described NeuroIS@Home scenario. Out of their points mentioned, the respondents were then asked to

---

[1] NeuroIS tools: Eye Tracking, Skin Conductance Response (SCR), Facial Electromyography (fEMG), Electrocardiogram (ECG), Functional Magnetic Resonance Imaging (fMRI), Positron Emission Tomography (PET), Electroencephalography (EEG), Magnetoencephalography (MEG).

choose the two most severe challenges as well as the two most promising opportunities. In total, we received 57 opportunities (M(SD) per respondent: 4.3 (1.4)) and 69 challenges (3.6 (1.5)), which were classified in an iterative procedure by three researchers. Overall, the procedure resulted in 16 categories (see Table 1) of which 5 are linked to opportunities, 7 to challenges, and 4 that are associated with both (challenges/opportunities not matching any category ended up in the category *other*).

Table 1. Summary of categorized challenges and opportunities.

Category	Challenges (most severe // induced by the scenario)	Opportunities (most promising // induced by the scenario)	Total
External/ecological validity		17 (14 // 9)	17
Convenience/ease of participation		7 (1 // 2)	7
Independence of lab infrastructure		4 (2 // 1)	4
New research questions		3 (1 // 1)	3
Longitudinal studies		2 (1 // 1)	2
Data quality	14 (9 // 5)		14
Lack of control	13 (7 // 5)		13
Equipment damage/loss	7 (3 // 3)		7
Logistics	6 (1 // 2)		6
Setup (installation, calibration, sensor attachment)	6 (3 // 0)		6
Hard-/software requirements	3 (2 // 1)		3
Remote troubleshooting/support	3 (1 // 0)		3
Data transfer, security/privacy, sovereignty	7 (4 // 4)	3 (0 // 0)	10
Scalability	1 (1 // 0)	9 (6 // 1)	10
Participant/sample diversity	1 (0 // 1)	5 (2 // 2)	6
Costs	1 (1 // 1)	1 (0 // 0)	2
Other	7 (1 // 4)	6 (2 // 2)	13
Total	69	57	126

Whereas most opportunities are covered by the categories *external/ecological validity*, *convenience/ease of participation*, and *independence of lab infrastructure*, most challenges refer to *data quality*, *lack of control*, and *equipment damage/loss*. However, the opportunities in connection with increased external/ecological validity were considered the most promising (e.g., experts mentioned "less artificial surrounding" or a "reduced interviewer bias" as benefits). The most severe challenges are seen in the *data quality* (e.g., due to "increased noise in data" or "increased loss of data") and *lack of experimental control*, including different kinds of "issues with participants' adherence to the experimental protocol" as well as "low control in terms of environmental conditions (light, temperature, noise, etc.)." Besides, four categories contained both, challenges and opportunities. For instance, the category *data transfer, security/privacy, and sovereignty* includes, on the one hand, challenges regarding privacy concerns or data transmission, but, on the other hand, opportunities regarding participants obtaining control and awareness of data being collected. Furthermore, the potential of *scalability* was rated as the second most promising opportunity, but also as a challenge by one expert if, for instance, an institution only owns a limited number of devices to send out. Additionally, several experts highlighted the opportunity to increase the *sample diversity* (e.g., recruit persons who would generally not come to the lab due to distance, special needs, disability, or time constraints), but one also stated that a self-selection bias might be increased in such a setting. The category *other* contains statements containing topics that only occurred once and could thus not be grouped with any other statements. Nevertheless, notable aspects were brought out, such as the challenge addressing the "necessity to simplify experimental procedures" as well as the opportunity of "higher hygiene, reuse of electrodes are avoided by definition" and "increased long-term trust in NeuroIS experiments," which certainly deserve attention.

Concerning the most severe challenges, we asked the respondents to think about measures that could be taken to remedy the challenge. All but three experts stated that they have ideas about such measures. These encompass very clear instructions, remote camera surveillance, insurances, or deposits, among others.

Since we are aware of that opportunities and challenges may be dependent on the respective NeuroIS tool, respondents should reflect on which NeuroIS tools (out of the tools relevant for NeuroIS@Home, see above) they apply or whether they are induced by the overall scenario. Table 1 reports the number of challenges and opportunities within each category explicitly labeled as induced by the scenario. Overall, this applies to 26 out of the 69 challenges (38%) and 19 out of the 57 opportunities (33%). We further find that 20 challenges (29%) and 13 opportunities (23%) were marked as being related to all tools. Table 3 in the appendix lists detailed results on how the categorized challenges and opportunities relate to individual tools.

Relevance: Table 2 shows further results indicating that the experts' stated likelihood to conduct NeuroIS@Home experiments within their own research ("How likely is it that you would apply the approach of conducting NeuroIS experiments remotely (e.g. at the participant's home) in the future?") increases in the next years and that the overall potential ("Do you see any potential in the approach of conducting NeuroIS experiments remotely (e.g. at the participant's home)?") is rated highest in case using a lab is impossible. The stated likelihood that the potential will change within the next five years ("How likely is it that the potential will change within the next five years?") is mixed among the surveyed experts. Most of them argue that technological advances affecting cost, availability, handling, etc., might influence their assessment in this regard.

Finally, we were interested in the respondent's view on a slightly modified scenario in which participants collect data via their own hardware (e.g., smartwatches, webcams). Expected advantages include the potential to collect more data, less administrative effort, and more realistic settings. However, they also raise concerns about the lower data quality of these devices compared to established NeuroIS tools.

Table 2. Descriptive statistics exploring the potential of NeuroIS@Home.

Question		Answers on 7-point Likert Scale							M	SD
		1	2	3	4	5	6	7		
Likelihood to apply NeuroIS@Home in future (1 = very unlikely l 7 = very likely)	<1 years	8	2	0	0	3	3	0	2.81	2.13
	1–5 years	6	1	0	1	3	3	2	3.69	2.36
	> 5 years	6	2	0	0	1	1	6	3.94	2.75
Potential of NeuroIS@Home (1 = very low potential l 7 = very high potential)	In case using a lab is impossible	1	1	1	3	1	3	6	5.19	1.91
	as complement to lab experiments	1	2	0	2	2	5	4	5.06	1.89
	as general alternative to lab experiments	2	3	1	2	1	4	3	4.31	2.14
Likelihood that potential will change within next five years? (1 = very unlikely l 7 = very likely)		1	4	1	3	2	3	2	4.13	1.90

4   Future Research

Even though the present study provides a number of intriguing considerations regarding the challenges and opportunities of NeuroIS@Home, further research is required to address the subject matter. As a next step, a follow-up survey could expand the insights of the present study by recruiting a larger sample of NeuroIS experts. Further, we argue that only if NeuroIS@Home study designs take into account the participants needs, researchers can benefit greatly from the new opportunities for their research purposes. Thus, it will be equally important to consider the participant perspective on the potential of NeuroIS@Home. By recruiting both experienced (i.e., participants of a previous NeuroIS study in the lab) *and* inexperienced subjects and contrasting these insights with the perspective of NeuroIS scholars, future research can develop a comprehensive and holistic understanding of the potential of NeuroIS@Home. The extracted knowledge may serve as a starting point to iteratively develop a set of guidelines and best practices for conducting NeuroIS@Home studies. As indicated by the experts within the survey, the alternative scenario of including participants' own sensor equipment instead of sending out the tools could bear high potential for future NeuroIS investigations. We see it as an integral part of NeuroIS@Home that we only touched slightly within the present paper.

5   Concluding Note

While at this stage the majority of empirical NeuroIS studies have been conducted in laboratory environments, recent technical and societal developments around sensor technologies, software interfaces, and social distancing have sparked a debate around how NeuroIS research can be conducted in the wild. In this study, we reported the results of an exploratory expert survey on the potential of conducting remote NeuroIS@Home experiments. Given this work's exploratory nature, we acknowledge that the identified challenges and opportunities are not necessarily mutually exclusive and collectively exhaustive. In this vein, we see this work as a starting point and hope that the identified challenges and opportunities will be valuable for researchers to iteratively develop methodological and technical approaches that reduce barriers for NeuroIS experiments in the wild. Accompanying other elements, this may contribute to addressing the "three-horned dilemma." Further, we acknowledge that with the current state of technology, there are still significant challenges that limit the reliability of neurophysiological measurements in home environments. Nevertheless, it is likely that further technological advances could soon help to alleviate these concerns.

Appendix: Further Results

Table 3. Relations between tools and challenges/opportunities.

Category	Eye Tracking	SCR	fEMG	ECG	EEG	$\sum$	Induced by the scenario
External/ecological validity	8	4	3	5	4	24	9
Convenience/ease of participation	5	3	1	4	4	17	2
Independence of lab infrastructure	2	1	1	1	1	6	1
New research questions	2	–	–	–	–	2	1
Longitudinal studies	–	1	–	1	–	2	1
Data quality	9	3	1	3	3	19	5
Lack of control	8	5	3	5	5	26	5
Equipment damage/loss	5	4	5	5	5	24	3
Logistics	4	3	3	4	3	17	2
Setup (installation, calibration, sensor attachment)	5	4	5	4	5	23	0
Hard-/software requirements	–	1	–	1	2	4	1
Remote troubleshooting/support	3	2	2	2	2	11	0
Data transfer, security/privacy, sovereignty	6	5	4	5	5	25	4
Scalability	9	5	3	6	5	28	1
Participant/sample diversity	3	2	1	2	2	10	3
Costs	1	–	–	1	–	2	1
Other	6	6	4	6	5	27	6
$\sum$	76	49	36	55	51		45

References

1. Dimoka, A., et al.: On the use of neurophysiological tools in IS research: developing a research agenda for NeuroIS. MIS Q. **36**, 679–702 (2012). https://doi.org/10.2307/41703475

2. Riedl, R., Fischer, T., Léger, P.-M., Davis, F.: A decade of neurois research: progress, challenges, and future directions. ACM SIGMIS Database DATABASE Adv. Inf. Syst. **51**, 13–54 (2020). https://doi.org/10.1145/3410977.3410980
3. Fischer, T., Davis, F.D., Riedl, R.: NeuroIS: a survey on the status of the field. In: Davis, F.D., Riedl, R., vom Brocke, J., Léger, P.-M., Randolph, A.B. (eds.) Information Systems and Neuroscience. LNISO, vol. 29, pp. 1–10. Springer, Cham (2019). https://doi.org/10.1007/978-3-030-01087-4_1
4. Karahanna, E., Benbasat, I., Bapna, R., Rai, A.: Editor's comments: opportunities and challenges for different types of online experiments. MIS Q. **42**, iii–x (2018)
5. Fischer, T., Riedl, R.: Lifelogging as a viable data source for NeuroIS researchers: a review of neurophysiological data types collected in the lifelogging literature. In: Davis, F.D., Riedl, R., vom Brocke, J., Léger, P.-M., Randolph, A.B. (eds.) Information Systems and Neuroscience. LNISO, vol. 16, pp. 165–174. Springer, Cham (2017). https://doi.org/10.1007/978-3-319-41402-7_21
6. Rouast, P.V., Adam, M.T.P., Chiong, R., Cornforth, D., Lux, E.: Remote heart rate measurement using low-cost RGB face video: a technical literature review. Front. Comp. Sci. **12**(5), 858–872 (2018). https://doi.org/10.1007/s11704-016-6243-6
7. van der Aalst, W., Hinz, O., Weinhardt, C.: Impact of COVID-19 on BISE research and education. Bus. Inf. Syst. Eng. **62**(6), 463–466 (2020). https://doi.org/10.1007/s12599-020-00666-9
8. González-Cabrera, J., Calvete, E., León-Mejía, A., Pérez-Sancho, C., Peinado, J.M.: Relationship between cyberbullying roles, cortisol secretion and psychological stress. Comput. Human Behav. **70**, 153–160 (2017). https://doi.org/10.1016/j.chb.2016.12.054
9. Caya, O., Léger, P.M., Brunelle, É., Grebot, T.: An empirical study on emotions, knowledge management processes, and performance within integrated business process teams. In: Proceedings of the Annual Hawaii International Conference on System Sciences, pp. 514–524. IEEE Computer Society (2012)
10. Baumgartner, D., Fischer, T., Riedl, R., Dreiseitl, S.: Analysis of heart rate variability (HRV) feature robustness for measuring technostress. In: Davis, F.D., Riedl, R., vom Brocke, J., Léger, P.-M., Randolph, A.B. (eds.) Information Systems and Neuroscience. LNISO, vol. 29, pp. 221–228. Springer, Cham (2019). https://doi.org/10.1007/978-3-030-01087-4_27
11. Langner, M., Toreini, P., Maedche, A.: AttentionBoard: a quantified-self dashboard for enhancing attention management with eye-tracking. In: Davis, F.D., Riedl, R., vom Brocke, J., Léger, P.-M., Randolph, A.B., Fischer, T. (eds.) NeuroIS 2020. LNISO, vol. 43, pp. 266–275. Springer, Cham (2020). https://doi.org/10.1007/978-3-030-60073-0_31
12. Toreini, P., Langner, M., Maedche, A.: Using eye-tracking for visual attention feedback. In: Davis, F.D., Riedl, R., vom Brocke, J., Léger, P.-M., Randolph, A., Fischer, T. (eds.) Information Systems and Neuroscience. LNISO, vol. 32, pp. 261–270. Springer, Cham (2020). https://doi.org/10.1007/978-3-030-28144-1_29
13. Riedl, R., Minas, R.K., Dennis, A.R., Müller-Putz, G.R.: Consumer-grade EEG instruments: insights on the measurement quality based on a literature review and implications for NeuroIS research. In: Davis, F.D., Riedl, R., vom Brocke, J., Léger, P.-M., Randolph, A.B., Fischer, T. (eds.) NeuroIS 2020. LNISO, vol. 43, pp. 350–361. Springer, Cham (2020). https://doi.org/10.1007/978-3-030-60073-0_41
14. Riedl, R., Davis, F.D., Hevner, A.R.: Towards a NeuroIS research methodology: intensifying the discussion on methods, tools, and measurement. J. Assoc. Inf. Syst. **15**, i–xxxv (2014). https://doi.org/10.17705/1jais.00377
15. vom Brocke, J., Liang, T.-P.: Guidelines for neuroscience studies in information systems research. J. Manag. Inf. Syst. **30**, 211–234 (2014). https://doi.org/10.2753/MIS0742-1222300408

Exploring the Recognition of Facial Activities Through Around-The-Ear Electrode Arrays (cEEGrids)

Michael T. Knierim[1]([✉]), Max Schemmer[1], and Monica Perusquía-Hernández[2]

[1] Karlsruhe Institute of Technology, Institute for Information Systems and Marketing, Karlsruhe, Germany
{michael.knierim,max.schemmer}@kit.edu
[2] University of Essex, Colchester, UK
perusquia@ieee.com

Abstract. NeuroIS scholars increasingly rely on more extensive and diverse sensor data to improve the understanding of information system (IS) use and to develop adaptive IS that foster individual and organizational productivity, growth, and well-being. Collecting such data often requires multiple recording devices, which leads to inflated study cost and decreased external validity due to greater intrusion in natural behavior. To overcome this problem, we investigated the potential of using an around-the-ear electrode array capable of capturing neural and cardiac activity for detecting an additional set of variables, namely facial muscle activity. We find that reading, speaking, chewing, jaw clenching, and six posed emotion expressions can be differentiated well by a Random Forest classifier. The results are complemented by the presentation of an open-source signal acquisition system. Thereby, an economical approach for naturalistic NeuroIS research and artefact development is provided.

Keywords: Face activity · Distal EMG · cEEGrid · OpenBCI · Random forest

1 Introduction

Observing neurophysiological and behavioral activity offers exciting possibilities like the support of productivity, personal and social growth, and general well-being [1, 2]. This potential is enabled by placing sensors on or near a person, which collect data in high temporal resolution and real-time. NeuroIS researchers leverage these data to (1) develop better understandings of IS use, acceptance and engagement [1]; or (2) for the development of adaptive IS for the regulation of states like mental workload, stress or flow [1, 2]. Despite the substantial advancements in the past decade [1, 2], the observation of such intangible phenomena remains a major challenge due to the lack of one-to-one relationships between physiological and psychological variables [3]. To address these challenges, there is an increasing consensus that data collected from multiple sensors is required [4, 5]. As NeuroIS research requires efficiency, mobility and low intrusiveness

© The Author(s), under exclusive license to Springer Nature Switzerland AG 2021
F. D. Davis et al. (Eds.): NeuroIS 2021, LNISO 52, pp. 47–55, 2021.
https://doi.org/10.1007/978-3-030-88900-5_6

to realize externally valid studies, a multi-sensor approach is, however, often difficult to implement.

To overcome this problem, we explore the potential of extending the known feature space of a recently developed wearable sensor called cEEGrid [6]. cEEGrids are placed around the ear and conveniently collect neural and cardiac activity [6, 7]. We argue that they have the additional capability of collecting other variables indicative of mental and physical states, namely facial muscle activity patterns, by capturing the spatio-temporal progressions of electrical activity generated from facial muscles. To assess this potential, we conducted a controlled laboratory experiment in which participants repeatedly posed 12 different facial activities (FA). We find that a Random Forest classifier can differentiate these activities to a promising degree with an average F1-score of 0.77. Speaking, jaw clenching, chewing, yawning, and posed smiling can all be detected with an average F1-score of ~0.85. These results highlight the added potential of using the cEEGrid sensors for the unobtrusive study of neural and behavioral (facial) activities and related phenomena of high interest to NeuroIS scholars. As an additional contribution, we show that the sensor is usable with a low-cost open-source biosignal acquisition system for which the reproduction materials are made available.

2   Related Work

The cEEGrids are flexible, printed Ag/AgCl electrodes arranged in a c-shaped array to fit around the ear [6] (see Fig. 1). These sensors have been developed to unobtrusively and comfortably collect EEG data in field settings, enabling high-quality and multiple hour EEG recordings [6, 8–10]. So far, the cEEGrids have demonstrated their ability to record typical EEG patterns related to visual stimulation [6], auditory stimulation [8, 11–13], sleep stage detection [7, 14, 15] and changes in mental workload [16]. Furthermore, the possibility to extract an ECG trace from the cEEGrid data has been demonstrated [7]. However, the research on cEEGrids has, up to now, focused on answering fundamental EEG methodology questions. The placement close to the face and the numerous dispersed electrodes led us to consider if the activity of muscles in the face can be collected through distal EMG measurement.

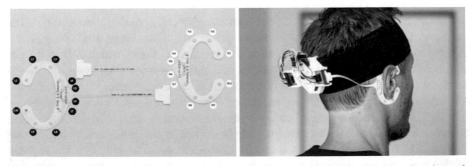

Fig. 1. The cEEGrid recording system. Left: Two cEEGrid electrodes showing the electrode positions. Right: The OpenBCI biosignal acquisition board with cEEGrids.

Distal EMG research has seen increasing interest for various applications due to the benefit of recording a phenomenon of interest from less visible or obtrusive recording sites. These approaches operate by the principle of volume conduction to pick up electrical potentials that propagate through the body. A prime example of distal EMG measurement is the research on affective facial expression recognition. To investigate the detection of emotion expression in real life, researchers have, for instance, employed the placement of electrodes on the side of the head to successfully differentiate real from posed smiles and even related micro-expressions [17]. Comparably, there has been increasing research on the ear-adjacent electrophysiological recording to pick up the heart's electrical activity [7], or to record neural activity from inside or around the ear [6, 18]. Here, we follow the same distal EMG principles to explore which facial muscle activity patterns can be reliably differentiated from each other using the cEEGrids.

3   Method

A controlled experiment was conducted to collect data for an FA classifier (see Fig. 2). Participants completed resting phases, performed maximum voluntary contraction (jaw clenching), read aloud text passages, chewed on gum, yawned, and mimicked one of six discrete emotion expressions shown on screen (see exemplary instructions in Fig. 3). Each task was performed multiple times for a few seconds with a three-second break between consecutive trials (see Fig. 2 for the trial counts and durations). Upon arrival, participants were informed about the recording and signed the consent form. Afterwards, the gelled cEEGrids were attached, and the signal quality assessed. Then, the experimenter left the room, and the experiment was completed autonomously by the participants. Data were collected for five healthy participants (2 female) in the age range of 25 to 37 (mean = 30).

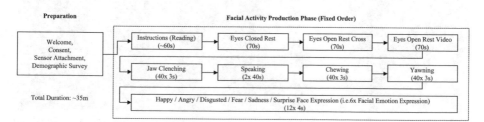

Fig. 2. Visualization of the experiment procedure with phase durations and repetitions

The cEEGrids were connected to an OpenBCI Cyton board with Daisy shield (see Fig. 1). The OpenBCI biosignal acquisition system has been introduced to the market in 2014. Since then, it has been increasingly used in scientific research[1] that seeks to leverage more accessible and flexible technology for cognitive-affective phenomena investigation and BCI development [19]. Thereby, it has so far been more readily

---

[1] At the time of writing, almost 200 research articles using the OpenBCI system are listed on the manufacturers website: https://docs.openbci.com/citations; Last retrieved 12.04.21.

employed in computer science research but is now also starting to appear in the IS discipline [20]. The combination of cEEGrids and OpenBCI amplifiers enables the low-cost mobile biosignal acquisition on 18 recording channels (two for reference and ground). For this purpose, electronics components and 3D-printed enclosures were designed. While this novel setup means that two electrodes had to be left out from the possible recording sites (here, channel L3 and R3), the use of this open-source platform allows for a much less expensive (~1.500 USD instead of ~ 8.000 USD) and flexible (i.e., prototyping friendly) recording solution. Both aspects are important when considering NeuroIS research setups and applications that should be usable in various settings and for various levels of scholar's methodological experience.

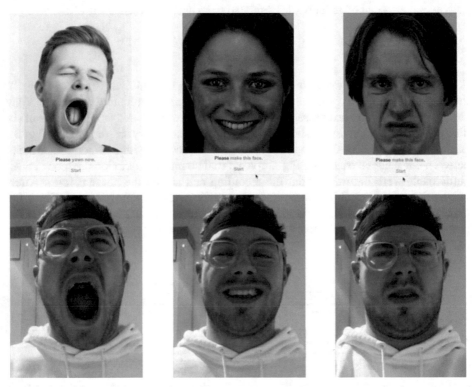

Fig. 3. Exemplary screenshots of the instructions and reactions during the posed yawning and emotion expressions (here: happy and disgust) experiment phases. Notice how the cEEGrids are not visible from this frontal view.

Each channel signal was first mean-centered, then band-pass (FIR 5–62 Hz) and notch (50 Hz) filtered. Following, the signals were epoched using non-overlapping one-second windows. We extracted a first feature set ($FS_{1\text{-Signal}}$) for each epoch comprising the sum, maximum, Hurst exponent [21], Petrosian and Higuchi Fractal Dimension [22, 23], and the Hjorth parameters Activity, Mobility, and Complexity [24]. These features were chosen due to their signal describing properties and their previous use in EEG-based event detection [25, 26]. Two additional feature sets were created that additionally include the same features computed for the first three independent components (ICs – considered to capture the main muscle activity of interest - FS2-Signal+), and the same features for all ICs together with median and SD statistics of the rectified and smoothed ICs (using a 100ms sliding-window median smoother on the absolute signal – similar to related distal fEMG work [17] – FS3-All). Features were computed for each channel resulting in datasets with 128, 152, and 248 features and between 1746 and 1966 samples.

4   Results

As classifiers, Random Forest, Support Vector Machine, AdaBoost, Logistic Regression, and a Multi-layer Perceptron were tested. Each classifier was trained for a single participant to explore how well FA events can be detected on a subject-specific level through this relatively short data collection period (~30 min). Overall, Random Forest classifiers with 100 trees showed the best performance. To evaluate the model results, stratified five-fold cross-sampling was used with Synthetic Minority Oversampling (SMOTE) to account for the dataset imbalance [27]. The performance metrics reported in Table 1 represent the average of all folds.

Table 1. Results of the classifier for twelve classes, measured with the F1-score. Values in rows are aggregated across participants. M = Mean, SD = Standard Deviation, MIN = Minimum, MAX = Maximum.

	Read	Rest	Clench	Speak	Chew	Yawn	Happy	Angry	Disgs.	Fear	Sad	Surpr.	Mean
FS1-Signal – Cross-FA Mean per Participant: .37, .53, .56, .63, .64													
M	.54	.64	.64	.68	.78	.72	.72	.35	.27	.43	.24	.48	**.54**
SD	.06	.12	0.36	.09	.08	.09	.15	.14	.16	.14	.16	.12	**.14**
MIN	.47	.50	.11	.57	.67	.60	.52	.12	.04	.29	.28	.32	**.37**
MAX	.59	.80	.91	.81	.87	.81	.88	.48	.40	.60	.37	.59	**.64**
FS2-Signal+ – Cross-FA Mean per Participant: .56, .62, .70, .70, .75													
M	.71	.63	.80	.72	.78	.76	.78	.61	.46	.61	.39	.60	**.65**
SD	.20	.14	.19	.12	.08	.04	.09	.13	.15	.09	.26	.10	**.13**
MIN	.49	.42	.47	.60	.65	.71	.67	.45	.25	.50	.49	.46	**.56**
MAX	.98	.79	.92	.90	.86	.81	.87	.78	.64	.71	.55	.70	**.75**
FS3-All – Cross-FA Mean per Participant: .70, .77, .78, .80, .80													
M	.91	.63	.87	.91	.83	.78	.83	.67	.63	.76	.66	.78	**.77**
SD	.06	.07	.05	.08	.04	.04	.04	.05	.08	.06	.18	.03	**.07**
MIN	.84	.53	.81	.77	.76	.72	.77	.60	.51	.68	.40	.75	**.70**
MAX	.97	.71	.93	.95	.88	.82	.89	.72	.70	.82	.78	.81	**.80**

The classification accuracy increases with the inclusion of the additional features with an improvement of the grand mean F1-score from 0.54 to 0.77. The FA poses that benefited most from the additional features are reading and emotion expressions. Overall, the FS_{3-All} classifier can differentiate the 12 different FA to a promising degree with an average F1-score of 0.77 (SD: 0.07). These scores still vary by FA type, with activities involving larger facial muscles strongly (e.g., muscles close to the jaw - masseter and zygomaticus major) showing better prediction scores (e.g., clenching, speaking, chewing, yawning, and smiling). An interesting finding is also, that reading shows high prediction accuracy, a finding that is likely based on the electrical activity generated from eye movements. It has been previously suggested that eye movements might be detectable with cEEGrids [7]. In contrast, FA that involve more and smaller facial muscles at the center of the face (e.g., corrugator supercilii involved primarily in expressions of anger) shows the weakest prediction scores.

5 Discussion and Outlook

The presented results highlight the added potential of using the cEEGrid sensors for the unobtrusive study of behavioral (facial) activities and related phenomena of high interest to the NeuroIS field. This potential is added to the already useful capabilities of the cEEGrids to capture neural and other physiological phenomena. With an average F1 score of 0.77, the classifier is able to predict one of 12 different posed FA events well above chance level. Presently, the detection appears most suitable for FA events based on large facial muscles such as chewing, smiling, clenching and yawning. NeuroIS scholars could, for instance, use the cEEGrids to investigate relationships between technostress [2] and bruxism or between flow experiences and IS enjoyment [28]. In contrast, the observation of FA events reliant on smaller muscles and muscles closer to the center of the face will require further investigation. It could be possible to extend the cEEGrid system to include electrodes on the side of the face (e.g., on the temporalis muscle closer to the face). These results are made possible by the novel combination of the OpenBCI biosignal acquisition board with the cEEGrids. The materials needed to reproduce the system are provided online[2]. Altogether, this system offers numerous advantages (flexibility in sensor use, low cost, open-source access to materials and APIs) that enable the development of NeuroIS studies and applications. Importantly, the system can be used in field research and allows for comfortable multi-modal data collection over the whole day [6, 7].

    To build on the latter point is considered the next most important step to further evaluate the FA diagnostic potential of the cEEGrids. So far, only the detection of controlled, posed FA activities (i.e., not naturally occurring FA events) has been investigated. In natural settings, FA events are likely to occur under the influence of various confounding effects, and likely with higher variance and complexity. Therefore, more investigation is needed to assess not only the detection potential in the presence of confounding factors, but to more comprehensively assess the system's sensing properties in terms of validity, reliability, sensitivity, diagnosticity, objectivity and intrusiveness [3]. Previous validation work on the comparable signal quality of the OpenBCI amplifier to medical grade EEG systems [29, 30], and the cEEGrid evaluation studies that demonstrate signal reliability and sensitivity to EEG phenomena (that represent more subtle electrophysiological

---

[2] https://github.com/MKnierim/openbci-ceegrids.

patterns than the large-amplitude EMG patterns investigated here) [6, 7] can already be seen as a first pillar for the usability of the system. As a first step, lab-based experiments that include the detectability of both posed and naturally occurring FA events (e.g., by induction of emotion expression using video clips – see [31]) should be pursued. Afterwards, for an even more complex field study evaluation, we propose to leverage an interactive machine learning approach [32] that builds on an initial controlled classifier and asks participants occasionally whether a particular FA was observed correctly or asks about what happened when an unknown FA instance was recognized. Through this approach, classifiers can be iteratively improved without having participants to complete diaries or work with additional sensors. As an incentive, participants could use the detection logs at the end of the study period to reflect on their experiences (e.g., what made them happy or what made them grind their teeth during the day? – see, e.g. [33]) and to evaluate whether or not the interactive process helped becoming more aware of a particular phenomenon (e.g., to avoid clenching their jaw). While the technology can improve further in terms of wearability, miniaturization advancements will make this technology genuinely wearable and usable in real life at some point (e.g., integrated into headphones or simply with less visible amplifier placements). For now, the cEEGrid-OpenBCI system already represents a promising novel approach for NeuroIS scholars to observe neural, physiological and behavioral activity patterns in a highly flexible, convenient, and accessible manner in the lab and the field.

References

1. Xiong, J., Zuo, M.: What does existing NeuroIS research focus on? Inf. Syst. **89**, 101462 (2020)
2. Riedl, R., Fischer, T., Léger, P.-M., Davis, F.D.: A decade of NeuroIS research: progress, challenges, and future directions. ACM SIGMIS Database DATABASE Adv. Inf. Syst. **51**, 13–54 (2020)
3. Riedl, R., Davis, F.D., Hevner, A.R.: Towards a NeuroIS research methodology: intensifying the discussion on methods, tools, and measurement. J. Assoc. Inf. Syst. **15**, i–xxxv (2014)
4. Ortiz de Guinea, A., Titah, R., Léger, P.-M.: Measure for Measure: A two study multi-trait multi-method investigation of construct validity in IS research. Comput. Human Behav. **29**, 833–844 (2013)
5. Léger, P.M., Davis, F.D., Cronan, T.P., Perret, J.: Neurophysiological correlates of cognitive absorption in an enactive training context. Comput. Human Behav. **34**, 273–283 (2014)
6. Debener, S., Emkes, R., De Vos, M., Bleichner, M.: Unobtrusive ambulatory EEG using a smartphone and flexible printed electrodes around the ear. Sci. Rep. **5**, 1–11 (2015)
7. Bleichner, M.G., Debener, S.: Concealed, unobtrusive ear-centered EEG acquisition: ceegrids for transparent EEG. Front. Hum. Neurosci. **11**, 1–14 (2017)
8. Mirkovic, B., Bleichner, M.G., Vos, M.D., Debener, S.: Target Speaker Detection with Concealed EEG Around the Ear. Front. Neurosci. **10**, 1–11 (2016)
9. Pacharra, M., Debener, S., Wascher, E.: Concealed around-the-ear EEG captures cognitive processing in a visual Simon task. Front. Hum. Neurosci. **11**, 1–11 (2017)
10. Bleichner, M.G., Kidmose, P., Voix, J.: Ear-centered sensing: from sensing principles to research and clinical devices. Front. Neurosci. **13**, 1437 (2019)
11. Nogueira, W., et al.: Decoding selective attention in normal hearing listeners and bilateral cochlear implant users with concealed ear EEG. Front. Neurosci. **13**, 1–15 (2019)

12. Garrett, M., Debener, S., Verhulst, S.: Acquisition of subcortical auditory potentials with around-the-ear CEEgrid technology in normal and hearing impaired listeners. Front. Neurosci. **13**, 1–15 (2019)
13. Jaeger, M., Mirkovic, B., Bleichner, M.G., Debener, S.: Decoding the attended speaker from EEG using adaptive evaluation intervals captures fluctuations in attentional listening. Front. Neurosci. **14**, 1–16 (2020)
14. Sterr, A., et al.: Sleep EEG derived from behind-the-ear electrodes (cEEGrid) compared to standard polysomnography: a proof of concept study. Front. Hum. Neurosci. **12**, 1–9 (2018)
15. Mikkelsen, K.B., et al.: Machine-learning-derived sleep–wake staging from around-the-ear electroencephalogram outperforms manual scoring and actigraphy. J. Sleep Res. **28**, e12786 (2019)
16. Wascher, E., et al.: Evaluating mental load during realistic driving simulations by means of round the ear electrodes. Front. Neurosci. **13**, 1–11 (2019)
17. Perusquía-Hernández, M., Hirokawa, M., Suzuki, K.: A wearable device for fast and subtle spontaneous smile recognition. IEEE Trans. Affect. Comput. **8**, 522–533 (2017)
18. Tabar, Y.R., Mikkelsen, K.B., Rank, M.L., Christian Hemmsen, M., Kidmose, P.: Muscle activity detection during sleep by ear-EEG. In: Proceedings of the Annual International Conference of the IEEE Engineering in Medicine and Biology Society, EMBS, pp. 1007–1010 (2020)
19. Sawangjai, P., Hompoonsup, S., Leelaarporn, P., Kongwudhikunakorn, S., Wilaiprasitporn, T.: Consumer grade EEG measuring sensors as research tools: a review. IEEE Sens. J. **20**, 3996–4024 (2020)
20. Riedl, R., Minas, R.K., Dennis, A.R., Müller-Putz, G.R.: Consumer-grade EEG instruments: insights on the measurement quality based on a literature review and implications for NeuroIS research. In: Davis, F.D., Riedl, R., vom Brocke, J., Léger, P.-M., Randolph, A.B., Fischer, T. (eds.) NeuroIS 2020. LNISO, vol. 43, pp. 350–361. Springer, Cham (2020). https://doi.org/10.1007/978-3-030-60073-0_41
21. Hurst, H.E.: Long-term storage capacity of reservoirs. Trans. Amer. Soc. Civ. Eng. **116**, 770–799 (1951)
22. Petrosian, A.: Kolmogorov complexity of finite sequences and recognition of different preictal EEG patterns. In: Proceedings of the IEEE Symposium on Computer-Based Medical Systems, pp. 212–217 (1995)
23. Higuchi, T.: Approach to an irregular time series on the basis of the fractal theory. Phys. D Nonlinear Phenom. **31**, 277–283 (1988)
24. Hjorth, B.: EEG analysis based on time domain properties. Electroencephalogr. Clin. Neurophysiol. **29**, 306–310 (1970)
25. Oh, S.-H., Lee, Y.-R., Kim, H.-N.: A novel EEG feature extraction method using Hjorth parameter. Int. J. Electron. Electr. Eng. **2**, 106–110 (2014)
26. Val-Calvo, M., Álvarez-Sánchez, J.R., Ferrández-Vicente, J.M., Fernández, E.: Optimization of real-time EEG artifact removal and emotion estimation for human-robot interaction applications. Front. Comput. Neurosci. **13**, 80 (2019)
27. Chawla, N.V., Bowyer, K.W., Hall, L.O., Kegelmeyer, W.P.: SMOTE: synthetic minority over-sampling technique. J. Artif. Intell. Res. **16**, 321–357 (2002)
28. Labonté-Lemoyne, É., et al.: Are we in Flow? Neurophysiological correlates of flow states in a collaborative game. In: Proceedings of the 2016 CHI Conference, pp. 1980–1988 (2016)
29. Frey, J.: Comparison of a consumer grade EEG amplifier with medical grade equipment in BCI applications. In: Proceedings from the International BCI Meeting, pp. 443–452 (2016)
30. Rashid, U., Niazi, I.K., Signal, N., Taylor, D.: An EEG experimental study evaluating the performance of Texas instruments ADS1299. Sensors **18**, 1–18 (2018)

31. Schaefer, A., Nils, F., Sanchez, X., Philippot, P.: Assessing the effectiveness of a large database of emotion-eliciting films: a new tool for emotion researchers. Cogn. Emot. **24**, 1153–1172 (2010)
32. Holzinger, A.: Interactive machine learning for health informatics: when do we need the human-in-the-loop? Brain Informatics **3**(2), 119–131 (2016). https://doi.org/10.1007/s40708-016-0042-6
33. McDuff, D., Karlson, A., Kapoor, A., Roseway, A., Czerwinski, M.: AffectAura: an intelligent system for emotional memory. In: Proceedings of the 2012 ACM Annual Conference on Human Factors in Computing Systems - CHI '12, pp. 849–858 (2012)

Leveraging NeuroIS Tools to Understand Consumer Interactions with Social Media Content

Jen Riley[1,2] and Adriane B. Randolph[3(✉)]

[1] Department of Marketing, Kansas State University, Manhattan, KS, USA
jenr@k-state.edu
[2] Department of Marketing and Professional Sales,
Kennesaw State University, Kennesaw, GA, USA
[3] Department of Information Systems and Security,
Kennesaw State University, Kennesaw, GA, USA
arandol3@kennesaw.edu

Abstract. Social media has risen as one of the leading budget allocations for advertising within many firms, demonstrating its increasing dominance of the marketing mix. As such, many corporate entities have increased their presence on social media platforms in recent years. We seek to better understand the impact of non-consumer generated content on the social media user experience. This study presents the application of electroencephalography to uncover mental activity by consumers when processing social media content. This research continues from a larger study exploring how consumers process content based on the author of social media content. While this extension focuses on understanding how consumers process social media content based on the author of the post, it has implications for further studies in human-computer interaction and content optimization.

Keywords: Social media · EEG · Neuromarketing · Content generation

1 Introduction

In recent years, marketing budgetary spending on advertising via social media platforms has drastically increased over traditional marketing mediums [1]. Over 97% of companies have shifted marketing strategies to take advantage of social media within their strategic approach [2]. As companies increase their application of social media tactics, departmental participation has also increased. The responsibility of reaching consumers via social media spans both marketers and salespeople producing content (e.g., salesperson generated content (SGC), and marketer generated content (MGC)) on behalf of the company [3]. While influencer and celebrity marketing have an established track record within marketing, advertising, and promotions, salesperson social media content presents a new paradigm [4].

Overall, the goals of this project are to: 1) determine if users are cognitively aware of the author of a post when engaging with social media content, 2) understand areas

© The Author(s), under exclusive license to Springer Nature Switzerland AG 2021
F. D. Davis et al. (Eds.): NeuroIS 2021, LNISO 52, pp. 56–62, 2021.
https://doi.org/10.1007/978-3-030-88900-5_7

of attention and cognitive load as it impacts user experiences with social media content posted by a variety of stakeholders, and 3) refine the process for developing and disseminating content more effectively via social media. This research is driven by the increase of social media content and alternative stakeholders leveraging content within social media platforms throughout the sales process [5].

This research is informed by a larger study conducted using the Q-sort method and semi-structured interviews. The larger study seeks to assess if consumers evaluate the author of social media content as a determination for engagement. Here, we look to neuroIS tools to provide insights about user cognition of who is posting social media content. As the increase in stakeholder participation evolves, the research question of *"Do customers differentiate social media content based on the author of the post?"* needs to be addressed. Although many insights were gathered, the order in which respondents saw various elements of content or the mental processing of the respondent was unable to be determined. Hence, we launched this extension to more closely examine consumer engagement with social media content using neuroIS tools as a complement to the traditional qualitative data.

Constructs such as emotion and cognitive load have received increased focus where neuroIS tools are most effective at extracting key insights [6, 7]. The use of neuroIS tools such as electroencephalography (EEG) has been established as a complementary tool to more traditional methods such as interviews and observation [8]. Continuing with this research and adding EEG will not only help us answer the above research question, but also provide additional understanding of the user experience and human behavior when engaging with social media content.

Historically, brain-computer interface (BCI) tools have been a means of providing communication and environmental control to people with severe motor disabilities [9]. However, in recent years, they have also been used to assess mental states [10]. In this study, we leverage passive BCI concepts [11] to gain a deeper understanding of participants within the Q-sort exercise as conducted in the larger study. In addition to understanding brain activations, we are interested in what elements of social media posts attract user attention (i.e., the profile name, the #hashtags, caption, or media). This paper represents an exploratory extension using neuroIS tools to understand the mental processing of consumers of social media content to help shed light on results from a larger study.

2 Social Media and Content Generation

Kaplan and Haenlien define social media as "a group of internet-based applications that build on the ideological and technical foundations of Web 2.0, and that allow the creation and exchange of user-generated content" (p. 61) [12]. However, this conceptualization of social media omits content types not generated by standard users. According to Kaplan and Haenlein for content to be considered user-generated it must meet these three criteria: 1) published publicly or to a select group of people, 2) creative, and 3) outside of professional purposes. To achieve all three of these criteria, ignores content created and published for professional purposes.

Based on the aforementioned narrative, it is clear that social media is no longer limited to a social, casual consumer activity. Social networking platforms also provide a community where consumers can interact with each other and companies [13, 14]. Corporate-level involvement within social media platforms requires content generation, posting, and engagement on behalf of the company. This content can consist of firm or company representatives posting as the company (i.e., firm-generated content (FGC) and company-generated content (CGC)) [3, 15]. Other options include marketer-generated content (MGC) where content is generated by a marketer on behalf of a company and salesperson-generated content (SGC) where a salesperson leverages their own social media profile to build connections with customers on behalf of their company's brand [14–17].

Due to the increase of information available to the modern consumer, salespeople must innovate and find new ways to influence the sales process. Harrigan et al. [18] notes that social media provides a nontraditional means to facilitate customer relationships and engage customers. Social media establishes an alternative connection with consumers and brings that relationship to the forefront of communication [19]. Thus, the purpose of this study is to evaluate consumer perceptions of social media content being used to sell to them.

3 Methodology

The objective of the study is to evaluate consumer engagement and cognitive processing of social media content in order to understand perceptions of content by author. In the preliminary data collection, seven (7) student-respondents (4 men, 3 women, ages 20–28) from a university in a southeastern metropolitan city who are business majors participated. Participants were gathered by offering in-class extra credit and the snowball effect.

Study respondents were recorded using EEG to measure their mental processes as they completed three rounds of a Q-sort activity [20, 21]. Eye tracking data was also recorded using Tobii eye tracking glasses (www.tobii.com) while respondents viewed and categorized social media posts to understand areas of focus and attention during sorts (i.e., if they looked at the authors name to determine where they would categorize that post). Participants of the study were asked to sort social media posts printed on 8.5 × 11 inch cards into categories over the course of three rounds, where round 1 was: they choose the categories, round 2: person or company, and round 3: salesperson or company. Students were then interviewed regarding their categorical choices.

Here, we present preliminary results based on just the EEG data indicating areas of activation during the three rounds. Respondents' electrical brain activity was measured using a non-invasive, 16-channel research-grade BioSemi ActiveTwo bioamplifier system on a laptop (http://www.cortechsolutions.com/Products/Physiological-dataacquisition/Systems/ActiveTwo.aspx). Active electrodes were placed on a standardly-configured electrode cap to allow for the recording of brain activations down-sampled to 256 Hz using a Common Average Reference (CAR). The sixteen recorded channels were: frontal-polar (Fp1, Fp2), frontal-central (FC3, FCz, FC4), central (C3, Cz, C4), temporal-parietal (TP7, TP8), parietal (P3, Pz, P4), and occipital (O1, Oz, O2).

4 Preliminary Results

Data was collected for an extension study of seven participants. After visual inspection of EEG, we analyzed the recordings from the sixteen channels of scalp electrodes using a previously-validated technique for brain localization called standardized low resolution brain electromagnetic tomography (sLORETA) [22]. These activations are presented on a fixed scale such that brighter areas with yellow indicate highest levels of activation. For each grouping of topological plots, the image on the top row in the center is a back-end view of the brain whereas the image on the bottom row in the center is a front-on view of the brain. Theory indicates that higher activation in the left hemisphere may indicate a stronger positive approach to the stimulus whereas higher activation in the right hemisphere may indicate a negative approach to the stimulus [23].

We may begin our understanding of participant experience while engaging in social media content using a qualitative lens by considering each participant as a case. For example, we can see from the activation maps in Fig. 1 that Participant 002's experience changed across the three rounds moving from a negative approach to a slightly more positive approach back to a negative approach with greatest levels of activation focused in the frontal lobe. The frontal lobe is most associated with conscious thinking, judgement, and complex reasoning. We may apply this approach to the remaining six participants.

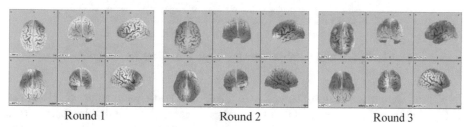

| Round 1 | Round 2 | Round 3 |

Fig. 1. Participant 002's averaged brain activations for each round of engagement with social media content.

We may then take a different view and examine a particular round and contrast experiences across participants. We found there were challenges in data collection where we are now only able to delineate brain activations across the second and third rounds for all participants and gathered cleanest results of the Q-sort during Round 3. Hence, in Fig. 2 we provide illustrations of averaged brain activations for all participants for Round 3. In Round 3, participants had to determine if the content was generated by an individual salesperson or a company. Participant 002 appeared to experience greater levels of mental processing along the midline of the cortex than what registered for other participants. However, Participant 002 and 004 both appeared to experience a negative approach to that round with highest activity in their frontal lobes. They are in contrast with Participant 003 and Participant 006 whose visual centers were instead most activated, Participant 007 who appeared to have a more neutral response in the frontal lobe, and Participant 008 whose sensorimotor area most associated with right-handed movement appeared to be most activated. We might then further dissect the data according to demographics or observations during the rounds.

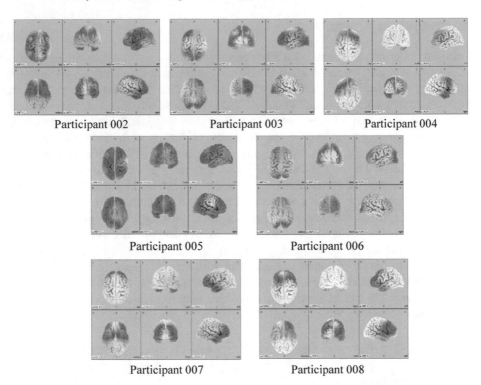

Participant 002   Participant 003   Participant 004

Participant 005   Participant 006

Participant 007   Participant 008

Fig. 2. Contrasting all participants' brain activations for Round 3.

Table 1 shows the results of Round 3 for the Q-sort. Participants categorized thirteen social media posts according to if the content was generated by an individual salesperson or a company. Only results for Participant 004 through 008 are reported due to inconsistencies in data capture for Participants 002 and 003. There are inconsistent results across participants for nearly half of the social media posts. Particularly, there are split results for posts numbered 3, 6, 9, and 11–13. These inconsistencies may be reflective of the differences in the brain activations for Participants 004 through 008. This encourages further investigation to understand if individuals who exhibit similar patterns of brain activation might also exhibit similar interpretations for social media categorizations.

Table 1. Round 3 Q-sort results (I = individual salesperson, C = company-generated post).

#	Description	Author type	004	005	006	007	008
1	Multiple pictures made collage, with caption	Salesperson	I	–	I	I	I
2	Multiple pictures made collage, with caption	Salesperson	C	C	I	I	I
3	Picture with watermark and caption	Salesperson	C	I	C	C	I
4	Post to group, photo with caption	Salesperson	C	C	C	C	C
5	Link with picture, teaser headline	Salesperson	I	I	I	I	I
6	Multiple pictures made collage, caption	Business	I	C	I	I	C
7	Photo with caption	Salesperson	–	C	C	C	C
8	Photo flyer with caption and #Hashtag block	Business	I	I	I	I	I
9	Multiple pictures made collage, with caption	Business	C	C	I	I	I
10	Photo with caption	Business	C	C	C	C	C
11	Photo with caption	Business	C	I/C	–	I	C
12	Two photos with caption	Business	I	I	–	I	C
13	Photo with caption	Business	–	I	C	C	C

5 Conclusion

Many corporate entities have increased their presence on social media platforms in recent years, impacting the social media user experience. This study extension illustrates how the use of neuroIS tools, such as EEG, may provide further insights into the complex thinking of consumers as they engage with social media content. This research continues from a larger study exploring how consumers process content based on the author of social media content. Future analysis will overlay insights from eye tracking data with qualitative research results.

References

1. Sweeney, E.K.: US ad spend reached $151B in 2018, a 4.1% jump, 24 January 2019. https://www.marketingdive.com/news/kantar-us-ad-spend-reached-151b-in-2018-a-41-jump/546725/
2. Stelzner, M.A.: How marketers are using social media to grow their businesses. Social Media Marketing Industry Report (2014)
3. Goh, K.-Y., Heng, C.-S., Lin, Z.: Social media brand community and consumer behavior: quantifying the relative impact of user-and marketer-generated content. Inf. Syst. Res. **24**(1), 88–107 (2013)

4. Stubb, C., Nyström, A.-G., Colliander, J.: Influencer marketing: the impact of disclosing sponsorship compensation justification on sponsored content effectiveness. J. Commun. Manage. **23**(2), 109–122 (2019)
5. Marshall, G.W., et al.: Revolution in sales: the impact of social media and related technology on the selling environment. J. Pers. Sell. Sales Manag. **32**(3), 349–363 (2012)
6. Riedl, R., Fischer, T., Léger, P.-M.: A decade of neurois research: status quo, challenges, and future directions. In: Thirty Eighth International Conference on Information Systems, South Korea (2017)
7. Fischer, T., Davis, F.D., Riedl, R.: NeuroIS: a survey on the status of the field. In: Davis, F.D., Riedl, R., vom Brocke, J., Léger, P.-M., Randolph, A.B. (eds.) Information Systems and Neuroscience. LNISO, vol. 29, pp. 1–10. Springer, Cham (2019). https://doi.org/10.1007/978-3-030-01087-4_1
8. Tams, S., et al.: NeuroIS – alternative or complement to existing methods? Illustrating the holistic effects of neuroscience and self-reported data in the context of technostress research. J. Assoc. Inf. Syst. **15**(10), 723–752 (2014)
9. McFarland, D.J., Wolpaw, J.R.: Brain-computer interfaces for communication and control. Commun. ACM **54**(5), 60–66 (2011)
10. Randolph, A.B., Labonté-LeMoyne, É., Léger, P.-M., Courtemanche, F., Sénécal, S., Fredette, M.: Proposal for the use of a passive BCI to develop a neurophysiological inference model of IS constructs. In: Davis, F.D., Riedl, R., vom Brocke, J., Léger, P.-M., Randolph, A.B. (eds.) Information Systems and Neuroscience. LNISO, vol. 10, pp. 175–180. Springer, Cham (2015). https://doi.org/10.1007/978-3-319-18702-0_23
11. Zander, T.O., Kothe, C.: Towards passive brain-computer interfaces: applying brain-computer interface technology to human-machine systems in general. J. Neural Eng. **8**(2), 025005 (2011)
12. Kaplan, A.M., Haenlein, M.: Users of the world, unite! The challenges and opportunities of Social Media. Bus. Horiz. **53**(1), 59–68 (2010)
13. Labrecque, L.I.: Fostering consumer–brand relationships in social media environments: the role of parasocial interaction. J. Interact. Mark. **28**(2), 134–148 (2014)
14. Riley, J.: Sustaining customer engagement through social media brand communities. J. Glob. Scholars Market. Sci. **30**(4), 344–357 (2020)
15. Kumar, A., et al.: From social to sale: The effects of firm-generated content in social media on customer behavior. J. Mark. **80**(1), 7–25 (2016)
16. Nunan, D., et al.: Reflections on "social media: Influencing customer satisfaction in B2B sales" and a research agenda. Ind. Mark. Manage. **75**, 31–36 (2018)
17. Salo, J.: Social media research in the industrial marketing field: review of literature and future research directions. Ind. Mark. Manage. **66**, 115–129 (2017)
18. Harrigan, P., et al.: Modelling CRM in a social media age. Australas. Mark. J. (AMJ) **23**(1), 27–37 (2015)
19. Ansari, A., Mela, C.F.: E-customization. J. Mark. Res. **40**(2), 131–145 (2003)
20. Shinebourne, P.: Using Q method in qualitative research. Int .J. Qual. Methods **8**(1), 93–97 (2009)
21. Brown, S.R.: Q methodology and qualitative research. Qual. Health Res. **6**(4), 561–567 (1996)
22. Pascual-Marqui, R.D.: Standardized low-resolution brain electromagnetic tomography (sLORETA): technical details. Methods Find. Exp. Clin. Pharmacol. **24**(Suppl D), 5–12 (2002)
23. Davidson, R.J.: anterior cerebral asymmetry and the nature of emotion. Brain Cogn. **20**, 125–151 (1992)

Optimizing Scatterplot-Matrices
for Decision-Support:
An Experimental Eye-Tracking Study Assessing Situational Cognitive Load

Lisa Perkhofer$^{(\boxtimes)}$ and Peter Hofer

University of Applied Sciences Upper Austria, Wels, Austria
{lisa.perkhofer,peter.hofer}@fh-steyr.at

Abstract. The scatterplot matrix is defined to be a standard method for multivariate data visualization; nonetheless, their use for decision-support in a corporate environment is scarce. Amongst others, longstanding criticism lies in the lack of empirical testing to investigate optimal design specifications as well as areas of application from a business related perspective. Thus, on the basis of an innovative approach to assess a visualization's fitness for efficient and effective decision-making given a user's situational cognitive load, this study investigates the usability of a scatterplot matrix while performing typical tasks associated with multidimensional datasets (correlation and distribution assessment). A laboratory experiment recording eye-tracking data investigates the design of the matrix and its influence on the decision-maker's ability to process the presented information. Especially, the information content presented in the diagonal as well as the size of the matrix are tested and linked to the user's individual processing capabilities. Results show that the design of the scatterplot as well as the size of the matrix influenced the decision-making greatly.

Keywords: Information visualization · Big data visualization · Decision-support · Cognitive load · Eye-tracking

1 Introduction

Information Visualization is a growing field, highly important for data analysis and decision-support [1, 2]. Many visualization options are available for widespread use; however, especially visualizations designed for larger and more complicated datasets are hardly tested and thus, limiting their scope and their impact in managerial decision-making [3, 4]. With respect to multidimensional data sets including multiple attributes, proposed and highly cited options range from visualizations such as the parallel coordinates plot and the heatmap visualization, to options for dimensional stacking or the scatterplot matrix (SPLOM) [5–7]. While the benefits of each particular visualization type are discussed at large, only a handful of authors discuss the visualization's optimal design [4, 8, 9]. This is somewhat surprising, as in traditional visualization use (e.g. bar, column, and pie charts) it has been shown that the design of a visualization largely influences the decision-makers ability to extract information [10, 11]. Thus, before comparing

© The Author(s), under exclusive license to Springer Nature Switzerland AG 2021
F. D. Davis et al. (Eds.): NeuroIS 2021, LNISO 52, pp. 63–76, 2021.
https://doi.org/10.1007/978-3-030-88900-5_8

the identified options for optimal decision-support, for each identified visualization type the best design needs to be located first.

In this paper, a promising visualization option—the SPLOM—is presented and its optimal design, given typical tasks—correlation estimation and analysis of the variables distribution—is empirically assessed, indicating a first step into the analysis of multidimensional data visualization. To this end, a latent variable, which is claimed to explain and predict decision-making outcome is analyzed. This variable distinguishing the analysis from others, is the situational cognitive load and it's measurement is on the basis of multiple eye-tracking related measures in order to consider the users individual processing abilities [12–14].

2   The Scatterplot Matrix

A simple scatterplot depicts two variables by presenting a collection of points on two continuous, orthogonal dimensions and is designed to emphasize the spatial distribution of the data presented [15]. Thus, the visualization uses positions to encode the respective values of two variables and allows for patterns and correlations to become visible [16]. The degree of the correlation is described by the correlation coefficient. Unfortunately, if the dataset increases in complexity the simple scatterplot rapidly becomes ineffective, as only a handful of dimensions can be presented at a time [17]. Thus, as an alternative, multiple scatterplots ordered in a matrix can be applied [18–20]. To this end, the scatterplots are positioned horizontally and vertically in the same order, and duplicated relative to the diagonal [18]. In doing so, one can cancel out the negative aspect of low dimensionality of the scatterplot but at the same time, draw on the readers ability to extract information from the well-known and widely used visualization type [15, 21].

By scanning proposed options in the literature and on open access libraries as well as matrixes implemented in BI tools, the following design choices become apparent [9]:

(1)   The design of the scatterplot itself (e.g. color use, trend line, scale)
(2)   The content of the diagonal (scatterplot vs. histogram) and if used, the design of the histogram (simple histogram, histogram including normal distribution curve, multiple histograms based on data clusters).
(3)   The content below the diagonal (the same scatterplot but with changed axis vs. presenting the respective correlation coefficient) and
(4)   the size of the matrix (the upper limit—due to readability issues, but also due to a decision-maker's cognitive constraints—is suspected to lie at 10 attribute [22]).

3   Study Design

The different design questions presented above could not be answered with just one study. Thus, two pre-studies have been conducted with the aim of answering some of the design questions a-priori.

Pre-Studies: The first pre-study assessed the design of a single scatterplot and is presented in Appendix A. Results indicate, that decision-support is optimal if the scatterplot is designed in multi-colors to distinguish clusters within the data set, and that a trend line increases task accuracy (TA) when assessing the correlation coefficient. The second pre-study is targeted towards identifying the optimal content as well as design of the diagonal. Further the content and the design of the information presented below the diagonal in the bottom left corner is assessed (for details please see Appendix B). It could be concluded that additional insight–next to presenting scatterplots–is helpful: a histogram presented in the diagonal for distribution assessment and the correlation coefficient including a heatmap design to highlight high and low scores increases TA greatly. With respect to the histogram, no evidence could be found to one of the two tested designs. Thus, it is evaluated again in the study presented in this paper.

Main Study: In this study, the size as well as the design of the histogram is further investigated and not only assessed by looking at decision-making outcome but also by analyzing the mental load it possesses on working memory through evaluating situational cognitive load. In all, 16 students educated in managerial accounting participated in the experiment. Tasks were targeted towards correlation and distribution assessment, which are presented in Table 1:

Table 1. Task types used

Task 1	Please assess the degree of the correlation highlighted in the matrix
Task 2	How many unique positive correlations above 0.6 are presented in the matrix?
Task 3	How many unique negative correlations below -0.6 are presented in the matrix?
Task 4	Please look at the company highlighted. In which year could the highest price be reached?
Task 5	Please look at the company highlighted. Tick all options that are true for the distribution over the period of 4 years. (multiple answers correct; answer options: skewed to the left, skewed to the right, two peaked, flat, peaked, nor-mal)

Procedure: Participants were asked to perform several tasks presented on a stationary computer screen equipped with an eye tracking recording device (SMI RED 250). A 5-point calibration as well as a 4-point validation was used at the beginning and the middle of the experiment to ensure high quality. Further, constant artificial lightning conditions were ensured and a noise-free room was used for data collection. For analysis, the tracking ratio was controlled per stimulus and only used if equal to or above 95%. The data set was on the basis of real life stock prices, but they are controlled to ensure comparable levels of distribution (in terms of kurtosis, skewness, and standard deviation), as large differences influence the user's ability to perform the tasks [9]. For creating the scatterplots and as well as the matrix MS Excel was used.

The experiment started with a trial run including explanations on how to assess correlations and how to derive information on the distribution of a variable by reading scatterplots and histograms. After informing participants on the content of the study and performing the trial run, the actual experimental tasks were presented in randomized order. Lastly, a self-assessment on previous experience and question on the demography of the participants were presented. Data collection lasted approximately 30 min. Assessment was based on TA (task accuracy), TT (task time), and situational cognitive load (sCL). TA measures whether a task is answered correct or incorrect, TT is measured from the time of the stimulus onset until its offset, and sCL assesses the current cognitive burden the user is under during task performance (Fig. 1).

Design Question 1: How should the histogram in the diagonal be designed?

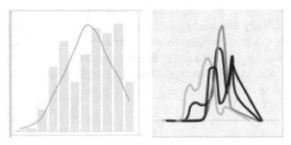

Fig. 1. Stimuli material (left – H1: histogram including normal distribution curve; right – H2: histogram on identified clusters within the dataset)

Design Question 2: When does the size of the matrix overburden the decision-maker? As. navigation in a scatterplot matrix is of high relevance for search efficacy, which is restricted to orthogonal movement along the same row or column given the nature of the visualization. Consequently, and to ensure high quality on eye-tracking data, static stimuli are used; however, interaction (highlighting) is simulated (Fig. 2 – left: highlight one scatterplot for analysis of its correlation; right: highlight of the row and column for the variable's distribution analysis) [21, 23].

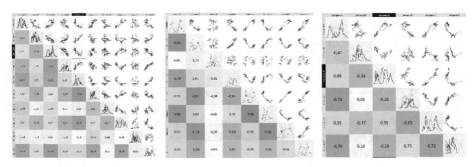

Fig. 2. Stimuli material design (left: size 10 × 10, middle: size 8 × 8, and right: size: 6 × 6)

4 Results

The assessment of the latent variable sCL is presented in Appendix C and data collection and analysis is analogous to Perkhofer and Lehner [13]. The variable is included in the assessment as sCL is supposed to be a reliable measure to explain and predict decision-making outcome, especially if differences on cognitive burden might occur [12]. As no clear evidence was visible for the type of histogram used, and the size of a matrix is always discussed in connection with overburdening the user, it seems to be necessary to check for CL as well as for its effect on decision-making outcome.

Results Design Question 1: The design of the histogram has no significant effect on the three variables of investigation (see Table 2); however, when analyzed in detail the two questions targeted towards distribution analysis (task 4 & 5) show opposing results with respect to TA. While for information acquisition on data clusters the second design of the histogram (H2) is better suited, for an overall assessment of the variable's distribution, the first design of the histogram (H1) is best. With respect to TT H2 is superior. Participants were less familiar with distribution analysis, thus the overall results on these two questions (task 4 & 5) in terms of accuracy were low, while the effect on task time was only visible in task 5 (see Table 3).

Table 2. Descriptive analysis design question 1

	Histogram	Task 1	2	3	4	5	
TA	1	0,96	0,93	0,93	0,64	0,72	**0,84**
	2	0,93	0,93	0,97	0,76	0,57	**0,83**
		0,95	**0,93**	**0,95**	**0,70**	**0,65**	
TT	1	8,49	5,77	7,14	9,97	15,27	**9,33**
	2	7,37	6,19	7,56	5,87	18,07	**9,01**
		7,93	**5,98**	**7,35**	**7,92**	**16,67**	
sCL	1	0,27	-0,53	-0,47	0,07	0,83	**0,03**
	2	-0,04	-0,46	-0,17	-0,48	0,85	**-0,06**
		0,12	**-0,50**	**-0,32**	**-0,20**	**0,84**	

Table 3. Overview repeated measures ANOVA on histogram design

	Hisotgram design	Task		Interaction	
TA	n.s.	-	0,000	0,391	0,026	0,192
TT	n.s.	-	0,000	0,695	0,029	0,178
sCL	n.s.	-	0,000	0,568	n.s	-

Results Design Question 2: With respect to TA, the higher the size the lower the outcome. However, this effect occurred only for the tasks targeted towards correlation assessment (tasks 1–3) while no differences appeared for distribution assessment (tasks 4–5). Interestingly, task 4 as well as the two tasks targeted towards counting correlations (task 2&3) are fastest, while identifying positive correlations is easier than identifying negative ones. With respect to the cognitive burden placed on the participants only the independent variable "task" is significant (task 5 is hardest). Nonetheless, the significant interaction effect reveals that task 5 is getting easier as the size of the matrix increases (as more possibilities to analyze the variables distribution become available). In contrast, tasks 1 and 2 are getting harder. For those tasks, an increase in size increased the cognitive burden, as it led to more squares needing to be scanned for analysis (Tables 4 and 5).

Table 4. Descriptive analysis design question 2

		Task					
	Size	1	2	3	4	5	
TA	6x6	1,00	1,00	1,00	0,50	0,60	**0,82**
	8x8	0,87	0,87	0,93	1,00	0,55	**0,84**
	10x10	0,87	0,88	0,63	0,50	0,86	**0,75**
		0,91	**0,91**	**0,85**	**0,67**	**0,67**	
TT	6x6	7,20	4,19	5,81	4,76	19,20	**8,23**
	8x8	7,54	8,07	9,42	6,90	17,03	**9,79**
	10x10	15,08	9,04	10,86	7,28	12,03	**10,86**
		9,94	**7,10**	**8,70**	**6,31**	**16,08**	
sCL	6x6	-0,14	-0,67	-0,33	-0,75	1,08	**-0,16**
	8x8	0,05	-0,27	0,00	-0,23	0,63	**0,04**
	10x10	0,97	0,06	0,16	-0,51	0,06	**0,15**
		0,30	**-0,30**	**-0,06**	**-0,50**	**0,59**	

Table 5. Overview repeated measures ANOVA on matrix size

	Size		Task		Interaction	
TA	0,027	0,401	0,016	0,431	0,005	0,422
TT	n.s	-	0,002	0,644	n.s	-
sCL	n.s	-	0,004	0,508	0,048	0,279

5 Discussion and Conclusion

The presented study investigates the optimal design of a scatterplot matrix, a visualization type designed to present multidimensional data sets. The inclusion of sCL, an identified variable to explain and predict decision-making outcome, helped in understanding in more detail why an increased size in some cases increases decision-making and in some deteriorates them. Further, it was again demonstrated how a task, especially if not frequently performed as is the case with distribution analysis, influences TA and TT.

Most importantly, and being the aim of this study, it could be shown that a good design can mitigate the effect of high mental demand of a task. Thus, based on the findings of the main study as well as the pre-studies, a decision-maker can be provided with a much better visualization for performing typical tasks in a managerial context. Deduced design recommendations are.

(1) to use a multi-color scheme and a regression line highlighting the type and strength of the correlation
(2) to use a histogram in the diagonal and choose the type of histogram depending on the relevant task
(3) to include correlation coefficients as well as a heatmap additional to the scatterplot itself, which are presented in the bottom left corner and
(4) to reduce the size of the scatterplot to only necessary dimensions, especially if used for correlation assessment to reduce the cognitive burden on the decision-maker.

Appendix

Appendix A: Pre-study 1 – Scatterplot Design

Study Design: Participants had to first assess the type of correlation (positive, negative of no correlation) and then the correlation's magnitude on a scale from 0 to 100 for positive correlations and a scale from -100 to 0 for negative correlations. Participants were recruited on Amazon Mechanical Turk (50 per design). For each design question, a different experiment was created keeping the experiment short and simple, as recommended for online experiments. Each experiment included scatterplots showing two strong positive ($r > 0,8$ and $r > 0,6$) as well as two strong negative correlations ($r < -0,6$, $r > -0,8$), two weak positive ($r > 0,2$) and two weak negative correlations ($r < -0,2$), as well as four scatterplots showing no correlation ($-0,2 < r > 0,2$).

Procedure: Participants were asked to perform an online experiment using Lime Survey. The experiment started by explaining how to assess correlations by reading scatterplots. To check for the participant's attention, they had to answer six questions on the presented content. Experimental tasks were presented in randomized order. Assessment was based on TA (task accuracy) which is measured by using the deviation of the participant's estimate to the true score of the presented correlation. Lastly, a self-assessment on previous experience and question on the demography of the participants were presented.

Design Question 1 and Stimuli Material: Should a regression-line be used to indicate the direction and the strength of the correlation coefficient? (Fig. 3).

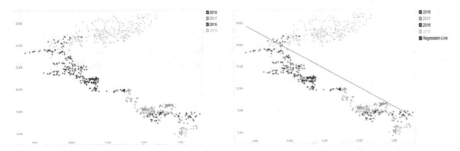

Fig. 3. Stimuli material design question 1 (left: default; right: trendline)

Design Question 2 and Stimuli Material: Should a monochrome or a multi-color scheme be used to differentiate identified data clusters? (Fig. 4).

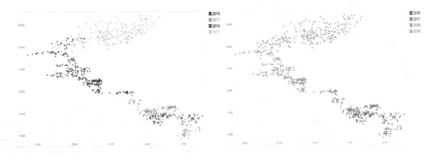

Fig. 4. Stimuli material design question 2 (left: default, right color scheme mono-color)

Results: Table 6 demonstrates, that assessing positive correlations is easier than assessing negative ones. Further, it is clearly indicated that a regression-line stressing the type (positive or negative) and the magnitude of the correlation increases its assessment.

Table 6. Overview results pre-study 1

		Ø	significance	eta²
Design 1:	Default	21,69		
Design 2:	Trend	15,64		
Design 3:	MonoColor	21,52		
Magnitude 1: <-0,8		20,42		
Magnitude 1: <-0,6		29,54		
Magnitude 1: <-0,2		21,42		
Magnitude 1: >-0,2<0,2		12,58		
Magnitude 1: >0,2:		20,65		
Magnitude 1: >0,6:		19,39		
Magnitude 1: >0,8:		14,43		
Design:			0,000	0,019
Magnitude:			0,000	0,070

With respect to the color scheme used, no clear evidence in terms of TA could be found, nonetheless, participants preferred the multi-color scheme.

The results of this study are applied to design the scatterplots used in the scatterplot matrix in pre-study 2 as well as the main experiment presented in this paper.

Appendix B: Pre-study 2 – Scatterplot Matrix Design

Study Design: The same five questions and the same data set were used in the pre-study as in the main experiment. Participants (N = 20), recruited at the University of Applied Sciences Upper Austria, educated in business administration were recruited but for this study a different group was used compared to the main experiment. Stimuli material was similar, but targeted towards answering two particular design question: (1) the design of the diagonal (a scatterplot vs. using a histogram) and (2) the design of the bottom left corner (scatterplots vs. the corresponding correlation coefficient).

Procedure: Participants were asked to perform an online experiment, which was sent out with a link and designed using Lime Survey. The experiment started by explaining how to assess correlations and how to derive information on the distribution of a variable by reading scatterplots and histograms. To check for the participant's attention, they had to answer questions on the presented content (six concerning a single scatterplot and how to assess correlations; four concerning the distribution of a variable; and four concerning the scatterplot matrix). After informing participants on the content of the study, experimental tasks were presented in randomized order. Assessment is based on TA (task accuracy) and TT (task time). Lastly, a self-assessment on previous experience and question on the demography of the participants were presented.

As experience is of utmost importance when analyzing information in visualized form and self-assessment has proven to be inefficient in the past [24], the experiment was repeated one week after the first trial to check for improvement through learning and repetition. One student did not participate in the second trial and one student did not pass the validity questions at the beginning of the experiment and thus, these two were excluded from analysis.

Design Question 1 and Stimuli Material: Should the area beneath the diagonal (bottom left) also include scatterplots or should the respective correlation coefficient be presented instead? In addition to the correlation coefficient, also a heatmap representing the strength of the correlation was included as it has proven to increase efficiency and effectiveness in tabular representations (Fig. 5).

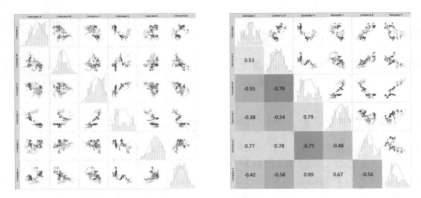

Fig. 5. Stimuli material design question 1

Results: Tasks targeted towards assessing one specific correlation as well as counting the amount of positive or negative correlations presented within the matrix (tasks 1–3) were significantly better supported by using the second design (correlation coefficient in the bottom left corner). TA was significantly higher when correlation coefficients were presented. Additionally, the effect size (eta^2) is stronger for the tasks targeted towards counting the amount of positive and negative correlations. Supporting previous research, showing correlation coefficients is more important for counting negative correlations, as those are harder to assess (see task 3).

Further, TA was higher in the second trial indicating participants did increase their effectiveness with repetition. As expected, no effects were visible for the tasks targeted towards assessing the variable's distribution (tasks 4–5). TT shows significance with respect to the repetition of the trial for all posed tasks, but no direct effect can be detected concerning the design of the area in the bottom left corner. However, when analyzing the significant interaction effects, design 2 increased the participant's efficiency in the second trial indicating better decision-support after getting used to its design (Table 7).

Table 7. Overview on repeated measures ANOVA on matrix design (bottom left)

		Bottom Left		Repetition/Time		Interaction effect	
Task1	TA	0,001	0,175	0,028	0,087	0,016	0,105
	TT	n.s.	-	0,000	0,485	n.s.	-
Task2	TA	0,000	0,247	0,000	0,261	0,006	0,135
	TT	n.s.	-	0.000	0,512	0,015	0,107
Task3	TA	0,000	0,322	0,042	0,076	0,013	0,111
	TT	n.s.	-	0,000	0,338	n.s.	-
Task4	TA	n.s.	-	n.s.	-	0,008	0,124
	TT	n.s.	-	0,000	0,234	n.s.	-
Task5	TA	n.s.	-	n.s.	-	n.s.	-
	TT	n.s.	-	0,0000	0,284	n.s.	-

Design Question 2 and Stimuli Material: The second design questions was targeted towards identifying the best information to be presented in the diagonal. Identified options form previous literature as well as published options on visualization platforms highlight three possibilities (Fig. 6):

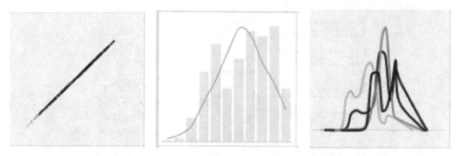

Fig. 6. Stimuli material design question 2 (left: scatterplot; middle: histogram including normal distribution curve; right: histogram on identified clusters within the dataset)

Results: For the tasks targeted towards assessing a variable's distribution, no direct effect on TA could be found. Only when examining interaction effects, differences on the improvement based on repetition can be found: performance using the first histogram design (hole dataset and normal distribution curve) improved, while performance using the second design decreased. Using a correlation coefficient shows the lowest TA. With respect to TT, the histogram representing cluster information significantly increased efficiency for task 4, which asked for an assessment of cluster information (Table 8).

Table 8. Overview on repeated measures ANOVA on design of the diagonal

		Diagonal		Repetition/Time		Interaction effect	
Task1	TA	n.s.	-	0,031	0,126	n.s.	-
	TT	n.s.	-	0,000	0,605	n.s.	-
Task2	TA	n.s.	-	0,001	0,259	n.s.	-
	TT	n.s.	-	0,000	0,531	n.s.	-
Task3	TA	n.s.	-	n.s	-	0,024	0,197
	TT	n.s.	-	0,000	0,370	n.s.	-
Task4	TA	n.s.	-	n.s	-	0,016	0,217
	TT	0,024	0,114	0,000	0,319	n.s.	-
Task5	TA	n.s.	-	n.s	-	n.s.	-
	TT	n.s.	-	0,0000	0,355	n.s.	-

Appendix C: Second-Order Formative Construct – Situational CL

For analysis of the second order formative construct we used SmartPLS. We applied the repeated measures approach to analyze the latent variable of situational cognitive load, while we consider balanced indicators for the first order constructs as suggested by Perkhofer and Lehner [13] (Fig. 7).

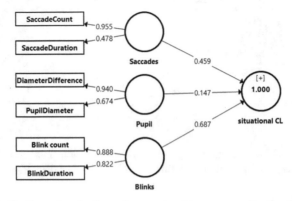

Fig. 7. Second-order formative construct to measure situational CL

Situational cognitive load is calculated during information processing, using an adaptive baseline to focus not only on the user and his/her capabilities but also on the presented stimulus. The variable is supposed to indicate a mismatch between the presented visualization and the needs of the decision maker, as it indicates increased cognitive resources by focusing on the duration of events [25]. Saccade and blink duration are suspected to be decelerated as the cognitive burden increases, and the size of pupil diameter is said to increase [13, 14, 26–28]. Further, by considering measures on blink and saccade count, also the need to search through bad design (not instantly locating the information) is accounted for [26, 29, 30]. Both, information on count measures (highly correlated with time) and information on the duration of events (correlated with the complexity of the material and task accuracy), explain situational cognitive load and thus increase understanding of good or bad design depending on the decision-maker and his/her capabilities.

References

1. Bačić, D., Fadlalla, A.: Business information visualization intellectual contributions: an integrative framework of visualization capabilities and dimensions of visual intelligence. Decis. Support Syst. **89**, 77–86 (2016)
2. Appelbaum, D., Kogan, A., Vasarhelyi, M., et al.: Impact of business analytics and enterprise systems on managerial accounting. Int. J. Account. Inf. Syst. **25**, 29–44 (2017)
3. Perkhofer, L.M., Hofer, P., Walchshofer, C., et al.: Interactive visualization of big data in the field of accounting. J. Appl. Account. Res. **5**, 78 (2019)

4. Perkhofer, L., Walchshofer, C., Hofer, P.: Does design matter when visualizing big data? An empirical study to investigate the effect of visualization type and interaction use. J. Manag. Control. **31**(1–2), 55–95 (2020). https://doi.org/10.1007/s00187-020-00294-0

5. Bertini, E., Tatu, A., Keim, D.A.: Quality metrics in high-dimensional data visualization: an overview and systematization. IEEE Trans. Vis Comput. Graph. **17**, 2203–2212 (2011)

6. Keim, D.A.: Information visualization and visual data mining. IEEE Trans. Vis. Comput. Graph. **8**, 1–8 (2002)

7. Inselberg, A.: The plane with parallel coordinates. Vis. Comput. **1**, 69–91 (1985). https://doi.org/10.1007/BF01898350

8. Isenberg, T., Isenberg, P., Chen, J., et al.: A systematic review on the practice of evaluating visualization. IEEE Trans. Vis. Comput. Graph. **19**, 2818–2827 (2013)

9. Perkhofer, L., Walchshofer, C., Hofer, P.: Designing visualizations to identify and assess correlations and trends: an experimental study based on price developments. In: Lehner, O. (ed.) Proceedings of the 17th Conference on Finance, Risk and Accounting Perspectives. ACRN Oxford, pp. 294–340 (2019)

10. Ware, C.: Information Visualization: Perception for Design , 3rd edn. Elsevier Ltd., Oxford (2012)

11. Tufte, E.R.: The Visual Display of Quantitative Information, 1st edn. Graphics Press, Connecticut (1983)

12. Perkhofer, L.: A cognitive load-theoretic framework for information visualization. In: Lehner, O. (ed.) Proceedings of the 17th Conference on Finance, Risk and Accounting Perspectives, in Print. ACRN Oxford, pp. 9–25 (2019)

13. Perkhofer, L., Lehner, O.: Using gaze behavior to measure cognitive load. In: Davis, F.D., Riedl, R., vom Brocke, J., Léger, P.-M., Randolph, A.B. (eds.) Information Systems and Neuroscience. LNISO, vol. 29, pp. 73–83. Springer, Cham (2019). https://doi.org/10.1007/978-3-030-01087-4_9

14. Brücken, R., Plass, J.L., Detlev, L.: Direct measurement of cognitive load in multimedia learning. Educ. Psychol. **38**, 53–61 (2003)

15. Sarikaya, A., Gleicher, M.: Scatterplots: tasks, data, and designs. IEEE Trans. Vis. Comput. Graph. **24**, 402–412 (2018)

16. Li, J., Martens, J.-B., van Wijk, J.J.: Judging correlation from scatterplots and parallel coordinate plots. Inf. Visual. **1**, 69 (2008)

17. Li, J., Martens, J.-B., van Wijk, J.J.: A model of symbol size discrimination in scatterplots. In: Mynatt, E.D., Hudson, S.E., Fitzpatrick, G. (eds.) CHI Conference, pp. 2553–2562. Association for Computing Machinery, New York (2010)

18. Kanjanabose, R., Abdul-Rahman, A., Chen, M.: A multi-task comparative study on scatter plots and parallel coordinates plots. Comput. Graph. Forum **34**, 261–270 (2015)

19. Netzel, R., Vuong, J., Engelke, U., et al.: Comparative eye-tracking evaluation of scatterplots and parallel coordinates. Vis. Inform. **1**, 118–131 (2017)

20. Nguyen, Q.V., Simoff, S., Qian, Y., et al.: Deep exploration of multidimensional data with linkable scatterplots. In: Zhang, K., Kerren, A. (eds.) Proceedings of the 9th International Symposium on Visual Information Communication and Interaction, pp. 43–50. ACM, New York (2016)

21. Elmqvist, N., Dragicevic, P., Fekete, J.-D.: Rolling the dice: multidimensional visual exploration using scatterplot matrix navigation. IEEE Trans. Vis. Comput. Graph. **14**, 1141–1148 (2008)

22. Urribarri, D.K., Castro, S.M.: Prediction of data visibility in two-dimensional scatterplots. Inf. Vis. **16**, 113–125 (2016)

23. Elmqvist, N., Moere, V.A., Jetter, H.-C., et al.: Fluid interaction for information visualization. Inf. Vis. **10**, 327–340 (2011)

24. Falschlunger, L., Lehner, O., Treiblmaier, H., et al.: Visual representation of information as an antecedent of perceptive efficiency: the effect of experience. In: Proceedings of the 49th Hawaii International Conference on System Sciences (HICSS), pp. 668–676. IEEE (2016)

25. Zagermann, J., Pfeil, U., Reiterer, H.: Measuring cognitive load using eye tracking technology in visual computing. In: Sedlmair, M., Isenberg, P., Isenberg, T., et al. (eds.) Proceedings of the Beyond Time and Errors on Novel Evaluation Methods for Visualization - BELIV 2016, pp. 78–85. ACM Press, New York (2016)

26. Toker, D., Conati, C., Steichen, B., et al.: Individual user characteristics and information visualization: connecting the dots through eye tracking. In: Proceedings of the 31st CHI - Changing Perspectives, pp. 295–304. ACM, New York (2013)

27. Granholm, E., Asarnow, R.F., Sarkin, A.J., et al.: Pupillary responses index cognitive resources limitations. Psychophysiology **33**, 457–461 (1996)

28. Hossain, G., Yeasin, M.: Understanding effects of cognitive load from pupillary responses using Hilbert analytic phase. In: IEEE Conference on Computer Vision and Pattern Recognition Workshops, pp. 381–386. IEEE (2014)

29. Renshaw, J.A., Finlay, J.E., Tyfa, D., et al.: Designing for visual influence: an eye tracking study of the usability of graphical management information. In: Rauterberg, M., Menozzi, M., Wesson. J. (eds.) Proceedings of the INTERACT 2003, pp. 144–151. IOS Press (2003)

30. Steichen, B., Carenini, G., Conati, C.: User-adaptive information visualization: Using eye gaze data to infer visualization tasks and user cognitive abilities. In: Proceedings of the IUI 2013, pp. 317–328. ACM, New York (2013)

"Overloading" Cognitive (Work)Load: What Are We Really Measuring?

Jacek Gwizdka[(✉)]

School of Information, University of Texas at Austin, Austin, USA
neurois2021@gwizdka.com

Abstract. Cognitive load is one of the most studied constructs in NeuroIS [1]. Not surprisingly, we have identified 27 papers presented at NeuroIS retreats between 2012 and 2020 which included measurement of cognitive load or related constructs. This paper reviews terminology used to refer to cognitive load, mental workload and its variations, as well as their operationalizations and measurements. All 27 papers employed physiological NeuroIS measures, while six of them additionally used subjective self-ratings. The wide range of measurements prompts us to question if we are measuring the same construct. We provide an overview and a summary of cognitive load terminology and measurement used in these 27 papers and conclude with recommendations for future research.

Keywords: Cognitive load · Mental workload · Workload · Attentional load · Executive function load · Working memory · Information overload · Measurement

1 Introduction

Cognitive load is one of the most studied constructs in NeuroIS [1]. As the field develops and matures it is useful to review our conceptualizations, operationalizations and measurements of this important construct. This paper starts by reviewing papers published in NeuroIS Retreat between the years 2012 and 2020. The specific years are defined by the online availability of papers (and, in the earlier years, abstracts). We provide an overview of cognitive load terminology and measurements used in these 27 papers and conclude with recommendations for future research.

2 Background

Historically, cognitive load (CL), mental workload (MW) and memory load (mL) originated mostly independently from different fields, CL from Educational Psychology, MW from Human Factors and Ergonomics [2], and mL from cognitive psychology (e.g., [3]). The constructs are explicated by three theories prominent in their respective fields, CL by Cognitive Load Theory (CLT) [4], MW by Multiple Resource Theory (MRT) [5], mL by Baddley's theory of working memory (Wm) [6]. The constructs share core theoretical assumptions [7] and are similar in their assumption of limited mental capacity and

© The Author(s), under exclusive license to Springer Nature Switzerland AG 2021
F. D. Davis et al. (Eds.): NeuroIS 2021, LNISO 52, pp. 77–89, 2021.
https://doi.org/10.1007/978-3-030-88900-5_9

competing task and environmental demands. The originating fields and many cognate disciplines continue to be interested in investigating this construct (and related). This continued trend is exemplified in a recently (2017) established series of International Symposia on Human Mental Workload (www.hworkload.org) [8]. In spite of the many decades long interest, the constructs and their measurement are still being disputed. For example, there is no common agreement on the three sub-constructs of CLT, the intrinsic, extraneous, and germane load and the new approaches to their measurement continued to be proposed [9]. As our review shows, the diversity of CL/MW/mL conceptualizations, operationalizations and measurements is also present in NeuroIS work which employed this construct (or shall we say these constructs?). In the text of this paper we will continue referring to it by Cognitive Load (CL).

Cognitive load measurement methods can be divided into 1) physiological (objective), 2) subjective, and 3) task performance measures. The physiological measures include signals from the central nervous systems (such as, brainwaves - EEG, oculography, pupillometry), and physiological signals from peripheral nervous system (such as heart rate variability - HRV, and electrodermal activity - EDA). The subjective measures rely on self-ratings scales (such as, NASA TLX, [40, 41]). The task performance measures include task completion time, number of errors, reaction times to secondary task. Another aspect of CL measurement is whether we measure instantaneous, average, accumulated, or peak load [41]. We review and classify NeuroIS Retreat papers using these types of measurements.

3 Method

We reviewed papers published at NeuroIS Retreat between 2012 and 2020. The papers were identified by performing search for "*load" (i.e. "load" with any prefix) on full-text of proceedings from 2012–2014 and on paper titles and abstracts from 2015–2020. The usage of "*load" was then manually checked. Only one paper was eliminated in this process, because it investigated a more general constructs of work overload and stress. We also searched for "mental effort" and did not identify any additional papers. The earliest identified work, was published as an abstract in NeuroIS Retreat 2012. To include measurements used in this research, we have identified a full conference paper where they were published [10]. In all other cases, we include information reported in the papers. We extracted from the papers all terms used to refer to some aspects of cognitive load, its descriptions, if any, as well as operationalizations and measures. In addition, whenever available we followed citations to the methods and measurement papers and reviewed them to include original terminology and further details of measurement described in these papers. Table 1 and Table 2 show a summary. Table 3 lists descriptions or definitions of cognitive load, if they were explicitly provided in the reviewed papers. Due to space limitations, we do not include all data extracted from the 27 papers.

Table 1. Summary of reviewed papers – part 1.

#	Year	Cit.	Paper type & N	Terms used	Subj. m.	NeuroIS measurements	Constructs in cited papers
1	2012	[11]	N=17, N=24. abs.	CL	Cam.	EEG: B-Alert workload index. Absolute/relative power spectra in 1-40Hz bands.	IO
2	2014	[12]	N=12	MW		Pupillary hippus (unrest)	MA
3	2015	[13]	N=13	MW		Absolute mean pupil diameter	MW
4	2015	[14]	WiP, pilot, N=2	CL;CW		Z-scores of pupil dilation. Mean of Z-scores per task.	CL (listening)
5	2015	[15]	WiP	CL	NASA	pupillometry and heart rate variability	CW; WmL;ME
6	2015	[16]	WiP	CL		pupillometry, EEG	CL
7	2015	[17]	WiP, pilot N=3	CL		Absolute pupil diameter	
8	2016	[18]	WiP	CL;ML		EEG "activations"; not specified further	Creation & retrieval of memories
9	2016	[19]	N=45 see 2015 WiP	MW	NASA TLX	Absolute mean pupil diameter	M engagement in automated task
10	2016	[20]	N=156	CW		EEG: spectral power ratio - engagement index	
11	2016	[21]	N=12; pilot	IO		EEG: Mindwave Neurosky headband + Neuro Experiment software output	CL
12	2016	[22]	N=10	MW; CL;CW	Cam.	EEG: Instantaneous, average, accumulated and peak load calculated from eXecutive Load Index (XLI) = ((delta + theta)/alpha) power ratio 2s compared with previous 20s.	Pupil response - Cognitive effects; Eye liveness
13	2017	[23]	N=12	MW		Stationary of absolute pupil size	MW
14	2017	[24]	N=22	CL;CW		Pupil diameter and square of pupil diameter. Most likely calculated from mean absolute pupil diameter per task (question)	
15	2017	[25]	WiP	CL; CO		no further detail	
16	2017	[26]	N=23	CL		Absolute mean pupil diameter	MA; CL
17	2017	[27]	WiP	CL		Pupial dilation to obtain continuous, average, and accumulated load. not specified in further detail	CL

Year-NeuroIS year (not pub. year); Cit.-citation to a paper in the list of references; N-number of participants; WiP-Work-in-Progress (otherwise completed research); Cam.-Cameron 6-item Likert scale; NASA-NASA TLX scale; CL-Cognitive Load; CW-Cognitive Workload; CO-Cognitive Overload; M-Mental; MW-Mental Workload; ME-Mental Effort; MA-Mental Activity m-memory; mL-memory Load; WmL-Working memory Load; IO-

Table 2. Summary of reviewed papers – part 2.

#	Year	Cit.	Paper type & N	Terms used	Subj. m.	NeuroIS measurements	Constructs in cited papers
18	2018	[28]	WiP; N=12	AL;AO; CL;ME;IL		Pupil dilation. EEG: event-related desynchronization in alpha band. not specified in further detail	WmL
19	2018	[29]	N=118; N=60;	CL; IO		Battery of eye-tracking derived measures: Fixation duration and count; Saccade duration and count; Pupil dilation - from baseline; pupil diameter diff; Eye blinks: count and duration. Establishes some reliability of these measures.	CL
20	2019	[30]	WiP	CL		pupil dilation	CL
21	2019	[31]	WiP (planned N=65)	CL;EFL		Three types of CL = three types of EFL 1. CL on inhibition: alpha event-related synchronization; 2. CL on updating: theta power; 3. CL on shifting: amplitude of posterior switch positivity (parietal electrodes)	WmL; m inhibition & updating; Attention shifting
22	2020	[32]	WiP	CL	NASA	Pupil dilation; EEG: theta, alpha and beta powers; No specific detail	M activity
23	2020	[33]	N=17	CL;CPL		Pupil: mean pupil diameter relative to mean pupil overall; EEG: relative low theta band power (10s post/10s pre)	Cognitive load
24	2020	[34]	N=10	CL		EEG: spectral power ratio (Beta/(Alpha + Theta)) - engagement index	M engagement in automated task
25	2020	[35]	N=10	CL		EEG: accumulated, average, peak load calculated from theta/beta power ratio; Neural signal complexity: fractal dim, multi-scale entropy, detrended fluctuation analysis.	Mind-wandering; CL during continuous cognitive task performance
26	2020	[36]	N=20	IO; CL; W	NASA	Electrodermal activity (EDA); Task completion time	CL
27	2020	[37]	N=60, prelim. results N=6	CL		EEG: spectral power ratio (Beta/(Alpha + Theta)) - engagement index	M. engagement in automated task

Year-NeuroIS year (not pub. year); Cit.-citation to a paper in the list of references; N-number of participants; WiP-Work-in-Progress (otherwise completed research); Cam.-Cameron 6-item Likert scale; NASA-NASA TLX scale; AL-Attentional Load; AO-Attentional Overload; CL-Cognitive Load; CW-Cognitive Workload; CO-Cognitive Overload; M-Mental; MW-Mental Workload; ME-Mental Effort; MA-Mental Activity m-memory; mL-memory Load; WmL-Working memory Load; IL-Information Load; IO-Information Overload; EFL-Executive Function Load.
Note, we provide citations to all cited papers on cognitive load measurement at the end of this paper's reference list.

Table 3. Cognitive Load descriptions provided in reviewed papers.

#	Year	Cit.	Terms used	Description or definition of Cognitive Load
5	2015	[15]	CL	"CL characterizes the demands of tasks imposed on the limited information processing capacity of the brain in the same way that physical workload characterizes the energy demands upon muscles. CL therefore represents an individual measure considering the individual amount of available resources and task-specific factors imposing CL. As independent construct, CL predicts performance for task execution, since high CL leads to poor task-performance and to wrong decisions"
7	2015	[17]	CL	"Cognitive load characterizes the demands of tasks imposed on the limited information processing capacity of the brain and constitutes an individual measure considering the individual amount of available resources"
8	2016	[18]	CL;ML	CLT. "Limited working memory with partly independent processing units for visual/spatial and auditory/verbal information, which interacts with a comparatively unlimited long-term memory"
12	2016	[22]	MW; CL;CW	"Mental workload can be defined as "the set of mental resources that people use to encode, activate, store, and manipulate information while they perform a cognitive task"
18	2018	[28]	AL;AO; CL;ME;IL	CLT "Cognitive load is the mental effort exerted by an individual to solve a problem or accomplish a task, during which information is retrieved from long term memory and temporarily stored in working memory for processing"
19	2018	[29]	CL; IO	CLT. Def: "the amount of working memory resources required in cognitive task execution"
20	2019	[30]	CL	"theoretical foundation grounded in the feature integration theory, dual processing, cognitive fit, cognitive load theory and works by Tversky and Kahneman"
21	2019	[31]	CL;EFL	Cognitive load as a mediator: between interruption characteristics and performance. Executive Function load. Cognitive load on inhibition; on updating; on shifting executive function (EF).
25	2020	[35]	CL	Cognitive load refers to the amount of working memory resources required to perform a particular task

Paper numbers (#) are from Table 1 and Table2.
Year-NeuroIS year (not pub. year); Cit.-citation to a paper in the list of references;
CL-Cognitive Load; CW-Cognitive Workload; CO-Cognitive Overload; M-Mental; MW-Mental Workload; ME-Mental Effort; MA-Mental Activity m-memory; mL-memory Load; WmL-Working memory Load; IO-Information Overload;

4   Observations

We observe a wide range of terminology used across papers. Eleven papers used CL, while four papers used MW. The remaining twelve papers used different terms interchangeably, presumably to refer to the same construct. For example, paper #12 [22] referred to MW, CL and CW. #18 [28] referred attentional load (AL), attentional overload (AO), CL, mental effort (ME), as well as to information load (IL). #19 [29] referred to CL and information overload (IO), while #21 [31] referred to CL and executive function load (EFL). We also observe differences in terminology between NeuroIS Retreat papers and cited by them papers which were used to inform measures and variables. For example, #9 [19] cites paper on a measure (pupil dilation) which reflects creation and retrieval of memories, but uses it to assess MW. Paper #25 [35] cites measures of mind-wandering and, separately, CL on continuous task performance and uses them to assess CL.

In Table 3 we provide a list of descriptions or definitions of CL, whenever they were provided in reviewed papers (#5, #7, #8, #12, #18, #19, #20, #21, #25). In addition, we observe that in six more papers (#14, #15, #16, #22, #26) Cognitive Load Theory provided an implicit definition of CL. A few papers (#8, #18, #25) refer in their definitions explicitly to working memory resources or to its limited capacity. Two describe mental demands imposed by a task (#5, #7) (which would roughly correspond to intrinsic load in CLT), while one (#18) refer to "mental effort exerted by an individual". We think that mental effort, while strongly related to CL, is a different construct and should not be equated with CL (e.g., see [38] for a recent review).

Six papers used a subjective measure using either Cameron self-reported 6-item Likert scale [39] and NASA TLX [40, 41]. All 27 papers used one or more NeuroIS physiological measures. In that nine papers used only EEG, ten used only pupil dilation, four pupil and EEG, one pupil and HRV, one a battery of eye-tracking measures (including pupil), one electrodermal activity & task completion time, and one did not specify any. As the NeuroIS field progresses, we observe measurements of more nuanced aspects of cognitive load. For example, papers #12, #17, #25 [22, 27, 35] introduce instantaneous, average, accumulated, and peak load [42]. A few earlier projects used absolute values of pupil dilation. Since this measure suffers from many drawbacks (sensitivity to external lightening and foreshortening errors), other projects used somewhat more advanced measures such as relative pupil dilation (or normalized Z-scores) or stationarity of pupil dilation signal. A few EEG spectral power ratios were used. One ((delta+ theta)/alpha) measures executive load (XLI [43]), and thus is closely related to CL, while another (beta/(alpha+ teta)) measures engagement [44] and thus shouldn't be simply equated with CL.

5   Comments and Recommendations

The wide variety of terms and measures used is somewhat concerning. Certainly, researchers may be interested in different constructs and sub-constructs related to CL, but they should be very explicit about what is being measured and avoid referring to general constructs like CL, while a more narrowly defined aspect is being measured. CL measures derived from the central (EEG, oculography) vs. peripheral nervous system (HRV, EDA) potentially tap into different aspects of CL.

Based on our review we offer the following recommendations:

- Terminology

  - Explicitly define constructs and how their measurement is operationalized.
  - Use terminology carefully and avoid referring to the same construct by different terms.

- Measurement

  - There is a wide variety of measurements used therefore it is difficult to offer a short list of specific recommendations. What follows are general guidelines.
  - Be explicit in defining a segment of time (or a unit of interaction) over which measures are aggregated (e.g., a task, a screen).
  - For *eye-tracking measures* - follow systematic approaches, such as [46, 47]. One of the reviewed papers [29] presented at NeuroIS'2018 provided initial demonstration of reliability of a battery of eye-tracking measures.
  - For *pupil-derived measurement* – use relative measures within subjects normalized to some baseline; incorporate lightning conditions or use measures not affected by lightning conditions; be aware of foreshortening errors. One possibility is to consider the Index of Pupillary Activity [48], a measure based on detecting frequency of periodic fluctuations of pupil diameter. It's insensitive to the lightning conditions and foreshortening errors, and offers a freely available algorithm.
  - For *EEG spectral power ratios* – there is a variation of employed measures. It is certainly warranted by measurement of different sub-constructs related to CL. However, let's consider other well established measures in assessing CL, such as alpha/theta ERD/ERS [49].

- Consider investigating reliability, validity and sensitivity of measures which we are used to assess CL. Recent examples of such work in NeuroIS Retreat include [29] and in other communities include for example [50, 51].

CL has been long of interest to many fields concerned with human performance. It is a complex construct and we still face difficulties in defining it, in understanding which factors influence it, and in measuring it. Such difficulties are not limited to NeuroIS, other cognate communities face them as well [9, 38, 45].

6 Limitations and Future Work

One limitation of our review is equal treatment of work-in-progress (WiP) and completed-research (CR) papers. The WiP papers report on research plans or pilot results. Consequently, the concepts and methods presented in them may not be fully developed yet, thus applying to them the same level of scrutiny as to the papers reporting on CR may be unjust. Furthermore, a few authors publish in NeuroIS Retreat sequential updates on their research projects. We have not attempted to group such papers, thus our summary may include "double counting".

This preliminary review was by design limited to NeuroIS Retreat, we plan to develop it in a more comprehensive (possibly systematic) review which includes publications from other conferences and journals. CL is just one of many constructs investigated in NeuroIS [52], we should consider other topics and constructs.

References

1. Fischer, T., Davis, F.D., Riedl, R.: NeuroIS: a survey on the status of the field. In: Davis, F.D., Riedl, R., vom Brocke, J., Léger, P.-M., Randolph, A.B. (eds.) Information Systems and Neuroscience. LNISO, vol. 29, pp. 1–10. Springer, Cham (2019). https://doi.org/10.1007/978-3-030-01087-4_1
2. Moray, N.: Models and measures of mental workload. In: Moray, N. (ed.) Mental Workload: Its Theory and Measurement, pp. 13–21. Springer US, Boston (1979). https://doi.org/10.1007/978-1-4757-0884-4_2
3. Baddeley, A.D.: Cognitive psychology and human memory. Trends Neurosci. **11**, 176–181 (1988). https://doi.org/10.1016/0166-2236(88)90145-2
4. Sweller, J.: Cognitive load theory, learning difficulty, and instructional design. Learn. Instr. **4**, 295–312 (1994)
5. Wickens, C.D.: Multiple resources and performance prediction. Theor. Issues Ergon. Sci. **3**, 159–177 (2002)
6. Baddeley, A.: Working memory: theories, models, and controversies. Annu. Rev. Psychol. **63**, 1–29 (2012). https://doi.org/10.1146/annurev-psych-120710-100422
7. Longo, L., Leva, M.C. (eds.): H-WORKLOAD 2017. CCIS, vol. 726. Springer, Cham (2017). https://doi.org/10.1007/978-3-319-61061-0
8. Longo, L., Leva, M.C. (eds.): H-WORKLOAD 2020. CCIS, vol. 1318. Springer, Cham (2020). https://doi.org/10.1007/978-3-030-62302-9
9. Orru, G., Longo, L.: The evolution of cognitive load theory and the measurement of its intrinsic, extraneous and germane loads: a review. In: Longo, L., Leva, M.C. (eds.) H-WORKLOAD 2018. CCIS, vol. 1012, pp. 23–48. Springer, Cham (2019). https://doi.org/10.1007/978-3-030-14273-5_3
10. de Guinea, A.O., Titah, R., Leger, P., Micheneau, T.: Neurophysiological correlates of information systems commonly used self-reported measures: a multitrait multimethod study. In: 2012 45th Hawaii International Conference on System Sciences, pp. 562–571 (2012). https://doi.org/10.1109/HICSS.2012.448
11. de Guinea, A.O., Titah, R., Leger, P.-M.: Neurophysiological correlates of information systems commonly used self-reported measures: a multitrait multimethod study. In: Proceedings Gmunden Retreat on NeuroIS 2012, Gmunden, Austria (2012)
12. Buettner, R.: Analyzing mental workload states on the basis of the pupillary hippus. In: Proceedings Gmunden Retreat on NeuroIS 2014, p. 1, Gmunden, Austria (2014)
13. Buettner, R.: Investigation of the relationship between visual website complexity and users' mental workload: a NeuroIS perspective. In: Davis, F.D., Riedl, R., vom Brocke, J., Léger, P.-M., Randolph, A.B. (eds.) Information Systems and Neuroscience. LNISO, vol. 10, pp. 123–128. Springer, Cham (2015). https://doi.org/10.1007/978-3-319-18702-0_16

14. Dumont, L., Chénier-Leduc, G., de Guise, É., de Guinea, A.O., Sénécal, S., Léger, P.-M.: Using a cognitive analysis grid to inform information systems design. In: Davis, F.D., Riedl, R., vom Brocke, J., Léger, P.-M., Randolph, A.B. (eds.) Information Systems and Neuroscience. LNISO, vol. 10, pp. 193–199. Springer, Cham (2015). https://doi.org/10.1007/978-3-319-18702-0_26

15. Neurauter, M., et al.: The influence of cognitive abilities and cognitive load on business process models and their creation. In: Davis, F.D., Riedl, R., vom Brocke, J., Léger, P.-M., Randolph, A.B. (eds.) Information Systems and Neuroscience. LNISO, vol. 10, pp. 107–115. Springer, Cham (2015). https://doi.org/10.1007/978-3-319-18702-0_14

16. Randolph, A.B., Labonté-LeMoyne, É., Léger, P.-M., Courtemanche, F., Sénécal, S., Fredette, M.: Proposal for the use of a passive BCI to develop a neurophysiological inference model of IS constructs. In: Davis, F.D., Riedl, R., vom Brocke, J., Léger, P.-M., Randolph, A.B. (eds.) Information Systems and Neuroscience. LNISO, vol. 10, pp. 175–180. Springer, Cham (2015). https://doi.org/10.1007/978-3-319-18702-0_23

17. Weber, B., et al.: Measuring cognitive load during process model creation. In: Davis, F.D., Riedl, R., vom Brocke, J., Léger, P.-M., Randolph, A.B. (eds.) Information Systems and Neuroscience. LNISO, vol. 10, pp. 129–136. Springer, Cham (2015). https://doi.org/10.1007/978-3-319-18702-0_17

18. Booyzen, T., Marsh, A., Randolph, A.B.: Exploring the mental load associated with switching smartphone operating systems. In: Davis, F.D., Riedl, R., vom Brocke, J., Léger, P.-M., Randolph, A.B. (eds.) Information Systems and Neuroscience. LNISO, vol. 16, pp. 67–71. Springer, Cham (2017). https://doi.org/10.1007/978-3-319-41402-7_9

19. Buettner, R.: The relationship between visual website complexity and a user's mental workload: a NeuroIS perspective. In: Davis, F.D., Riedl, R., vom Brocke, J., Léger, P.-M., Randolph, A.B. (eds.) Information Systems and Neuroscience. LNISO, vol. 16, pp. 107–113. Springer, Cham (2017). https://doi.org/10.1007/978-3-319-41402-7_14

20. Hariharan, A., Dorner, V., Adam, M.T.P.: Impact of cognitive workload and emotional arousal on performance in cooperative and competitive interactions. In: Davis, F.D., Riedl, R., vom Brocke, J., Léger, P.-M., Randolph, A.B. (eds.) Information Systems and Neuroscience. LNISO, vol. 16, pp. 35–42. Springer, Cham (2017). https://doi.org/10.1007/978-3-319-41402-7_5

21. Milic, N.(N.): Consumer grade brain-computer interfaces: an entry path into NeuroIS domains. In: Davis, F.D., Riedl, R., vom Brocke, J., Léger, P.-M., Randolph, A.B. (eds.) Information Systems and Neuroscience. LNISO, vol. 16, pp. 185–193. Springer, Cham (2017). https://doi.org/10.1007/978-3-319-41402-7_23

22. Mirhoseini, S.M.M., Léger, P.-M., Sénécal, S.: The influence of task characteristics on multiple objective and subjective cognitive load measures. In: Davis, F.D., Riedl, R., vom Brocke, J., Léger, P.-M., Randolph, A.B. (eds.) Information Systems and Neuroscience. LNISO, vol. 16, pp. 149–156. Springer, Cham (2017). https://doi.org/10.1007/978-3-319-41402-7_19

23. Buettner, R., Scheuermann, I.F., Koot, C., Rössle, M., Timm, I.J.: Stationarity of a user's pupil size signal as a precondition of pupillary-based mental workload evaluation. In: Davis, F.D., Riedl, R., vom Brocke, J., Léger, P.-M., Randolph, A.B. (eds.) Information Systems and Neuroscience. LNISO, vol. 25, pp. 195–200. Springer, Cham (2018). https://doi.org/10.1007/978-3-319-67431-5_22

24. Léger, P.-M., Charland, P., Sénécal, S., Cyr, S.: Predicting properties of cognitive pupillometry in human–computer interaction: a preliminary investigation. In: Davis, F.D., Riedl, R., vom Brocke, J., Léger, P.-M., Randolph, A.B. (eds.) Information Systems and Neuroscience. LNISO, vol. 25, pp. 121–127. Springer, Cham (2018). https://doi.org/10.1007/978-3-319-67431-5_14

25. Seeber, I., Weber, B., Maier, R., de Vreede, G.-J.: The choice is yours: the role of cognitive processes for IT-supported idea selection. In: Davis, F.D., Riedl, R., vom Brocke, J., Léger, P.-M., Randolph, A.B. (eds.) Information Systems and Neuroscience. LNISO, vol. 25, pp. 17–24. Springer, Cham (2018). https://doi.org/10.1007/978-3-319-67431-5_3

26. Sénécal, S., Léger, P.-M., Riedl, R., Davis, F.D.: How product decision characteristics interact to influence cognitive load: an exploratory study. In: Davis, F.D., Riedl, R., vom Brocke, J., Léger, P.-M., Randolph, A.B. (eds.) Information Systems and Neuroscience. LNISO, vol. 25, pp. 55–63. Springer, Cham (2018). https://doi.org/10.1007/978-3-319-67431-5_7

27. Weber, B., Neurauter, M., Burattin, A., Pinggera, J., Davis, C.: Measuring and explaining cognitive load during design activities: a fine-grained approach. In: Davis, F.D., Riedl, R., vom Brocke, J., Léger, P.-M., Randolph, A.B. (eds.) Information Systems and Neuroscience. LNISO, vol. 25, pp. 47–53. Springer, Cham (2018). https://doi.org/10.1007/978-3-319-674 31-5_6

28. Calic, G., Shamy, N.E., Hassanein, K., Watter, S.: Paying attention doesn't always pay off: the effects of high attention load on evaluations of ideas. In: Davis, F.D., Riedl, R., vom Brocke, J., Léger, P.-M., Randolph, A.B. (eds.) Information Systems and Neuroscience. LNISO, vol. 29, pp. 65–72. Springer, Cham (2019). https://doi.org/10.1007/978-3-030-01087-4_8

29. Perkhofer, L., Lehner, O.: Using gaze behavior to measure cognitive load. In: Davis, F.D., Riedl, R., vom Brocke, J., Léger, P.-M., Randolph, A.B. (eds.) Information Systems and Neuroscience. LNISO, vol. 29, pp. 73–83. Springer, Cham (2019). https://doi.org/10.1007/978-3-030-01087-4_9

30. Djurica, D., Mendling, J., Figl, K.: The impact of associative coloring and representational formats on decision-making: an eye-tracking study. In: Davis, F.D., Riedl, R., vom Brocke, J., Léger, P.-M., Randolph, A., Fischer, T. (eds.) Information Systems and Neuroscience. LNISO, vol. 32, pp. 305–313. Springer, Cham (2020). https://doi.org/10.1007/978-3-030-28144-1_34

31. Mirhoseini, S., Hassanein, K., Head, M., Watter, S.: User performance in the face of IT interruptions: the role of executive functions. In: Davis, F.D., Riedl, R., vom Brocke, J., Léger, P.-M., Randolph, A., Fischer, T. (eds.) Information Systems and Neuroscience. LNISO, vol. 32, pp. 41–51. Springer, Cham (2020). https://doi.org/10.1007/978-3-030-28144-1_5

32. Abbad Andaloussi, A., Soffer, P., Slaats, T., Burattin, A., Weber, B.: The impact of modularization on the understandability of declarative process models: a research model. In: Davis, F.D., Riedl, R., vom Brocke, J., Léger, P.-M., Randolph, A.B., Fischer, T. (eds.) NeuroIS 2020. LNISO, vol. 43, pp. 133–144. Springer, Cham (2020). https://doi.org/10.1007/978-3-030-60073-0_15

33. Giroux, F., Boasen, J., Sénécal, S., Léger, P.-M.: Hedonic multitasking: the effects of instrumental subtitles during video watching. In: Davis, F.D., Riedl, R., vom Brocke, J., Léger, P.-M., Randolph, A.B., Fischer, T. (eds.) NeuroIS 2020. LNISO, vol. 43, pp. 330–336. Springer, Cham (2020). https://doi.org/10.1007/978-3-030-60073-0_38

34. Jones, T., Randolph, A.B., Cortes, K., Terrell, C.: Using NeuroIS tools to understand how individual characteristics relate to cognitive behaviors of students. In: Davis, F.D., Riedl, R., vom Brocke, J., Léger, P.-M., Randolph, A.B., Fischer, T. (eds.) NeuroIS 2020. LNISO, vol. 43, pp. 181–184. Springer, Cham (2020). https://doi.org/10.1007/978-3-030-60073-0_20

35. Mizrahi, D., Laufer, I., Zuckerman, I.: The effect of individual coordination ability on cognitive-load in tacit coordination games. In: Davis, F.D., Riedl, R., vom Brocke, J., Léger, P.-M., Randolph, A.B., Fischer, T. (eds.) NeuroIS 2020. LNISO, vol. 43, pp. 244–252. Springer, Cham (2020). https://doi.org/10.1007/978-3-030-60073-0_28

36. Ocón Palma, M.D.C., Seeger, A.-M., Heinzl, A.: Mitigating information overload in e-commerce interactions with conversational agents. In: Davis, F.D., Riedl, R., vom Brocke, J., Léger, P.-M., Randolph, A., Fischer, T. (eds.) Information Systems and Neuroscience. LNISO, vol. 32, pp. 221–228. Springer, Cham (2020). https://doi.org/10.1007/978-3-030-28144-1_24

37. Randolph, A., Mekbib, S., Calvert, J., Cortes, K., Terrell, C.: Application of NeuroIS tools to understand cognitive behaviors of student learners in biochemistry. In: Davis, F.D., Riedl, R., vom Brocke, J., Léger, P.-M., Randolph, A., Fischer, T. (eds.) Information Systems and Neuroscience. LNISO, vol. 32, pp. 239–243. Springer, Cham (2020). https://doi.org/10.1007/978-3-030-28144-1_26

38. McGregor, M., Azzopardi, L., Halvey, M.: Untangling cost, effort, and load in information seeking and retrieval. In: CHIIR 2021, p. 11 (2021)

39. Cameron, A.-F., Webster, J.: Multicommunicating: juggling multiple conversations in the workplace. Inf. Syst. Res. **24**, 352–371 (2012). https://doi.org/10.1287/isre.1120.0446

40. Hart, S.G.: Nasa-task load index (NASA-TLX); 20 years later. Proc. Hum. Factors Ergon. Soc. Annu. Meet. **50**, 904–908 (2006). https://doi.org/10.1177/154193120605000909

41. Hart, S.G., Staveland, L.E.: Development of NASA-TLX (Task Load Index): results of empirical and theoretical research (1988)

42. Xie, B., Salvendy, G.: Prediction of metal workload in single and multiple task environments. Int. J. Cogn. Ergon. **4**, 213–242 (2000)

43. Coyne, J.T., Baldwin, C., Cole, A., Sibley, C., Roberts, D.M.: Applying real time physiological measures of cognitive load to improve training. In: Schmorrow, D.D., Estabrooke, I.V., Grootjen, M. (eds.) FAC 2009. LNCS (LNAI), vol. 5638, pp. 469–478. Springer, Heidelberg (2009). https://doi.org/10.1007/978-3-642-02812-0_55

44. Pope, A.T.: Biocybernetic system evaluates indices of operator engagement in automated task. Biol. Psychol. **40**, 187–195 (1995). https://doi.org/10.1016/0301-0511(95)05116-3

45. Moustafa, K., Luz, S., Longo, L.: Assessment of mental workload: a comparison of machine learning methods and subjective assessment techniques. In: Longo, L., Leva, M.C. (eds.) H-WORKLOAD 2017. CCIS, vol. 726, pp. 30–50. Springer, Cham (2017). https://doi.org/10.1007/978-3-319-61061-0_3

46. Zagermann, J., Pfeil, U., Reiterer, H.: Measuring cognitive load using eye tracking technology in visual computing. Presented at the Proceedings of the Sixth Workshop on Beyond Time and Errors on Novel Evaluation Methods for Visualization, 24 October 2016 (2016). https://doi.org/10.1145/2993901.2993908

47. Zagermann, J., Pfeil, U., Reiterer, H.: Studying eye movements as a basis for measuring cognitive load. In: Extended Abstracts of the 2018 CHI Conference on Human Factors in Computing Systems, New York, NY, USA, pp. 1–6. Association for Computing Machinery (2018). https://doi.org/10.1145/3170427.3188628

48. Duchowski, A.T., et al.: The index of pupillary activity: measuring cognitive load vis-à-vis task difficulty with pupil oscillation. In: Proceedings of the 2018 CHI Conference on Human Factors in Computing Systems, New York, NY, USA, pp. 282:1–282:13. ACM (2018). https://doi.org/10.1145/3173574.3173856

49. Antonenko, P., Paas, F., Grabner, R., van Gog, T.: Using electroencephalography to measure cognitive load. Educ. Psychol. Rev. **22**, 425–438 (2010). https://doi.org/10.1007/s10648-010-9130-y

50. Longo, L., Orru, G.: An evaluation of the reliability, validity and sensitivity of three human mental workload measures under different instructional conditions in third-level education. In: McLaren, B.M., Reilly, R., Zvacek, S., Uhomoibhi, J. (eds.) CSEDU 2018. CCIS, vol. 1022, pp. 384–413. Springer, Cham (2019). https://doi.org/10.1007/978-3-030-21151-6_19

51. Bracken, B., et al.: Validation of a physiological approach to measure cognitive workload: CAPT PICARD. In: Longo, L., Leva, M.C. (eds.) H-WORKLOAD 2019. CCIS, vol. 1107, pp. 66–84. Springer, Cham (2019). https://doi.org/10.1007/978-3-030-32423-0_5

52. Riedl, R., Léger, P.-M.: Topics in NeuroIS and a taxonomy of neuroscience theories in NeuroIS. In: Riedl, R., Léger, P.-M. (eds.) Fundamentals of NeuroIS. SNPBE, pp. 73–98. Springer, Heidelberg (2016). https://doi.org/10.1007/978-3-662-45091-8_4

Papers on CL Measurement Cited in the Reviewed 27 Papers (Note: May Somewhat Overlap with the References Above)

53. Antonenko, P., Paas, F., Grabner, R., van Gog, T.: Using electroencephalography to measure cognitive load. Educ. Psychol. Rev. **22**, 425–438 (2010). https://doi.org/10.1007/s10648-010-9130-y

54. Bagyaraj, S., Ravindran, G., Devi, S.S.: Analysis of spectral features of EEG during four different cognitive tasks. Int. J. Eng. Technol. **6**, 10 (2014)

55. Berka, C., et al.: EEG correlates of task engagement and mental workload in vigilance, learning, and memory tasks. Aviat. Space Environ. Med. **78**, B231-244 (2007)

56. Bouma, H., Baghuis, L.C.: Hippus of the pupil: periods of slow oscillations of unknown origin. Vision. Res. **11**, 1345–1351 (1971). https://doi.org/10.1016/0042-6989(71)90016-2

57. Brünken, R., Plass, J.L., Leutner, D.: Direct measurement of cognitive load in multimedia learning. Educ. Psychol. **38**, 53–61 (2003)

58. Campbell, F.W., Robson, J.G., Westheimer, G.: Fluctuations of accommodation under steady viewing conditions. J. Physiol. **145**, 579–594 (1959). https://doi.org/10.1113/jphysiol.1959.sp006164

59. Chen, F., et al.: Eye-based measures. In: Chen, F., et al. (eds.) Robust Multimodal Cognitive Load Measurement. HIS, pp. 75–85. Springer, Cham (2016). https://doi.org/10.1007/978-3-319-31700-7_4

60. Czajka, A.: Pupil dynamics for iris liveness detection. IEEE Trans. Inf. Forensics Secur. **10**, 726–735 (2015). https://doi.org/10.1109/TIFS.2015.2398815

61. Elchlepp, H., Best, M., Lavric, A., Monsell, S.: Shifting attention between visual dimensions as a source of switch costs. Psychol. Sci. **28**, 470–481 (2017). https://doi.org/10.1177/095679 7616686855

62. Friedman, N., Fekete, T., Gal, K., Shriki, O.: EEG-based prediction of cognitive load in intelligence tests. Front. Hum. Neurosci. **13** (2019). https://doi.org/10.3389/fnhum.2019.00191

63. Gärtner, M., Grimm, S., Bajbouj, M.: Frontal midline theta oscillations during mental arithmetic: effects of stress. Front. Behav. Neurosci. **9** (2015). https://doi.org/10.3389/fnbeh.2015.00096

64. Goldinger, S.D., Papesh, M.H.: Pupil dilation reflects the creation and retrieval of memories. Curr. Dir. Psychol. Sci. **21**, 90–95 (2012). https://doi.org/10.1177/0963721412436811

65. Haapalainen, E., Kim, S., Forlizzi, J.F., Dey, A.K.: Psycho-physiological measures for assessing cognitive load. In: Proceedings of the 12th ACM International Conference on Ubiquitous Computing, New York, NY, USA, pp. 301–310. Association for Computing Machinery (2010). https://doi.org/10.1145/1864349.1864395

66. Hess, E.H., Polt, J.M.: Pupil size in relation to mental activity during simple problem-solving. Science **143**, 1190–1192 (1964). https://doi.org/10.1126/science.143.3611.1190

67. Klimesch, W., Schimke, H., Schwaiger, J.: Episodic and semantic memory: an analysis in the EEG theta and alpha band. Electroencephalogr. Clin. Neurophysiol. **91**, 428–441 (1994). https://doi.org/10.1016/0013-4694(94)90164-3

68. Klimesch, W.: EEG alpha and theta oscillations reflect cognitive and memory performance: a review and analysis. Brain Res. Rev. **29**, 169–195 (1999). https://doi.org/10.1016/S0165-0173(98)00056-3

69. Klimesch, W., Sauseng, P., Hanslmayr, S.: EEG alpha oscillations: the inhibition-timing hypothesis. Brain Res Rev. **53**, 63–88 (2007). https://doi.org/10.1016/j.brainresrev.2006.06.003

70. Knapen, T., de Gee, J.W., Brascamp, J., Nuiten, S., Hoppenbrouwers, S., Theeuwes, J.: Cognitive and ocular factors jointly determine pupil responses under equiluminance. PLoS ONE **11**, e0155574 (2016). https://doi.org/10.1371/journal.pone.0155574

71. Kruger, J.-L., Hefer, E., Matthew, G.: Measuring the impact of subtitles on cognitive load: eye tracking and dynamic audiovisual texts. In: Proceedings of the 2013 Conference on Eye Tracking South Africa, New York, NY, USA, pp. 62–66. Association for Computing Machinery (2013). https://doi.org/10.1145/2509315.2509331

72. Laeng, B., Sirois, S., Gredebäck, G.: Pupillometry a window to the preconscious? Perspect. Psychol. Sci. **7**, 18–27 (2012). https://doi.org/10.1177/1745691611427305

73. Lange, F., et al.: Neural correlates of cognitive set shifting in amyotrophic lateral sclerosis. Clin. Neurophysiol. **127**, 3537–3545 (2016). https://doi.org/10.1016/j.clinph.2016.09.019

74. Niedermeyer: Niedermeyer's Electroencephalography: Basic Principles, Clinical Applications, and Related Fields. Oxford University Press (2017)

75. Paas, F.G., Van Merriënboer, J.J., Adam, J.J.: Measurement of cognitive load in instructional research. Percept Motor Skills **79**, 419–430 (1994). https://doi.org/10.2466/pms.1994.79.1.419

76. Paas, F., Tuovinen, J.E., Tabbers, H., Gerven, P.W.M.V.: Cognitive load measurement as a means to advance cognitive load theory. Educ. Psychol. **38**, 63–71 (2003). https://doi.org/10.1207/S15326985EP3801_8

77. Pope, A.T.: Biocybernetic system evaluates indices of operator engagement in automated task. Biol. Psychol. **40**, 187–195 (1995). https://doi.org/10.1016/0301-0511(95)05116-3

78. Roux, F., Uhlhaas, P.J.: Working memory and neural oscillations: α-γ versus θ-γ codes for distinct WM information? Trends Cogn. Sci. **18**, 16–25 (2014). https://doi.org/10.1016/j.tics.2013.10.010

79. Shi, Y., Ruiz, N., Taib, R., Choi, E., Chen, F.: Galvanic skin response (GSR) as an index of cognitive load. In: CHI 2007 Extended Abstracts on Human Factors in Computing Systems, New York, NY, USA, pp. 2651–2656. Association for Computing Machinery (2007). https://doi.org/10.1145/1240866.1241057

80. Siegle, G.J., Ichikawa, N., Steinhauer, S.: Blink before and after you think: blinks occur prior to and following cognitive load indexed by pupillary responses. Psychophysiology **45**, 679–687 (2008). https://doi.org/10.1111/j.1469-8986.2008.00681.x

81. Speier, C., Valacich, J.S., Vessey, I.: The influence of task interruption on individual decision making: an information overload perspective. Decis. Sci. **30**, 337–360 (1999). https://doi.org/10.1111/j.1540-5915.1999.tb01613.x

82. van Son, D., de Rover, M., De Blasio, F.M., van der Does, W., Barry, R.J., Putman, P.: Electroencephalography theta/beta ratio covaries with mind wandering and functional connectivity in the executive control network. Ann. N. Y. Acad. Sci. **1452**, 52–64 (2019). https://doi.org/10.1111/nyas.14180

83. Wang, Q.: An eye-tracking study of website complexity from cognitive load perspective. Decis. Support Syst. **62**, 1–10 (2014)

84. Zekveld, A.A., Heslenfeld, D.J., Johnsrude, I.S., Versfeld, N.J., Kramer, S.E.: The eye as a window to the listening brain: neural correlates of pupil size as a measure of cognitive listening load. NeuroImage **101**, 76–86 (2014). https://doi.org/10.1016/j.neuroimage.2014.06.069

On Electrode Layout in EEG Studies: A Limitation of Consumer-Grade EEG Instruments

Gernot R. Müller-Putz[1](✉), Ursula Tunkowitsch[1], Randall K. Minas[2],
Alan R. Dennis[3], and René Riedl[4,5]

[1] Institute of Neural Engineering, Graz University of Technology, Graz, Austria
gernot.mueller@tugraz.at
[2] Shidler College of Business, University of Hawaii, Manoa, HI, USA
rminas@hawaii.edu
[3] Kelley School of Business, Indiana University, Bloomington, IN, USA
ardennis@iu.edu
[4] University of Applied Sciences Upper Austria, Steyr, Austria
rene.riedl@fh-steyr.at
[5] Johannes Kepler University, Linz, Austria

Abstract. There is an ongoing discussion in the NeuroIS (Neuro-Information-Systems) discipline on whether consumer-grade EEG instruments are as suitable for scientific research as research-grade instruments. Considering the increasing adoption of consumer-grade instruments along with the fact that many NeuroIS EEG papers used such tools, this debate is fundamental. We report on a study in which we contrasted a 61-channel EEG recording with a 14-channel recording that should simulate the electrode layout of the EPOC headset, the presumably worldwide most widely used consumer-grade tool. The contrast was carried out based on topographic mapping, because this kind of EEG data analysis does not only play a significant role in cognitive neuroscience, but also in NeuroIS research. Our findings show noticeable differences in the topoplots between both conditions. The current research results are limited by the fact that our task context is a non-IS context (i.e., upper limb movements). Hence, future research should validate our results based on IS tasks and situations in order to confirm, revise, or falsify the present results.

Keywords: Brain · Consumer-grade EEG · Electroencephalography · EEG · EPOC · NeuroIS · Research-grade EEG

1 Introduction

An important question which has recently been raised in cognitive neuroscience research in general, and also in the NeuroIS literature, is whether the number and placement of sensors (referred to as electrode layout) used by electroencephalography (EEG) instruments affect the conclusions we can draw. Research-grade EEG instruments usually have many more sensors than cheaper consumer-grade EEG instruments. From an Information Systems (IS) perspective, this question is important because much NeuroIS research uses consumer-grade instruments that often have only 14 sensors. A recent review of the NeuroIS literature [1] found that EEG is the dominant method, and that out of all 27 completed empirical EEG studies, 11 used Emotiv's EPOC headset, a 14-channel wireless EEG headset, the presumably worldwide most frequently used consumer-grade EEG system. Major reasons why researchers in various disciplines, including NeuroIS, use consumer-grade tools is that typically these instruments are wireless, portable, cheap, and easy to use.

Riedl, Minas, Dennis, and Müller-Putz [2] published a review on the measurement quality of consumer-grade instruments in 2020. In total, they reviewed 16 studies from various scientific fields in which the measurement quality of consumer-grade EEG devices was assessed (tools included were: EPOC, Emotiv, USA; ThinkGear, Neurosky, USA; MUSE brain sensing technology headband, InteraXon, Canada; OpenBCI, OpenBCI, USA). In essence, 14 out of the 16 studies concluded that use of consumer-grade EEG was acceptable (based on assessment of reliability, concurrent validity, and comparative validity). However, one study analyzed in this review, Duvinage et al. [3], write that "the Emotiv headset performs significantly worse" than a research-grade system (p. 1). Therefore, consumer-grade systems with fewer sensors appear to be appropriate in most cases [2], but there appear to be some cases where fewer sensors may cause problems.

The goal of this paper is to investigate one potential boundary condition; that is, one situation where a lower number of sensors may lead to different conclusions. Our intention was to compare two devices with a different number and placement of sensors, but due to the COVID-19 crisis and resulting lockdowns, it was not possible to conduct this study. Therefore, we analyzed existing EEG data from the context of upper limb movements.

The logic of our approach is to contrast EEG recordings based on many channels with a subset of the *same* data that contains many fewer channels. In this paper, we report on a study in which we contrasted a 61-channel EEG recording with a 14-channel subset of the same data with electrode positions similar to a commonly used consumer-grade instrument (Emotiv EPOC). The contrast was carried out based on *topographic mapping*, because this kind of EEG data analysis plays a significant role in cognitive neuroscience, and also in NeuroIS research; for details, please see a paper by Müller-Putz and colleagues [4, pp. 926–927].

2 Methods

2.1 Data Set

The dataset used in this work was previously recorded in a study by Ofner et al. 2017 [5] and downloaded from the BNCI-Horizon-2020 Database[1]. It offers 61-channel EEG recordings during several separate upper limb movements, including: elbow extension, elbow flexion, wrist pronation, wrist supination, hand open, and hand close of 15 healthy

---

[1] http://bnci-horizon-2020.eu/database/data-sets.

participants. Additionally, the sensor data of a data glove and an exoskeleton are provided. For the analysis in this work, we applied the same preprocessing and biosignal analysis as in the original paper to (i) prepare the data for analysis and (ii) identify the time points of movement onset.

The entire data set was used as the 61-channel data set. The reduced channel setup was created by removing channels from the full 61-channel dataset to produce a reduced channel data set of 14 channel positioned as with the EPOC headset. We note that the 14 channels of the EPOC headset are placed to best capture cognition, not motion, so the EPOC headset has no sensors on the top of the head near the sensorimotor region, which is associated with movement [5, 7]. However, we note that in the final preparation stages of this paper, Emotiv indicated on its website (https://www.emotiv.com/epoc/) that "… EMOTIV EPOC+ is designed for scalable and contextual human brain research and provides access to professional grade brain data with a quick and easy to use design".

Because of the form of the dataset some electrodes needed to be adjusted to fit the EPOC setup (see Table 1). The same preprocessing and processing steps were performed once with the entire EEG data (61 electrodes) and once with the reduced channel setup (14 electrodes).

Table 1. Original EEG electrodes adapted to the reduced channel electrodes

Reduced	AF3	AF4	F3	F4	F7	F8	FC5
Original	F1	F2	F3	F4	FFC5h	FFC6h	FC6
Reduced	FC6	T7	T8	P7	P8	O1	O2
Original	FC6	C5	C6	P3	P4	PPO1	PPO2

2.2 Movement Onset Detection

First, the detection of the movement onset was done, according to [5]. Therefore, the data glove sensor and the exoskeleton sensors were used. In order to detect elbow flexion/extension as well as forearm pronation/supination, the elbow and wrist sensors from the exoskeleton were used. The data glove sensor data was used to detect hand opening/closing, by performing principal component analysis (PCA) on the data glove sensor data; the first principal component was used for further processing. A movement was detected when the absolute difference between the sensor data and the average of the data in a preceding time window (from -1 s to -0.5 s) crossed a threshold. Thresholds were chosen separately for each sensor to ensure timely detection of movement onsets.

2.3 EEG Data Processing

EEGLAB was used for all preprocessing and processing steps. Preprocessing was done according to [5]. The data was downsampled to 256 Hz to reduce computation time and then bandpass filtered between 0.3 to 70 Hz (4-th order zero-phase butterworth filter). Artifacts were marked when values were above or below $\pm 200\,\mu$V or trials with abnormal joint probability (threshold 5x SD) or trials with abnormal kurtosis (threshold 5x SD). Trials with these artifacts were marked, but not yet removed. Afterwards, the original (unfiltered) 256 Hz EEG data was filtered with a 4-th order butterworth filter between 0.3 to 3 Hz. Thereafter the before marked trials containing artifacts were rejected. For each condition, we merged the data of all participants. By applying channel positions for both data sets (original (61-channel) and reduced (14-channel)) based on EEGLAB topoplot function we averaged topoplots for each condition. To get a temporal overview, these topoplots were calculated every 250 ms in an interval of −1000 ms to 1000 ms relative to the movement onset. This window shows typically the movement-related cortical potential (MRCP) which occurs before (Bereitschaftspotential) and during executed, attempted or imagined movements [6, 7].

3   Results

In Fig. 1 we show example topoplots of brain activity during elbow flexion. Figure 1A shows the original data with 61 channels and Fig. 1B shows the topolots with the reduced set of electrodes (14-channel). Both plots were calculated without any re-referencing (e.g., common average reference, CAR), because in the reduced electrode set this is not allowed since the electrode positions are not well distributed over the whole head and the number of electrodes is too small. Applying CAR in the 61-channel set, would lead to even more pronounced focal activity around the movement onset.

The plots in Fig. 1 show some similarities and some differences. Our focus is on the movement around time 0 ms. This appears as a focal negativity at time −500 ms in Fig. 1A, is at its maximum around 0 ms, and disappears at 500 ms and is followed by a positive activity in the same region. The pattern for this focal activity in Fig. 1B is different. The focal pattern (Bereitschaftspotential and MRCP) is visible in the full channel set,

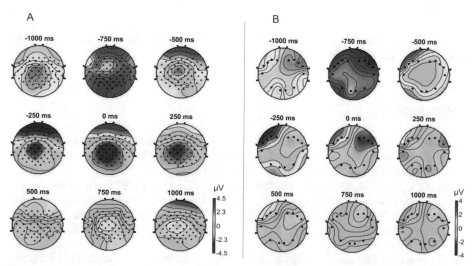

Fig. 1. Grand average topoplots for "elbow extension" performed from second 0. A) Topoplots for the original 61-channel data set. B) Topoplots for a reduced 14-channel subset for the same data.

whereas the reduced set shows arbitrary patterns without any clear neurophysiological meaning.

4 Conclusion

In this paper, we examined an existing data set from the context of upper limb movements and found noticeable differences in topoplots between 61 channel EEG recordings and 14 channel EEG recordings. For the 14-channel data, we selected electrode positions that are similar to the electrode positions used by the EPOC headset, which has no sensors on the top of the head, near the sensorimotor region which controls movement. Our analysis context was chosen deliberately as topographic mapping plays an important role in NeuroIS research.

Thus, we conclude that one boundary condition to the general conclusions of Riedl, et al. [2] that consumer-grade systems are appropriate for NeuroIS research, is the number and placement of sensors. When equipment is used that has a reduced channel set (e.g., the Emotiv EPOC), researchers should be careful to ensure that the placement of sensors is appropriate for their research questions. More details on the theory of the necessary number of channels and even source imaging can be found here [8–10]. The current research focused on upper limb movements and examined a region of the brain where there were no nearby sensors in the reduced channel set. We found a pattern of results that was not consistent with the full data set when sensors close to this focal region were included. A similar result would have been achieved with a classical ERP data set, where usually the main activity is in the midline around positions Fz, Cz, and Pz. Future research should validate our results based on IS tasks and situations in order to arrive at more definitive conclusions. Moreover, future research should also investigate systems with a low number of electrodes, yet with a more appropriate electrode placement on the skull.

Acknowledgements. This work was partly supported by the European ERC Consolidator "Feel Your Reach" (ERC-Cog-2015 681231).

References

1. Riedl, R., Fischer, T., Léger, P.-M., Davis, F.D.: A decade of NeuroIS research: progress, challenges, and future directions. ACM SIGMIS Database DATABASE Adv. Inf. Syst. **51**(3), 13–54 (2020). https://doi.org/10.1145/3410977.3410980
2. Riedl, R., Minas, R., Dennis, A., Müller-Putz, G.: Consumer-grade EEG instruments: insights on the measurement quality based on a literature review and implications for NeuroIS research. In: Davis, F.D., Riedl, R., vom Brocke, J., Léger, P.-M., Randolph, A.B., Fischer, T. (eds.) NeuroIS 2020. LNISO, vol. 43, pp. 350–361. Springer, Cham (2020). https://doi.org/10.1007/978-3-030-60073-0_41
3. Duvinage, M., Castermans, T., Petieau, M., Hoellinger, T., Cheron, G., Dutoit, T.: Performance of the Emotiv Epoc headset for P300-based applications. Biomed. Eng. Online **12**(1), 56 (2013). https://doi.org/10.1186/1475-925X-12-56

4. Müller-Putz, G.R., Riedl, R., Wriessnegger, S.C.: Electroencephalography (EEG) as a research tool in the information systems discipline: foundations, measurement, and applications. Commun. Assoc. Inf. Syst. **37**(46), 911–948 (2015). https://doi.org/10.17705/1CAIS.03746

5. Ofner, P., Schwarz, A., Pereira, J., Müller-Putz, G.R.: Upper limb movements can be decoded from the time-domain of low-frequency EEG. PLoS ONE **12**(8), e0182578 (2017). https://doi.org/10.1371/journal.pone.0182578

6. Deecke, L., Grözinger, B., Kornhuber, H.H.: Voluntary finger movement in man: cerebral potentials and theory. Biol. Cybern. **23**(2), 99–119 (1976). https://doi.org/10.1007/BF00336013

7. Kobler, R.J., Kolesnichenko, E., Sburlea, A.I., Müller-Putz, G.R.: Distinct cortical networks for hand movement initiation and directional processing: an EEG study. Neuroimage **220**, 117076 (2020). https://doi.org/10.1016/j.neuroimage.2020.117076

8. Sullivan, L.R., Davis, S.F.: Electroencephalography. In: Kaye, A.D., Davis, S.F. (eds.) Principles of Neurophysiological Assessment, Mapping, and Monitoring, pp. 145–158. Springer, New York (2014). https://doi.org/10.1007/978-1-4614-8942-9_10

9. Srinivasan, R., Tucker, D.M., Murias, M.: Estimating the spatial Nyquist of the human EEG. Behav. Res. Methods Instrum. Comput. **30**, 8–19 (1998). https://doi.org/10.3758/BF03209412

10. Pascual-Marqui, R.D., Michel, C.M., Lehmann, D.: Low resolution electromagnetic tomography: a new method for localizing electrical activity in the brain. Int. J. Psychophysiol. **18**(1), 49–65 (1994). https://doi.org/10.1016/0167-8760(84)90014-X

Predicting In-Field Flow Experiences Over Two Weeks from ECG Data: A Case Study

Michael T. Knierim[1]([✉]), Victor Pieper[1], Max Schemmer[1], Nico Loewe[1], and Pierluigi Reali[2]

[1] Institute of Information Systems and Marketing, Karlsruhe Institute of Technology (KIT), Karlsruhe, Germany
{michael.knierim,victor.pieper,max.schemmer,nico.loewe}@kit.edu
[2] Department of Electronics, Information, and Bioengineering, Politecnico di Milano, Milan, Italy
pierluigi.reali@polimi.it

Abstract. Predicting flow intensities from unobtrusively collected sensor data is considered an important yet challenging endeavor for NeuroIS scholars aiming to understand and support flow during IS use. In this direction, a limitation has been the focus on cross-subject models built on data collected in controlled laboratory settings. We investigate the potential of predicting flow in the field through personalized models by collecting report and ECG data from a clerical worker over the course of two weeks. Results indicate that a lack of variation in flow experiences during this time likely diminished these potentials. Through pre-training feature selection methods, model accuracies could be achieved that nonetheless approach related cross-subject flow prediction work. Novel recommendations are developed that could introduce more flow variation in future flow field studies to further investigate the within-subject predictability of flow based on wearable physiological sensor data.

Keywords: Flow experience · Field study · ECG · LASSO · Random forest

1 Introduction

Flow, the experience of complete task involvement, is linked to improved task performance, growth and well-being, on the individual and organizational level [1]. Therefore, the flow support has been a growing research stream in the NeuroIS community [2–4]. With the ultimate aim of enabling flow-supporting neuro-adaptive systems that can, for example, reduce work interruptions to maintain flow [4], or provide self-regulation advice to enable the emergence of flow [6], an increasing amount of research has been put forward that uses heart rate variability (HRV) patterns to classify the presence of weaker or stronger flow [4, 7–10]. The advantage of relying on HRV features is that they can be collected conveniently through chest-belt electrocardiography (ECG), smartwatch photoplethysmography (PPG) or even contact-free through camera-based PPG [11]. Therefore, an increased potential to detect flow is given by the ubiquity of these sensing

© The Author(s), under exclusive license to Springer Nature Switzerland AG 2021
F. D. Davis et al. (Eds.): NeuroIS 2021, LNISO 52, pp. 96–102, 2021.
https://doi.org/10.1007/978-3-030-88900-5_11

approaches in daily life [1]. However, only two studies have so far investigated the potential of predicting flow occurrence in daily life with HRV data [7, 12]. In both studies, it is suggested that more extensive data collection, and importantly, within-subject flow prediction (i.e. personalized flow prediction models), might be a promising direction to further assess the accuracy and robustness of HRV-based flow detection. Here, we follow up on this proposition by implementing an experience sampling method (ESM) case study observing a clerical worker's flow experiences over the course of eleven workdays with self-report and ECG recordings. Leveraging data-driven modeling methods, the possibility of predicting the flow intensities of this individual are investigated. Results from a LASSO regression and a Random Forest (RF) classifier indicate that sparse models achieve better prediction performance (MAE of 1.18 for the regression, F1-Score of 0.65 for a binary RF). A central limitation emerged in the lack of flow experience variation. Thus, beyond the methodological contributions of the prediction modeling, we contribute to the advancement of flow-supporting neuro-adaptive systems with recommendations for future field studies on how to record more varied flow and enable better prediction through increased experiential and physiological contrasts.

2 Related Work

Flow is a construct from psychological literature and comprised of six characteristics: (1) merging of action and awareness, (2) sense of control, (3) loss of self-conscious thought, (4) transformation of time perception, (5) concentration on the task, and (6) intrinsic reward [13]. Flow is said to be possible whenever active engagement and three pre-conditions are required: (1) balance of perceived task difficulty and skill, (2) clear goals, and (3) unambiguous feedback [13]. To observe flow, researchers typically follow either controlled approaches (e.g. manipulating the difficulty of a task) or field studies using the ESM method explicitly designed to observe natural flow occurrence [5, 17]. To detect flow, self-reports are predominantly used [5], but automatic and specifically physiological methods have seen increased research interest in recent years. Thereby, most of this work has focused on identifying correlates of flow (for reviews see [18, 19]) or to classify flow levels (e.g. low/high or low/moderate/high) using machine learning. (for reviews see [7, 8]). Especially HRV data has been used for this purpose as it can be robustly and conveniently recorded. However, most of this previous flow prediction work has been conducted in the lab (achieving accuracies of up to 70% for binary classifications), and only two studies have observed HRV data in the field [7, 12]. In both cases, several participants were observed for a week and between-subject modeling allowed to identify interesting correlates (the LF/HF ratio in both studies) and a classification accuracy of up to 74% in [7]. To extend these works, we chose a case study approach to enable a more extensive data collection that would allow more nuanced insight into within-subject variations. In the young history of flow physiology research, this approach has been common to develop initial understandings – for example, about cardiac activity patterns during flow in orchestra conductors [20] or concert musicians [21]. However, no previous research has pursued building a personalized (i.e. single subject) flow prediction model based on continuous ECG recordings.

3   Method

An ESM study was conducted over eleven workdays with a 28-year-old male clerical worker. The participant had an average body mass index, exercised one to two times per week, did not smoke, take medication and had no allergies. After arriving in the office, the participant attached a MoviSens EcgMove4 sensor with adhesive disposable electrodes to the chest. The sensor allows raw ECG measurement with a sampling rate of 1024 Hz. During the day, flow and task type questionnaire interruptions were issued through the MoviSensXS ESM smartphone app. These interruptions were scheduled to be repeated at four random times from morning to noon and three random times throughout the afternoon. The minimum time between two questionnaires was at least 20 min. In the survey, to keep the intrusiveness into natural behavior low, the flow short scale (FKS) [22] was included that measures flow experience with ten items rated on a 7p Likert scale. Task categories were determined in alignment with related work [23] and a pre-study interview with the participant.

Removing incomplete or dismissed questionnaires and removing instances were ECG data was not available, overall 66 measurement instances of flow reports and five-minute-long ECG recordings before an interruption were available. Flow scores were created by mean averaging reports for all ten items. Cronbach's Alpha indicated good internal consistency (0.90). ECG data were processed according to the Pan-Tompkins method to perform R peaks detection [24]. Based on the computed RR-periods, HRV features were extracted using the pyHRV Python library [25]. Specifically, time-domain and frequency-domain features were extracted in alignment with related work [4, 7, 8]. Also, non-linear HRV features were extracted as they have not yet been investigated in relation to flow. In total, there were 66 samples and 24 HRV features available (see Fig. 2C for an overview).

4   Results

Figure 1 shows the flow reports with values mainly in the range from 3 to 5 (mean 4.38). An ANOVA based on a linear mixed model with day, interruption, and task type as fixed effects and day as random intercept effect indicated no significant effects for either day $(F(10, 48) = 1.33, p = 0.24)$ or daily interruption $(F(7, 48) = 1.65, p = 0.15)$ but for task $(F(6, 42) = 6.83, p < 0.001)$. The latter effect seems driven by the high flow values elicited primarily when not working. Therefore, when considering the work situations alone, within- and across day measurements do not seem to have elicited strong flow experience variation. Nonetheless, as some variation is present, the investigation of flow relationships with HRV features was pursued further.

To account for the high degree of intercorrelation of the HRV features (16 out of the 24 HRV features showed average correlations with other HRV features over 0.4 - see also [26, 27]), two embedded feature selection methods were used to create sparse flow prediction models. On the one hand, given the aggregated flow reports' interval-like nature, a LASSO regression model was created that penalizes excessive beta coefficients [28]. On the other hand, in alignment with previous flow classification research, a RF classifier with binary outcome labels (lower vs higher flow) was created with ANOVA-based feature selection in the cross-validation stage (see [9] with a similar approach for

EEG data from a flow laboratory experiment). For both models, the HRV features were z-standardized to account for the diversity in their scales.

For the LASSO regression, 10-fold cross-validation (CV) and grid search with lambda values from 0.001 to 1 was performed. The mean absolute error (MAE) was used as minimization criterion (see Fig. 2A&B). To extract a sparse model, a lambda value of 0.1 was selected, for which three HRV features were retained, namely LF_{peak}, LF/HF (both negative sign), and SampEn (positive sign). The MAE of this model is 1.18 and the R^2 is 0.17. These results indicate, that flow intensities of this individual can be predicted with an accuracy of a bit more than 1p on the used 7p Likert scale.

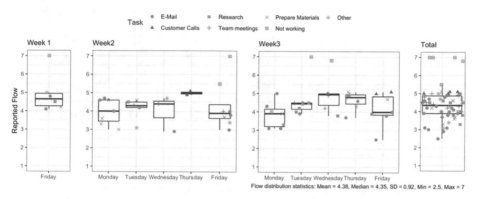

Fig. 1. Flow experience report distributions over the course of the study.

Splitting the flow report variable at the midpoint ($<4p$ was set to "lower flow") to create a binary classification task and using 10-fold CV with all HRV features as predictors, the average F-test values from an ANOVA with each predictor were calculated. Figure 2C shows that one feature stands out for the classification, namely LF_{peak}. The second most important feature is NNI_{min}, followed by LF_{log}. LF_{abs}, and LF_{rel} represent the fourth and fifth important feature that form a possible set of most useful variables. Given the high similarity of the LF features, the latter two features were not considered further. These results show similar feature importance as identified in the LASSO regression. Training a RF classifier with 100 trees using either all, the three most important or the single most important features resulted in weighted F1-scores of 0.54, 0.55, and 0.65, respectively (using 10-fold CV and oversampling to account for dataset imbalances). A subsequent grid search (100 to 1000 trees, 1 to 24 maximum leaf depth) showed slight improvements for each model with F1-scores of 0.59 with all features (400 trees, 8 leaves), 0.61 with the three best features (500 trees, 1 leaf), and 0.69 (200 trees, 8 leaves). Thereby, the best classification for this single subject is realized using the LF_{peak} feature alone, with accuracies approaching similar precision as previous between-subject classifiers in related flow field studies [4].

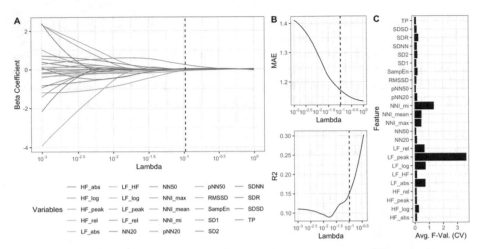

Fig. 2. HRV feature selection results. A-B: MAE minimization in the LASSO regression across lambda values. The dashed line shows the selected lambda. C: ANOVA-based feature selection in a 10-Fold Random Forest classifier. Average F-test statistics across all folds are shown.

5 Discussion and Outlook

In this study, we explored the potential of improving flow detection from HRV data through a case study. In contrast to related work, the focus on within-subject classification and a more extended collection period was considered a promising approach to test the merits of personalized models. Altogether, the results suggest that using pre-training feature selection, prediction performances of magnitudes approaching previous between-subject models could be achieved. On the one hand, the model development method represents a useful option for future flow field studies to improve prediction accuracies and identify important HRV features. The herein identified important features further confirm previous observations. LF and LF/HF ratios have been found as most important predictor variables in both previous field studies on flow [7, 12]. On the other hand, the case study approach showed a central limitation through narrow variation of recorded flow. This observation is considered the main reason why prediction performances could not be improved over previous work. Therefore, it is considered of central importance to devise future case studies that include broader flow variation. One solution could be to extend the data collection period further. However, previous work has shown more efficient approaches by either pre-selecting study participants or observation periods based on their likeliness for flow variation and by including at least a few participants. For example, [20] observed practice and competition events of two domain experts and [21] observe known, challenging and novel, stressful tasks performed by three domain experts. Thus, a future extension of this work could be improved by, for example, focusing on two to three workers with challenging tasks that are prone to experience flow (e.g. knowledge workers like software developers, engineers or designers – see [7, 23]). An interesting alternative could also be to include an element of manipulation into the field study. Thereby, participants could be either observed during two selected daily occupations (e.g. work and a hobby – as the results here showed higher flow during non-work situations) or perform a controlled task (e.g. play a game in manipulated difficulty) at some point in the day. Through this approach, the two main flow research paradigms of difficulty manipulation and experience sampling [5, 29] could be combined to collect a rich and extensive data set with both increased internal and external validity. In summary, the herein presented results and arguments contribute to the NeuroIS community by highlighting both potentials and limitations for personalized flow classification together with novel study design recommendations that further enable the exploration of in-field unobtrusive and automatic flow detection.

References

1. Spurlin, S., Csikszentmihalyi, M.: Will work ever be fun again? In: Fullagar, C.J., Delle Fave, A. (eds.) Flow at Work: Measurement and Implications, pp. 176–187 (2017)
2. Riedl, R., Fischer, T., Léger, P.-M., Davis, F.D.: A decade of NeuroIS research: progress, challenges, and future directions. ACM SIGMIS Database DATABASE Adv. Inf. Syst. **51**, 13–54 (2020)
3. Knierim, M.T., Rissler, R., Hariharan, A., Nadj, M., Weinhardt, C.: Exploring flow psychophysiology in knowledge work. In: Davis, F.D., Riedl, R., vom Brocke, J., Léger, P.-M., Randolph, A.B. (eds.) Information Systems and Neuroscience. LNISO, vol. 29, pp. 239–249. Springer, Cham (2019). https://doi.org/10.1007/978-3-030-01087-4_29
4. Rissler, R., Nadj, M., Li, M.X., Knierim, M.T., Maedche, A.: Got flow? Using machine learning on physiological data to classify flow. In: Extended Abstracts of the 2018 CHI Conference on Human Factors in Computing Systems, pp. 1–6 (2018)
5. Moneta, G.B.: On the measurement and conceptualization of flow. In: Engeser, S. (ed.) Advances in Flow Research, pp. 23–50 (2012)
6. Szegletes, L., Köles, M., Forstner, B.: The design of a biofeedback framework for dynamic difficulty adjustment in games. In: Proceedings of 5th IEEE International Conference on Cognitive Infocommunications, CogInfoCom 2014, pp. 295–299 (2014)
7. Rissler, R., Nadj, M., Li, M.X., Loewe, N., Knierim, M.T., Maedche, A.: To be or not to be in flow at work: physiological classification of flow using machine learning. IEEE Trans. Affect. Comput. 1–12 (2020)
8. Maier, M., Elsner, D., Marouane, C., Zehnle, M., Fuchs, C.: DeepFlow: detecting optimal user experience from physiological data using deep neural networks. In: IJCAI International Joint Conference on Artificial Intelligence, pp. 1415–1421 (2019)
9. Chanel, G., Rebetez, C., Bétrancourt, M., Pun, T.: Emotion assessment from physiological signals for adaptation of game difficulty. IEEE Trans. Syst. Man, Cybern. Part A Syst. Hum. **41**, 1052–1063 (2011)
10. Berta, R., Bellotti, F., Gloria, A.D., Pranantha, D., Schatten, C.: Electroencephalogram and physiological signal analysis for assessing flow in games. IEEE Trans. Comput. Intell. AI Games **5**, 164–175 (2013)
11. Rouast, P.V., Adam, M.T.P., Chiong, R., Cornforth, D.J., Lux, E.: Remote heart rate measurement using low-cost RGB face video: a technical literature review. Front. Comput. Sci. **12**, 858–872 (2018)
12. Gaggioli, A., Cipresso, P., Serino, S., Riva, G.: Psychophysiological correlates of flow during daily activities. In: Wiederhold, B.K., Riva, G. (eds.) Annual Review of Cybertherapy and Telemedicine, pp. 65–69. IOS Press (2013)

13. Nakamura, J., Csikszentmihalyi, M.: Flow theory and research. In: Lopez, S., Snyder, C.R. (eds.) Oxford Handbook of Positive Psychology, pp. 195–206. Oxford University Press, New York (2009)
14. Léger, P.M., Davis, F.D., Cronan, T.P., Perret, J.: Neurophysiological correlates of cognitive absorption in an enactive training context. Comput. Hum. Behav. **34**, 273–283 (2014)
15. Peifer, C., Schulz, A., Schächinger, H., Baumann, N., Antoni, C.H.: The relation of flow-experience and physiological arousal under stress - can u shape it? J. Exp. Soc. Psychol. **53**, 62–69 (2014)
16. Labonté-Lemoyne, É., et al.: Are we in flow? Neurophysiological correlates of flow states in a collaborative game. In: Proceedings of the 2016 CHI Conference, pp. 1980–1988 (2016)
17. Csikszentmihalyi, M., Hunter, J.: Happiness in everyday life: the uses of experience sampling. J. Happiness Stud. **4**, 185–199 (2003)
18. Knierim, M.T., Rissler, R., Dorner, V., Maedche, A., Weinhardt, C.: The psychophysiology of flow: a systematic review of peripheral nervous system features. In: Proceedings of the 9th Retreat on NeuroIS, pp. 109–120 (2017)
19. Peifer, C.: Psychophysiological correlates of flow-experience. In: Engeser, S. (ed.) Advances in Flow Research, pp. 139–164. Springer, New York (2012). https://doi.org/10.1007/978-1-4614-2359-1_8
20. Jaque, S.V., Karamanukyan, I.H., Thomson, P.: A psychophysiological case study of orchestra conductors. Med. Probl. Perform. Art. **30**, 189–196 (2015)
21. Harmat, L., et al.: Heart rate variability during piano playing: a case study of three professional solo pianists playing a self-selected and a difficult prima vista piece. Music Med. **3**, 102–107 (2011)
22. Engeser, S., Rheinberg, F.: Flow, performance and moderators of challenge-skill balance. Motiv. Emot. **32**, 158–172 (2008)
23. Quinn, R.W.: Flow in knowledge performance experience. Adm. Sci. Q. **50**, 610–641 (2005)
24. Pan, J., Tompkins, W.J.: A real-time QRS detection algorithm. IEEE Trans. Biomed. Eng. **32**, 230–236 (1985)
25. Gomes, P., Margaritoff, P., Silva, H.: Development of an open-source Python toolbox for heart rate variability (pyHRV). In: Proceedings of International Conference on Electrical, Electronic and Computing Engineering (IcETRAN), pp. 822–828 (2019)
26. Shaffer, F., Ginsberg, J.P.: An overview of heart rate variability metrics and norms. Front. Public Heal. **5**, 1–17 (2017)
27. Berntson, G.G., Quigley, K.S., Lozano, D.: Cardiovascular psychophysiology. In: Cacioppo, J.T., Tassinary, L.G., Berntson, G.G. (eds.) Handbook of Psychophysiology, pp. 182–210. Cambridge University Press, Cambridge (2007)
28. Tibshirani, R.: Regression shrinkage and selection via the lasso. J. R. Stat. Soc. **58**, 267–288 (1996)
29. Keller, J.: The flow experience revisited: the influence of skills-demands-compatibility on experiential and physiological indicators. In: Harmat, L., Ørsted Andersen, F., Ullén, F., Wright, J., Sadlo, G. (eds.) Flow Experience, pp. 351–374. Springer, Cham (2016). https://doi.org/10.1007/978-3-319-28634-1_21

An Inward Focus of Attention During Information Security Decision Making: Electrophysiological Evidence

Robert West[(✉)] and Kate Cowger

Department of Psychology and Neuroscience, DePauw University, Greencastle, USA
{robertwest,katecowger_2022}@depauw.edu

Abstract. Insider threat represents a significant source of violations of information security. Our previous research using event-related potentials (ERPs) has revealed patterns of neural activity that distinguish ethical decision making from decisions that do not involve an ethical component. In the current study, we sought to gain insight into the locus of the effect of ethical decision making on the posterior N2 component of the ERPs. The ERP data revealed that the N2 was greater in amplitude for control trials relative to ethical violation trials, and time-frequency analyses revealed that this resulted from a reduction in phase-locked activity across trials rather than a decrease in EEG power. These findings may indicate that ethical decision making related to information security is associated with a greater inward focus of attention than is the case for decision making on control trials.

Keywords: Information security · Insider threat · Ethical decision making · ERPs

1 Introduction

Violations of information security represent a significant threat to the well-being of individual citizens, corporations, and governments. Insider threats (i.e., violations of information security that result from the actions of individuals that are part of an organization) may account for as much as 50% of instances of violations of information security [1]. This has led to the intense investigation of system and person level variables such as personal norms and ethics, self-efficacy, rewards, detection certainty, etc. that moderate insider threat using a variety of methodologies [2]. In our own research, we have incorporated scalp recorded EEG to examine neural activity related to ethical decision making in the domain of information security [3, 4]. These studies reveal modulations of the ERPs that consistently distinguish ethical dilemmas from neutral dilemmas, and are sensitive to individual differences in variables known to predict instances of insider threat (e.g., moral belief and self-control) [5]. In the current study, we build upon our recent research using the Information Security Paradigm (ISP) [3] by examining one possible locus of the effect of ethical violations on the posterior N2 component of the ERPs. Specifically, we considered whether an inward focus of attention driven by the

F. D. Davis et al. (Eds.): NeuroIS 2021, LNISO 52, pp. 103–111, 2021.
https://doi.org/10.1007/978-3-030-88900-5_12

sustained consideration of an ethical dilemma presented in a scenario may lead to a reduction in the allocation of attention to visual stimuli including the decision prompt may account for the finding that the N2 is greater in amplitude for control trials than for ethical violation trials.

In the ISP [3], individuals read a set of scenarios and are then presented with a decision prompt. The scenarios differ in terms of whether or not a hypothetical information system specialist is faced with a possible violation of information security (e.g., the unauthorized access of a secure server) or a decision that does not have an ethical component (e.g., assisting a colleague in retrieving a client list). Research incorporating the ISP provides support for a Dual Process Theory of decision making wherein controlled strategic processes related to self-control and more rule-based processes related to moral convictions are associated with different patterns of neural activity (i.e., modulations of the event-related potentials or ERPs) [5]. For instance, low self-control is associated with a reduction in the amplitude of slow wave activity over the lateral frontal region of scalp between 500–1500 ms after the onset of the decision prompt in the ISP. This finding has been interpreted as reflecting a reduced tendency for individuals with low self-control to engage in effortful frontal processes during decision making. In contrast, high moral potency is associated with reduced slow wave activity over the frontal region. A finding that may reflect individuals with high moral potency rejecting violations of information security based upon rule-based moral standards that do not require the utilization of effortful processes.

ERPs recorded after the onset of the decision prompt in the ISP consistently reveal that the amplitude of the posterior N2 component is greater in amplitude for control trials than for violation trials [3–5]. This finding is interesting, as the perceptual characteristics (i.e., intensity, color, number of characters or words) of the prompts is similar across the two types of trials, so it seems unlikely that this effect is stimulus driven. One explanation for the effect on the posterior N2 is that the consideration of ethical violations is associated with a greater inward focus of attention that is be driven by the need to resolve conflict between the benefits (i.e., financial gain) and costs (i.e., being discovered or violating a personal or social norm) of the unethical act relative to when individuals are considering control scenarios that do not involve unethical behavior. This idea is consistent the literature examining the effects of spatial and feature attention on the amplitude of the posterior N2 [6]. This research reveals that the amplitude of the N2 component is greater when attention is directed to a feature of the external world (e.g., a spatial location or stimulus color) relative to when attention is not directed toward a stimulus. Furthermore, this enhancement effect is also observed in the steady state visual evoked potential. For instance, Anderson and Müller [7] report a selective increase in power at the stimulation frequency for an attended versus unattended color, but not at other frequencies. Work examining the effect of cognitive or memory load is also consistent with this idea. Specifically, the amplitude of the posterior N2 is reduced as working memory load increases and when individuals maintain a delayed intention in the context of prospective memory [8].

Here we report data from a new study designed to examine the locus of the effect of an ethical violation on the posterior N2 using ERP and time-frequency methods. The time-frequency analyses allowed us to explore the locus of the N2 effect in the ERPs, determining whether the effect was differentially related to the magnitude of the neural response (i.e., EEG power) or to phase-locking of the EEG across trials to onset of the prompt (i.e., intertrial coherence (ITC)). We also sought to examine the role that emotion may play in decision making in the ISP by having participants rate the intensity of their emotional response to the scenarios and prompts after each trial. Three hypotheses were considered based upon previous ERP studies using the ISP [3–5], the broader literature relating information processing to measures of EEG using time-frequency methods [6, 7], and the literature examining the role of emotion in moral decision making [9, 10].

H1: The amplitude of the posterior N2 will be greater for control trials than for violation trials, and the amplitude of sustained ERP activity over the right hemisphere will be greater for violation trials than for control trials.
H2: ITC at the time of the posterior N2 will be greater for control than for violation trials, if this effect reflects the inward focus of attention for violation trials.
H3: Violation trials will be associated with a stronger emotional response than control trials.

2   Method

2.1   Participants

Sixty-three students from DePauw University enrolled in introductory and intermediate level Psychology courses participated in the study. Participants were 18–22 years of age; 47 participants identified as female and 15 identified as male; the racial and ethnic distribution was 42 white, 8 black or African American, 8 Asian, 1 Hispanic/Laninx, 2 other, and 1 unidentified; and one participant's demographic information was lost due to an Internet connectivity error.

2.2   Materials

The ISP was adapted from Kirby et al. [4]. The task included two types of scenarios (i.e., control and violation). For each scenario a hypothetical situation was posed involving Josh, an information technology specialist with extensive knowledge of his company's IT systems. For control scenarios, the situation involved a decision to engage in some activity that did not involve an unethical behavior (e.g., assist a colleague with a project). For violation scenarios, an action was described wherein Josh would need to violate the company's information security policy in order to complete the task (e.g., unauthorized access of a secure server). Each trial began with a blank screen (500 ms) that was followed by the scenario that remained on the screen until participants pressed the space bar. The scenario was replaced by a fixation cross at the center of the screen for 500 ms followed by the onset of the prompt that remained on the screen until a response key was pressed. Individuals responded to the prompt by pressing C (left middle finger), V (left index finger), B (right index finger), or N (right middle finger) to indicate their choice (No, Likely No, Likely Yes, Yes, respectively). The prompt was followed by a screen indicating that the individual should enter their "Emotional Response" to the scenario and prompt on a 1–4 scale (Not at all strong, Somewhat Strong, Very Strong, Extremely Strong) using the same keys as used to respond to the prompt. A TTL pulse was aligned with the onset of the prompt and delivered to the amplifier to mark the onset and meaning of the prompt in the EEG data stream. There were two practice trials and 16 test trials for each type of scenario. Behavior (choice and response time) and ERPs were measured from onset of the prompt. Prior to beginning the trials, participants were told that Josh was under a lot of financial strain and work-related stress. They were also instructed to go through the task imagining that they were Josh and to answer the prompts from Josh's perspective.

2.3 Procedure

Once individuals had arrived at the lab, each was given a descriptive summary of the procedure that included an introduction to the EEG cap, electrodes, and conductive gel to be placed in their hair. Individuals provided signed informed consent for the study that was approved by the Institutional Review Board of DePauw University. The participants then completed a demographic survey reporting their age, gender, racial and ethnic background, and years of college completed followed by questionnaires that measured individual differences in self-control [3], grit [11], depression [12], Internet addiction [13], and smartphone addiction [14]. These measures were collected to examine how individual differences are related to neural recruitment elicited during ethical decision making. The participants completed the ISP, counting Stroop task, and doors task while EEG was recorded. After these tasks, individuals were debriefed and paid $5 in winning for the doors task and received research credit for an introductory or intermediate level psychology as compensation for participation.

2.4 EEG Recording and Analysis

The EEG were recorded with a 32 channel actiCAP and actiCHamp active AG/AGCL electrode system using the Brain Vision Recorder Software (Brain Vision, LLC). The data were sampled at 500 Hz, from DC-150 Hz and digitized at 24 bits. Thirty electrodes were placed in the standard Brain Vision 32 electrode cap configuration (CP5-CP6 were replaced with the ocular electrodes) and two were placed below the eyes to monitor blinks and vertical eye movements. During recording the electrodes were referenced to electrode Cz, and impedances were maintained below 20 KΩ. For analysis, the average reference was calculated after the ocular artifacts were corrected using ICA.

The EEG data were processed using EEGLAB [15] and ERPLAB [16]. A .1–30 Hz IIR filter was applied to the EEG; 1–2 bad electrodes were interpolated as needed; ocular artifacts associated with blinks and saccades were corrected using ICA and visual inspection of the data. ERPs were averaged for −200 to 2000 ms and the time-frequency analysis included −500 to 2000 ms around onset of the prompt. For the ERP and time-frequency analyses a permutation t-test was used with 1000 iterations for statistical inference with an uncorrected p = .01. The time-frequency analyses used a 500 ms pre- and post-stimulus buffer, so the epochs for these analyses represent onset of the prompt (time 0) to 1500 ms.

3 Results

3.1 Psychometric and Behavioral Data

To establish the reliability of the ISP, we examined internal consistency for the control and violation items using Chronbach's α. Both sets of items demonstrated clear internal consistency (control $\alpha = .90$, 95% CI [.86–.93]; violation $\alpha = .93$, 95% CI [.91–.96]) and there was no evidence that dropping any of the items would result in an improvement in reliability. These data indicate that the ISP has good psychometric properties for a research instrument.

The choice data revealed that on average individuals responded between likely no and likely yes for both control (M = 2.37, SD = .61) and violation (M = 2.51, SD = .69) items, and choices responses did not significantly differ between items, t(61) = 1.81, p = .075. Response time was also similar for control (M = 1.3 s, SD = .50) and violation (M = 1.3 s, SD = .47) items, t(61) = .30, p = .77. Relative to previous research with this task [3–5], choice responses are higher for violation items than would be expected and response times are noticeably faster. The reason for the difference in choice behavior between the current sample and prior studies is unclear.

The emotional response data revealed that on average individuals found the control items (M = 2.82, SD = .26) to elicit a stronger emotional response than the violation items (M = 1.89, SD = .55), t = 12.77, p < .001. Individuals did take longer to rate the violation items (M = 2.5 s, SD = .80) than the control item (M = 2.3 s, SD = .70), t = 3.77, p < .001. These data diverge for the moral decision-making literature in demonstrating that at least within the ISP ethical decision-making is not associated with heightened emotion.

3.2 ERP and Time-Frequency Data

Figure 1a includes the ERPs averaged over electrodes for the occipital (O1-Oz-O2) and frontal-central-parietal (FC2-CP2-Cz-Pz) regions. For the occipital region, the ERP data revealed that the amplitude of the posterior N2 was greater for control trials than for violation trials. This finding is consistent with our previous research and provides evidence supporting Hypothesis 1. The sustained ERP activity over the right frontal-central-parietal regions reflecting greater negativity for violation trials than for control trials between 200–1000 ms after onset of the prompt is also consistent with the findings of our previous research [3–5], and may reflect longer lasting deliberative processing related to ethical decision making in the task.

For the occipital region, the time-frequency analysis revealed an increase in power around 200 ms after onset of the prompt in the Theta/Alpha band (5–10 Hz) that was followed by suppression in the high Alpha band (Fig. 1b). This increase in Theta/Alpha power did not differ between the control and violation trials in the time-frequency data (Fig. 1b). The ITC analysis revealed increased coherence around 200 ms in the Theta/Alpha bands that was greater for control trials than for violation trials (Fig. 1c). These findings provide support for Hypothesis 2, indicating that the effect of trial on the posterior N2 arises from differences in phase locking to the prompt across trials rather than differences in the magnitude of the neural response (i.e., power) to the prompt.

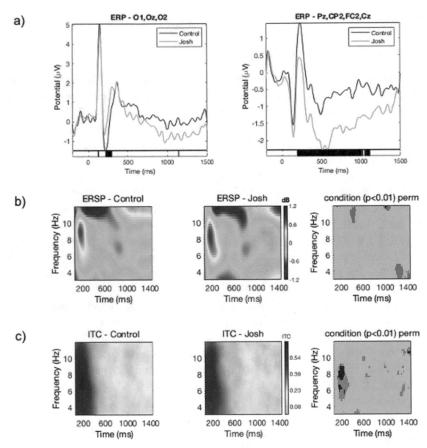

Fig. 1. a) Grand-averaged ERPs over the occipital region and right hemisphere demonstrating the effect of trial on the N2 and sustained ERP activity over the right hemisphere, b) time-frequency plots for control and violation (Josh) trials, c) ITC plots for control and violation (Josh) trials.

4   Discussion

The current data contribute to a growing body of literature that serves to refine our understanding of the psychological and neural correlates of ethical decision making in the ISP [3, 4]. The emotion rating data reveal that violation scenarios in the ISP may not elicit strong conscious emotion responses that shape decision making. This finding is interesting within the context of the broader literature related to moral and ethical decision making demonstrating the emotion or affective processes can drive decision making in these domains [9, 10]. However, the contribution of emotion to decision making is often considered a System 1 input that may not operate at the level of conscious awareness [9]. Given this, further efforts to assess the implicit influence of emotion in the ISP may be worth exploring. The ERP data also revealed slow wave activity extending from the frontal to parietal region over the right hemisphere. The is consistent with previous research using the ISP [3–5] and indicates that ethical decision making in the task is associated with slow deliberative processes that engage a broadly distributed cortical network.

The findings of the current study provide clear support for our hypotheses related to the locus of the effect of an ethical violation on the amplitude of the posterior N2. We replicated the effect of trial type on the posterior N2 and demonstrating that this effect is associated with a decrease in phase-locked activity rather than a reduction in EEG power associated with the onset of the prompt for violation trials relative to control trials. These data provide evidence for the idea that considering violations of information security may be associated with a greater inward focus of attention than considering scenarios that do not involve an ethical component. The effect of trial on ITC related to the posterior N2 may provide a foundation to further explore the nature of the effect of individual difference variables such as self-control and moral belief on decision-making processes related to information security [17, 18]. For instance, if greater self-control affords one increased attentional resources to support ongoing processing of the ethical dilemma and prompt, then the effect of trial type on ITC might decrease as self-control increases. The findings related to ITC may also provide a means to investigate the influence of state variables such as fatigue or stress on the efficiency of processes underpinning decision making in the context of information security [19], as both of these might alter attentional allocation.

The data for the N2 and ITC around 200 ms after onset of the prompt are consistent with the hypothesis that ethical decision making may be associated with an inward focus of attention. Our previous research reveals that individual differences in self-control, but not moral potency, are related to the amplitude of the N2 in the ISP with the N2 effect being larger in those individuals with low self-control [5]. These and the current findings together lead to the suggestion that reduced attentional resources related to low self-control may be one factor that could result in poor ethical decision making in the ISP. The individual differences measures related to self-control, grit, and pathological technology use should provide us the opportunity to further examine the relationship between these variables and neural recruitment related to ethical decision making in the ISP.

There are some limitations of our study that must be considered. First, as is true of all our work with the ISP, we are at best measuring an abstract intention related to hypothetical decision making in the laboratory rather than decision making tied to natural contexts with real outcomes. This limitation might be addressed with advances in data collection supported by low-cost reliable EEG systems that could facilitate data collection in quasi-realistic settings or field studies. Second, the sample included undergraduate students and that were primarily female, limiting the generalizability of the findings. Our initial work using ERPs with the ISP only included males in the studies [3], and the physiology observed in that study is quite similar to what we have reported with samples including primarily females [4, 5]. Given this, we do not believe that the gender of the sample undermines our findings, however, it may be worth directly considering gender effects in future studies. Also related to the gender imbalance of the study, one might wonder whether female participants are able to empathize with the male protagonist (i.e., Josh) in the ISP. We have sought to address this issue in a recent study using the ISP wherein a gender-neutral name was used for the protagonist along with gender neutral pronouns related to others mentioned in the scenarios. The new study also includes a larger sample (i.e., >200 participants) and roughly equal numbers of males and females, so we should be able to explore this issue at least at the behavioral level.

In conclusion, the current findings build upon prior research using ERPs with the ISP. We demonstrate that violation items are not associated with stronger emotion than control items when assessed with a conscious thought probe. We also replicated the effect of violation trials on the amplitude of the posterior N2 and demonstrated that this effect likely results from differences in phase locking in the Theta/Alpha band rather than differences in overall power between violation and control trials. These findings provide support for the idea that ethical decision making in the ISP may be associated with an inward focus of attention. We believe that the current findings provide a foundation to examine the interplay between individual differences and attentional allocation in the ISP that may serve to bolster or impede ethical decision making.

References

1. Richardson, R.: CSI computer crime and security survey (2011). http://www.GoSCI.com
2. Cram, W.A., D'Arcy, J., Proudfoot, J.G.: Seeing the forest and the trees: a meta-analysis of the antecedents to information security policy compliance. Manag. Inf. Syst. Q. **43**, 525–554 (2019)
3. Hu, Q., West, R., Smarandescu, L.: The role of self-control in information security violations: insights from a cognitive neuroscience perspective. J. Manage. Inform. Syst. **31**, 6–48 (2015)
4. Kirby, B., Malley, K., West, R.: Neural activity related to information security decision making: effects of who is rewarded and when the reward is received. In: Davis, F.D., Riedl, R., vom Brocke, J., Léger, P.-M., Randolph, A.B. (eds.) Information Systems and Neuroscience. LNISO, vol. 29, pp. 19–27. Springer, Cham (2019). https://doi.org/10.1007/978-3-030-010 87-4_3
5. West, R., Budde, E., Hu, Q.: Neural correlates of decision making related to information security: self-control and moral potency. PLoS ONE **14**(9), e0221808 (2019)
6. Luck, S., Hillyard, S.: Electrophysiology of visual attention in humans. In: The Cognitive Neurosciences, 5th edn, pp. 187–196 (2014)
7. Andersen, S., Müller, M.: Behavioral performance follows the time course of neural facilitation and suppression during cued shifts of feature-selective attention. Proc. Natl. Acad. Sci. USA **107**, 13878–13882 (2010)
8. West, R., Bowry, R., Krompinger, J.: The effects of working memory demands on the neural correlates of prospective memory. Neuropsychologia **44**, 197–207 (2006)
9. Greene, J., Sommerville, R., Nystrom, L., Darley, J., Cohen, J.: An fMRI investigation of emotional engagement in moral judgment. Science **293**, 2105–2108 (2001)
10. Greene, J., Nystrom, L., Engell, A., Darley, J., Cohen, J.: The neural bases of cognitive conflict and control in moral judgment. Neuron **44**, 389–400 (2004)
11. Duckworth, A., Quinn, P.: Development and validation of the Short Grit Scale (Grit-S). J. Pers. Assess. **91**, 166–174 (2009)
12. Easton, W., Smith, C., Ybarra, M., Muntaner, C., Tien, A.: Center for epidemiologic studies depression scale: review and revision (CESD and CESD-R). In: Maruish, M.E. (ed.) The Use of Psychological Testing for Treatment Planning and Outcomes Assessment: Instruments for Adults, pp. 363–377. Lawrence Erlbaum Associates Publishers (2004)
13. Pawlikowski, M., Alstötter-Gleich, C., Brand, M.: Validation and psychometric properties of a short version of Young's Internet Addiction Test. Comput. Hum. Behav. **29**, 1212–1223 (2013)

14. Kwon, M., Kim, D., Cho, H., Yang, S.: The smartphone addiction scale: development and validation of a short version for adolescents. PLoS One **8**(12), e83558 (2013)
15. Delorme, A., Makeig, S.: EEGLAB: an open source toolbox for analysis of single-trial EEG dynamics. J. Neurosci. Meth. **143**, 9–21 (2004)
16. Lopez-Calderon, J., Luck, S.J.: ERPLAB: an open-source toolbox for the analysis of event-related potentials. Front. Hum. Neurosci. **8**, 213 (2014)
17. Xu, Z., Hu, Q., Zhang, C.: Why computer talents become computer hackers. Commun. ACM. **56**, 64–74 (2013)
18. Hu, Q., Zhang, C., Xu, Z.: Moral beliefs, self-control, and sports: effective antidotes to the youth computer hacking epidemic. Paper Presented at 45th Hawaii International Conference on Systems Science (2012)
19. Riedl, R.: On the biology of technostress: literature review and research agenda. DATABASE Adv. Inf. Syst. **44**, 18–55 (2012)

EyeTC: Attentive Terms and Conditions of Internet-Based Services with Webcam-Based Eye Tracking

Peyman Toreini[1]($\boxtimes$), Moritz Langner[1], Tobias Vogel[2,3], and Alexander Maedche[1]

[1] Karlsruhe Institute of Technology (KIT), Karlsruhe, Germany
{peyman.toreini,moritz.langner,alexander.maedche}@kit.edu
[2] University of Mannheim, Mannheim, Germany
vogel@uni-mannheim.de, tobias.vogel@h-da.de
[3] Darmstadt University of Applied Sciences, Darmstadt, Germany

Abstract. Now and then, users are asked to accept terms and conditions (T&C) before using Internet-based services. Previous studies show that users ignore reading T&C most of the time and accept them tacitly without reading, while they may include critical information. This study targets solving this problem by designing an innovative NeuroIS application called EyeTC. EyeTC uses webcam-based eye tracking technology to track users' eye movement data in real-time and provide attention feedback when users do not read T&C of Internet-based services. We tested the effectiveness of using EyeTC to change users' behavior for reading T&C. The results show that when users receive EyeTC-based attention feedback, they allocate more attention to the T&C, leading to a higher text comprehension. However, participants articulated privacy concerns about providing eye movement data in a real-world setup.

Keywords: Eye tracking · Attentive user interface · Attention feedback · NeuroIS

1 Introduction

Internet users are confronted with legally binding documents such as terms and conditions (T&C) on a daily basis. However, almost no one reads them before agreeing on the content, which is also named as "the biggest lie on the internet" [1]. Nevertheless, such documents may include critical information that allow third parties to benefit from users' information while they do not truly agree on that. Users often give the provider permission to keep, analyze and sell their data when accepting T&Cs of Internet-based services. Previous studies show that when users signed up for a fictitious social network service, 98% of them missed clauses to allow data sharing with the NSA and employers [1]. Besides, not reading important legal texts has also been analyzed for computer usage policies [2], security warnings for downloads [3], or when connecting to public Wi-Fi [4]. One of the reasons users accept such information without reading it is that they consider it an interruption of their primary task like finishing an online purchase transaction

© The Author(s), under exclusive license to Springer Nature Switzerland AG 2021
F. D. Davis et al. (Eds.): NeuroIS 2021, LNISO 52, pp. 112–119, 2021.
https://doi.org/10.1007/978-3-030-88900-5_13

or signing up for a new Internet-based service [5]. Attitude, social trust, and apathy are also found to explain partially why users elect not to read such legal documents [2]. Also, habituation might explain such behavior, while the design of T&C can create this habituation and lead to fewer people reading and cognitively processing what they agree to [3, 5].

Apart from the reason why people do not read T&C, there is a need to increase user's awareness about their failure and to guide them in reading missed parts of T&C, especially when it includes critical information. Existing approaches focus on forcing users to stay on the T&C page for a specific time or force them to scroll until the end of the T&C before accepting them to inspire users to read them. However, these approaches do not guarantee that users properly read the document, and there is a need to design more intelligent approaches. One solution is to "convert" T&C to attentive documents in order to track of how documents are really read [6]. Attentive user interfaces (AUI) are known as user interfaces that are aware of the user's attention and support them to allocate their limited attention [7–10]. Eye tracking technology is the primary device for designing such AUIs as it allows to retrieve information about visual attention [11, 12]. NeuroIS researchers also suggested using this technology to design innovative applications [13–16] and AUIs [17–21]. However, there is a lack of research on using eye tracking devices for designing attention feedback [22]. Therefore, in this study, we suggest designing an AUI that focuses on T&C. We name this application EyeTC. EyeTC refers to an attentive T&C that tracks users' eye movement in real-time and provides attention feedback when users ignore reading the content of T&C. We especially focus on using webcam-based eye trackers since they are cheap and available for users, and they do not need to buy extra tools to use EyeTC. Therefore, in this study, we focus on answering the following research question (RQ):

RQ: How to design attentive T&Cs with webcam-based eye tracking to enhance user's attention to T&Cs and their comprehension?

To answer this question, we investigated webcam-based eye trackers' usage for designing attentive T&C within a design science research (DSR) project. Scholars have emphasized the need for the integration of the DSR and NeuroIS fields in order to designing innovative applications [14, 15]. In this project, we propose the EyeTC application that can track users eye movement via webcams in real-time and use this information to provide attention feedback while processing T&C. In this study, we focus on the development and evaluation phase of the first design cycle. After instantiating the suggested design, we evaluated it in a laboratory experiment. Our results show that using attentive T&C improves users attention allocation on T&C as well as their text comprehension. However, they articulated privacy concerns for sharing their eye movement data in a real-world scenario. We contribute to the field of NeuroIS by providing evidence of how eye tracking technology can be used for designing AUIs that support users to read T&C.

2 The EyeTC Prototype

To conceptualize and implement EyeTC, we followed the eighth and ninth-contribution types of the NeuroIS field suggested by [12]. Specifically, we defined two main components of EyeTC: an attentive T&C, which is considered as a neuro-adaptive IS, and attention feedback, in the form of live biofeedback. Figure 1 depicts an overview of the instantiation of these two dimensions in EyeTC.

For developing the attentive T&C component, we used webcam-based eye tracking technology. Using low-cost eye trackers is suggested for information system research [20, 23], and one of the options is using webcam-based eye trackers [24]. We converted webcams to eye trackers by integrating WebGazer JavaScript[1] [25]. Next, the eye tracking system retrieves gaze data using the webcam recording and stores the information about the predicted gaze position (sensing attention). After the user agrees to the T&C, the reading detector system of the attentive T&C analyses the user's reading intensity (reasoning about attention), and if visual attention does not pass a certain threshold, users will receive feedback on the lack of attention. Later, if the user agreed to the T&C without reading the text, the attention feedback system is activated (regulating interactions). First, in the attention feedback component, users are informed by a pop-up warning message stating the importance of reading legal documents and upcoming attention feedback design. Next, the attention feedback system uses the information about the reading activity of the user to highlight the specific AOIs that were not read yet by users sufficiently while accepting the T&C.

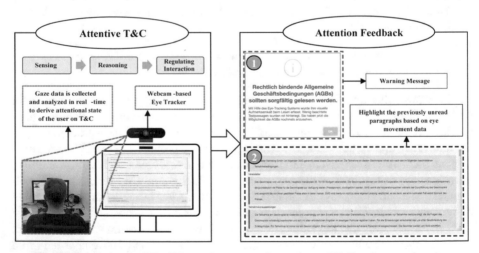

Fig. 1. Components of EyeTC to enhance users' attention to T&C and comprehension

---

[1] https://webgazer.cs.brown.edu/.

3   Experimental Design

To evaluate EyeTC, we executed a controlled laboratory experiment with two groups in which attention feedback types were manipulated between subjects. As apparatus, we used Logitech Brio 4K Ultra HD webcam on all laboratory computers and the WebGazer. In the following we discuss the two presented attention feedback types as well as the experimental procedure.

3.1   Attention Feedback Types

In this study, we designed two different types of attention feedback for T&C readers distributed in the control and treatment groups. Both groups received feedback types after being forced to read T&C by scrolling until the end of the T&C and choose to the agreement on the provided content (similar to existing approaches on the internet when facing T&C). After users click on the continue, the treatment group received EyeTC and the corresponding attention feedback with both warning message and highlighting option discussed in the previous section. The control group received general attention feedback in the form of only a warning message. This warning message aimed to create bottom-up attention and reminded participants about reading the legal text carefully. Both groups received the same warning message with the same primary text. The only difference is that the treatment group users were informed about receiving the highlighted passage in the next step. Therefore, with this design, we argue that both groups experienced the same situation except the personalized highlighted passage provided by EyeTC to the participants in the treatment group.

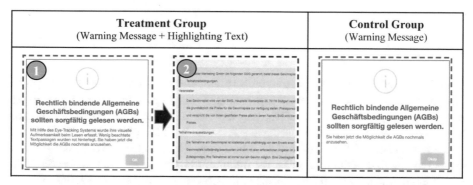

Fig. 2. Two types of attention feedback used in this study to investigate the EyeTC

3.2   Experimental Procedure

Figure 3 shows the experimental steps to evaluate EyeTC. After reading the experimental instruction and performing the calibration, we started a bogus experiment. In this experiment, we asked users to choose their favorite pictures among two options while we track their eyes to find the relationship between their choice and the visual behavior. After performing the bogus experiment, we offered the users to participate in a lottery to win an extra 20 euros besides the compensation for the experiment participation. For that, they had to read and accept our designed T&C. Both groups were forced to scroll down the T&C before the accept button got activated. In this phase, the attentive T&C started to record and analyze the user's eye movements while reading the T&C. After users accepted the T&C, the treatment group received a warning message and attention feedback, and the control group received only a warning message. Next, the participants from both groups were forced to check the T&C again, which was considered as their revisit phase. During all these steps, the user's interaction data is recorded, and users' exploration time on each step is considered as the duration of their allocated attention. After they were done with the experiment, we measured the T&C text comprehension of the participants with a declarative knowledge test in the form of 15 multiple-choice questions. Last, participants joined a survey for demographic questions, perceived usefulness of the attention feedback types, and the ability to articulate privacy concerns using webcam-based eye trackers.

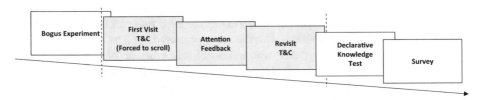

Fig. 3. Experiment steps used for evaluating EyeTC

4   Results

In total 62 university students (32 female, 30 male) with an average age of 22.82 (SD = 2.61) participated in this laboratory experiment. Users were assigned randomly to one of the two groups. Furthermore, all the participants in both groups visited the T&C for two times and system did not detect anyone that read the T&C precisely in the first visit.

First, we checked the users' first visit duration. Executing Wilcoxon rank-sum test shows that the first visit duration for the treatment group (M = 122 s, SD = 87 s) did not differ significantly from participants in the control group (M = 125 s, SD = 81 s) , W = 463, p = .81, r = −.03. It shows that both groups had similar behavior regarding reading T&C. However, in the revisit phase, participants in the treatment group (M = 144 s, SD = 97 s) had a significantly higher reading duration than the control group (M = 38 s, SD = 42 s), W = 814, p < .001, r = −.595. It shows that users that received attention feedback changed their behavior and spend more time on T&C. The provided T&C includes 914 words and with an assumed reading speed of 250 words/minute [26] the reader might need around of 195.85 s to read the text. By comparing the total reading time (first visit and revisit) of both groups we argue that a reader might read the T&C in the treatment condition (total time spent M = 266 s, SD = 130 s), but not in the control condition (with a total time spent M = M = 163 s, SD = 84 s). Also, comparing the total duration time on T&C shows that the treatment group spent significantly more seconds on T&C than participants in the control group, t = 3.69, p < .001.

Furthermore, the performance in a declarative knowledge test as measured by the number of correct answers was higher for users in the treatment group (M = 10, SD = 2.8) than for users in the control group (M = 8.8, SD = 2.1), W = 650, p < .05, r = − .305. Despite, the results of the survey show that both groups have high privacy concerns about using eye tracking technology and there is no difference between the treatment group (M = 5.74, SD = 1.1) and the control group (M = 5.55, SD = 1.12), W = 527.5, p = 0.511, r = −.083 (Fig. 4).

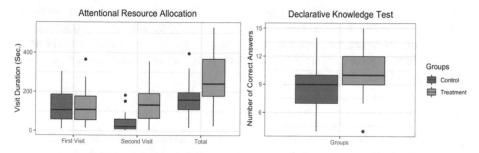

Fig. 4. The influence of EyeTC on attention allocation and text comprehension

5 Discussion

Our experimental results show a positive effect of EyeTC on the users to read T&C. The personalized highlighting of passages that have not been read was significantly more effective than a simple reminder in the form of a prompt. In conclusion, EyeTC caused a higher reading duration on the T&C and better text comprehension. Tracking tools are often known to decrease privacy, but we show that eye tracking can be used to increase privacy by supporting people in reading T&C and understanding them. Based on the DSR contribution types provided by Gregor and Hevner [27] this project is considered as "improvement" type since we could provide as the solution (EyeTC) for a known problem (ignore reading T&C). Furthermore, by implementing EyeTC as a trustable eye tracking software [28], users can decide to use eye tracking in a way to help them not to miss out important content.

However, this research also has some limitations that should be covered in the future. Using webcam-based eye tracking was beneficial for designing EyeTC as they are integrated into most personal computers and are more available than using eye trackers. However, they are less accurate and precise compared to the infrared eye trackers. Also, they are very sensitive to movements, and we controlled for the steady posture of the participants during the experiment. However, there is a chance that the EyeTC did not provide accurate highlighting visualization for some participants during the experiment. However, as people typically ignore reading T&C, it was not reported by any participants. Furthermore, we did not consider the user's eye movement data in the evaluation section to control data noise regarding the webcam-based eye trackers. For the evaluation, we focused on the users' mouse clicks as interaction data as well as the survey results. As future work, we suggest general highlighting of typical passages that people do not read and investigate the users' reaction and the need for personalized adaption of the system. Also, to validate the results, we suggest designing and evaluating EyeTC with accurate eye trackers in the future. A more accurate eye tracker can also help to better understand how users process T&C. Also can support EyeTC to distinguish between skimming, reading, and non-reading behavior, etc. [6, 29, 30]. Furthermore, the results are based on a controlled lab environment, and there is a need to check the effectiveness of EyeTC in the field and as long-term studies. Also, the future agenda is to establish standards for integrating EyeTC either by T&C providers or in a way that users can install it to receive support. Also, the findings from this study may be further developed to create applications beyond attentive T&C. For example, this system could be used in e-learning courses to motivate learners to read factual texts; companies might find it helpful to implement a reading enhancing system for certain documents, reading other legal documents like a contract, etc.

References

1. Obar, J.A., Oeldorf-Hirsch, A.: The biggest lie on the Internet: ignoring the privacy policies and terms of service policies of social networking services. Inf. Commun. Soc. **23**, 128–147 (2020)
2. Bryan Foltz, C., Schwager, P.H., Anderson, J.E.: Why users (fail to) read computer usage policies. Ind. Manag. Data Syst. **108**, 701–712 (2008)
3. Anderson, B.B., Jenkins, J.L., Vance, A., Kirwan, C.B., Eargle, D.: Your memory is working against you: How eye tracking and memory explain habituation to security warnings. Decis. Support Syst. **92**, 3–13 (2016)
4. Fox-Brewster, T.: Londoners give up eldest children in public Wi-Fi security horror show. https://www.theguardian.com/technology/2014/sep/29/londoners-wi-fi-security-herod-clause#:~:text=Londoners. Give up eldest children in public Wi-Fi security horror show,-This article is&text=When people connected to the, for the duration of eternity
5. Böhme, R., Köpsell, S.: Trained to accept? A field experiment on consent dialogs. Conf. Hum. Factors Comput. Syst. - Proc. **4**, 2403–2406 (2010)
6. Buscher, G., Dengel, A., Biedert, R., Elst, L.V.: Attentive documents: eye tracking as implicit feedback for information retrieval and beyond attentive documents: eye tracking as implicit feedback for information retrieval and beyond. ACM Trans. Interact. Intell. Syst. **1**, 1–30 (2012)
7. Vertegaal, R.: Attentive User Interfaces. Commun. ACM. **46**, 30–33 (2003)
8. Anderson, C., Hübener, I., Seipp, A.-K., Ohly, S., David, K., Pejovic, V.: A survey of attention management systems in ubiquitous computing environments. Proc. ACM Interact. Mob. Wearable Ubiquit. Technol. **2**, 1–27 (2018)
9. Bulling, A.: Pervasive Attentive User Interfaces. Comput. (Long. Beach. Calif.) **49**, 94–98 (2016)
10. Roda, C., Thomas, J.: Attention aware systems: theories, applications, and research agenda. Comput. Human Behav. **22**, 557–587 (2006)
11. Duchowski, A.T.: Eye Tracking Methodology: Theory and Practice. Springer, Cham (2017)
12. Holmqvist, K., Nyström, M., Andersson, R., Dewhurst, R., Jarodzka, H., Van De Weijer, J.: Eye Tracking: A Comprehensive Guide to Methods and Measures. Oxford University Press, Oxford (2011)

13. Davis, F.D., Riedl, R., Hevner, A.R.: Towards a NeuroIS research methodology: intensifying the discussion on methods, tools, and measurement. J. Assoc. Inf. Syst. **15**, I–XXXV (2014)
14. Riedl, R., Léger, P.-M.: Fundamentals of NeuroIS: Information Systems and the Brain. Springer, Heidelberg (2016)
15. vom Brocke, J., Riedl, R., Léger, P.-M.: Application strategies for neuroscience in information systems design science research. J. Comput. Inf. Syst. **53**, 1–13 (2013)
16. Dimoka, A., Davis, F.D., Pavlou, P.A., Dennis, A.R.: On the use of neurophysiological tools in IS research: developing a research agenda for NeuroIS. MIS Q. **36**, 679–702 (2012)
17. Hummel, D., Toreini, P., Maedche, A.: Improving digital nudging using attentive user interfaces: theory development and experiment design using eye-tracking. In: Research in Progress Proceedings of the 13th International Conference on Design Science Research in Information Systems and Technology (DESRIST), Chennai, India, pp. 1–8 (2018)
18. Toreini, P., Langner, M.: Desiginig user-adaptive information dashboards: considering limited attention and working memory. In: Proceedings of the 27th European Conference on Information Systems (ECIS2019), Stockholm-Uppsala, Sweden (2019)
19. Langner, M., Toreini, P., Maedche, A.: AttentionBoard: a quantified-self dashboard for enhancing attention management with eye-tracking (in press). In: Davis, F., Riedl, R., vom Brocke, J., Léger, P., Randolph, A., Fischer, T. (eds.) Information Systems and Neuroscience (NeuroIS Retreat 2020). Virtual Conference (2020)
20. Toreini, P., Langner, M., Maedche, A.: Using eye-tracking for visual attention feedback. In: Davis, F.D., Riedl, R., vom Brocke, J., Léger, P.-M., Randolph, A., Fischer, T. (eds.) Information Systems and Neuroscience. LNISO, vol. 32, pp. 261–270. Springer, Cham (2020). https://doi.org/10.1007/978-3-030-28144-1_29
21. Toreini, P., Langner, M., Maedche, A.: Use of attentive information dashboards to support task resumption in working environments. In: Proceedings of the 2018 ACM Symposium on Eye Tracking Research & Applications, pp. 1–3. ACM Press (2018)
22. Lux, E., Adam, M.T.P., Dorner, V., Helming, S., Knierim, M.T., Weinhardt, C.: Live biofeedback as a user interface design element: a review of the literature. Commun. Assoc. Inf. Syst. **43**, 257–296 (2018)
23. Zugal, S., Pinggera, J.: Low–cost eye–trackers: useful for information systems research? In: Iliadis, L., Papazoglou, M., Pohl, K. (eds.) Advanced Information Systems Engineering Workshops, CAiSE 2014. Lecture Notes in Business Information Processing, vol. 178, pp. 159–170. Springer, Cham (2014)
24. Burton, L., Albert, W., Flynn, M.: A comparison of the performance of webcam vs. infrared eye tracking technology. In: Proceedings of the Human Factors and Ergonomics Society Annual Meeting, vol. 58, pp. 1437–1441 (2014)
25. Papoutsaki, A.: Scalable webcam eye tracking by learning from user interactions. In: Proceedings of the 33rd Annual ACM Conference Extended Abstracts on Human Factors in Computing Systems, pp. 219–222. ACM, New York (2015)
26. Rayner, K.: Eye movements in reading and information processing: 20 years of research. Psychol. Bull. **124**, 372–422 (1998)
27. Gregor, S., Hevner, A.R.: Positioning and presenting design science research for maximum impact. MIS Q. **37**, 337–355 (2013)
28. Steil, J., Hagestedt, I., Huang, M.X., Bulling, A.: Privacy-aware eye tracking using differential privacy. In: Proceedings of the 11th ACM Symposium on Eye Tracking Research & Applications, pp. 1–9. ACM, New York (2019)
29. Biedert, R., Buscher, G., Schwarz, S., Hees, J., Dengel, A.: Text 2.0. In: Proceedings of the 28th of the International Conference Extended Abstracts on Human Factors in Computing Systems - CHI EA 2010, p. 4003. ACM Press, New York (2010)
30. Gwizdka, J.: Differences in reading between word search and information relevance decisions: evidence from eye-tracking. Lect. Notes Inf. Syst. Organ. **16**, 141–147 (2017)

Detecting Flow Experiences in the Field Using Video-Based Head and Face Activity Recognition: A Pilot Study

Christoph Berger[⊠], Michael T. Knierim, and Christof Weinhardt

Institute for Information Systems and Marketing (IISM), Karlsruhe Institute of Technology (KIT), Karlsruhe, Germany

{christoph.berger9,michael.knierim,christof.weinhardt}@kit.edu

Abstract. Flow represents a valuable daily life experience as it is linked to performance, growth, and well-being. As flow support is still a major challenge due to a lack of automatic and unobtrusive detection methods, NeuroIS scholars face the opportunity to devise measurement approaches for flow experience during IS use and, moreover, flow supporting, adaptive NeuroIS. This work presents the first results from a controlled experience sampling field study in which experiences were observed using video recordings during a week of scientific writing. Novel behavioral features (face and head activity) with negative flow-report correlations are identified. Together, the results contribute to the NeuroIS community by providing an extended concept of flow as a state of behavioral efficiency, the identification of novel correlates, and recommendations for economical and feasible extensions of the study approach.

Keywords: Flow experience · Field study · cESM · FaceReader · Emotion

1 Introduction

The experience of flow is linked to improved performance, growth, and well-being, on individual, organizational and social levels [1]. Flow support has been a growing NeuroIS research topic [2–4] that is still facing challenges in terms of unobtrusive, real-time measurement options [5]. Previous flow detection research was focused on automatic measures of the human central and peripheral nervous system [6, 7]. For real-world observation, these approaches are limited by requiring sensor placement on individuals [4, 8]. In contrast, the investigation of remote flow sensing is missing. Specifically, video recordings have so far not been used to detect flow, but gain unique relevance, especially as social distancing necessitates a stronger focus on field research. Presently available in most ubiquitous interaction devices, cameras are increasingly able to capture various information (e.g. face and head activity, and even facial blood flow observation – enabling e.g. remote HR recordings [9]). Especially as flow experiences might be related to positive emotions (or a lack of negative emotion) [10–12], video recordings of facial activity provide a promising opportunity for NeuroIS flow research. To pursue this

© The Author(s), under exclusive license to Springer Nature Switzerland AG 2021
F. D. Davis et al. (Eds.): NeuroIS 2021, LNISO 52, pp. 120–127, 2021.
https://doi.org/10.1007/978-3-030-88900-5_14

potential, we conducted a pilot study for video-based observation of flow in a highly ecological valid setting, observing students continuing their work on a personally relevant thesis project over a week. The first results show that the investigation of flow-emotion relationships through this method might be limited as emotion expression barely occurred in this non-social context. However, data exploration revealed a novel potential for video-based flow detection through activity variance observation. These findings represent a promising new direction and contribute to an improved understanding of flow detection during IS use.

2 Related Work

The flow experience is described by six characteristics: (1) merging of action and aware-ness, (2) sense of control, (3) loss of self-conscious thought, (4) transformation of time perception, (5) sense of complete control, and (6) intrinsic reward [13]. Flow is possible in tasks that require active engagement and fulfill three pre-conditions: (1) balance of perceived task difficulty and skill, (2) clear goals, and (3) unambiguous feedback [13]. In the NeuroIS community, this is for example confirmed in studies on flow experi-ences during IT-mediated learning [7, 14], electronic gaming [15], or digital knowledge work [3, 4]. While flow research already builds on a robust theory with high levels of researcher agreement [16], considerable debate still exists about conceptual aspects and flow measurement. This is partly because some integrate while others discriminate emo-tional experiences and flow. For example, some argue, flow is experienced as a positive affect [17]. Others posit flow as a state of high arousal and positive valence [10, 18, 19]. Yet again, others argue, that during flow, as self-reflective thinking is disabled, a neutral, non-affective experience must be present [16]. Much past research on the relationship between flow and emotion relied on self-reports [5, 12]. In reviews on the emerging neurophysiological and behavioral flow observation [6, 12], only three studies were identified using facial electromyography (fEMG) to observe relationships of flow with emotion expressions automatically and continuously [10–12]. These fEMG findings are mixed, which might be attributed to contextual differences (e.g., emotion expression as a social phenomenon [20] is possibly attenuated in isolation settings). Thus, more research is needed on the flow-emotion (expression) relationship and video recordings represent an economical and increasingly precise approach for it [21, 22]. While human observers are still the gold standard in expression recognition [23], the majority of currently avail-able software for automatic facial expression detection can analyze video recordings as well as real-time data with up to 80% agreement with human observers [24]. Building on these promising advancements, we considered exploring the relationship of video-based emotion expression and other head activity with flow experiences in the field.

3 Method

To capture naturally occurring flow, a controlled experience sampling (cESM) study was implemented [3, 25]. Students working on advanced thesis stages were observed at home during writing with repeated interruption to capture flow. The study was split into the preparation and a main stage. In the preparation stage, participants received required recording equipment together with setup instructions and a questionnaire for participant background information. In case of setup complications, remote assistance was provided. The main stage comprised five writing sessions (one per day) including three epochs with randomized durations (10, 15, or 20 min). Audio interruptions notified the participants to complete a survey (including the 10-item flow short-scale by [26] - 7p Likert). All participants received a Logitech BRIO 4k UHD webcam to record videos (25 fps and 1280 × 720 resolution) with Logitech Capture software (V2.04.13). The writing software was standardized to Microsoft Word. Four students (female from 21 to 30 years) participated voluntarily. One student did not complete the study and was excluded from the analyses. Flow scores were created by mean averaging reports for all ten items. Cronbach's Alpha indicated good internal consistency (overall: 0.93, across participants: 0.88–0.96). Video feature extraction was realized using Noldus FaceReader 8.1. The extracted features were limited to five minutes before each audio interruption. Two feature spaces were defined for subsequent analyses and distribution variables extracted for each feature: (1) the median –accounting for outliers, and (2) the standard deviation – as a measure of activity variation. The latter was considered interesting given previous reports on flow being related to a more consistent (i.e. less volatile) physical and mental state [7, 27, 28]. The feature spaces of interest are (1) the facial emotion expressions, which comprised dimensional emotion variables (arousal and valence) and discrete emotion expressions (neutrality, anger, disgust, happiness, sadness, fear, surprise) and (2) generic face and head behavior, gaze direction, eye and mouth activity, and head position.

4   Results

First, the reports were plotted to gain an understanding of flow experiences during the study. Figure 1 shows flow with moderate to high levels and average intensities paralleling related work on flow in scientific writing [3] and other tasks [29, 30]. The variation pattern indicates at least some degree of contrasts that would lend themselves to studying relationships of flow experiences with other variables. To understand the variation, an ANOVA was calculated on a linear mixed model (LMM) with participants as random intercept effects and session and epoch as fixed effects. A significant main effect was found for session ($F(4, 36) = 3.083$, $p = 0.028$), but not for epoch ($F(2, 36) = 0.586$, $p = 0.562$). This indicates a variation in flow experiences between sessions but not within. This finding parallels previous work on flow in scientific writing where no intensity variation was found in a single session [3].

Further follow-up tests indicated no main effects of epoch length (i.e. if the interruption came after 10, 15, or 20 min) on flow intensity calculating an ANOVA on an LMM with epoch duration as fixed and participant as random intercept effect ($F(2, 41.56) = 0.920$, $p = 0.406$). Also, the cumulative elapsed time of writing at each interruption within a session showed no significant relationship with flow intensities in an LMM with cumulative epoch duration as fixed and participant as random intercept effect (Beta coefficient: 0.004, $p = 0.739$). Therefore, while the counts of flow intensities in Fig. 1 suggest a trend towards more intense flow during later stages of a writing session (in 10 out of 15 sessions the most intense flow is reported in the last epoch - a trend that would be in line with related work that finds more intense flow in later task stages – see [25]), this effect might too weak to reach statistical significance.

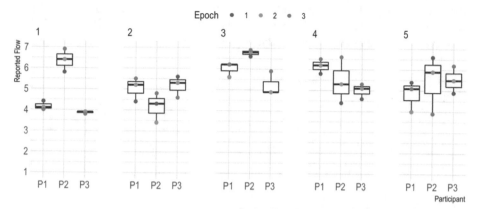

Fig. 1. Flow report distributions across participants and sessions

Next, the variable distribution plots for emotion expressions and general activity (examples see Fig. 2) revealed that hardly any emotion expression occurred before a survey interruption (almost exclusive neutral facial expressions). For the general activity, more variation is present, in particular for the head movement variables. Across all features, the variance metrics (i.e., SDs) showed superior suitability for relationship analysis given their more dispersed variation.

Following these descriptive analyses, the extracted features were entered into a selection process to identify possible flow predictors. To account for the within-subject variation, repeated measures correlation coefficients [31] were calculated for each flow report-feature variable pair mentioned above. This resulted in the computation of 34 correlation coefficients. At this stage, given the small number of participants and the exploratory nature of the analysis, it was decided to not include Bonferroni corrections in favor of identifying possibly novel relationships. Two significant relationships were identified for flow reports with Surprise SD ($r = -0.31$, $p = 0.042$) and Z-Axis Head Orientation SD ($r = -0.39$, $p = 0.009$). Accounting for the possibility of outliers causing these relationships, the analyses were repeated with values omitted that were more than three standard deviations distant from the construct mean (on the within-subject level). The relationship of Surprise SD remained tentative ($r = -0.28$, $p = 0.068$). No outliers were removed for Z-Axis Head Orientation SD and the coefficient, therefore, remained the same.

To assess sensitivity, the same analyses were conducted using video features computed for three- and ten-minute segments. No robust changes in identified relationships emerged, with both relationships always being significant (below the .05 level). Exploring the predictive potential of these variables, an LMM was set up with flow as a dependent variable, z-standardized Surprise SD and Z-Axis Head Orientation SD as fixed effects, and participants as random intercept effects (results see Fig. 2). Both predictors are significantly related to flow reports, albeit Surprise SD only at trend level. A marginal R^2 value of 0.204 indicates a useful explanation of flow report variance from this identified variable set. To further assess the predictive abilities of the regression model, leave-one-participant-out cross-validation was conducted. Mean absolute error (MAE) and root mean squared error (RMSE) averaged across all folds (MAE: 0.79, RMSE: 0.95) indicate promising flow prediction potential with these video-based features.

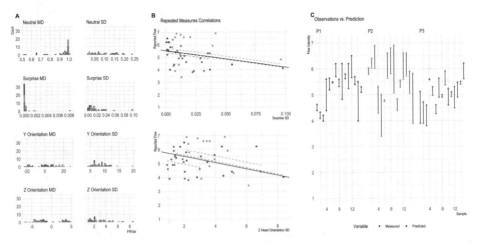

Fig. 2. Face and head activity variables and relationships with flow reports. A: Distributions of two feature examples per feature space. B: Significant repeated measures correlations with random intercepts. C: Reported and predicted flow intensities from an LMM.

5   Discussion and Outlook

In this work, we explored a novel approach for remote flow detection in the field. While the sample is still small with three participants, the data comprises 45 data samples already through the observation over the course of a week. Through this approach, we find first of all that the cESM approach is efficiently eliciting intensified flow episodes while providing some intensity variation. Importantly, this extends related work that showed flow variation limitations when using single task sessions [3]. Extending observation periods to several days overcomes this problem. However, limiting the study context to writing in isolation, revealed an interesting challenge for the investigation of flow-emotion expression relationships due to the absence of emotional expressions. Therefore, the absence of flow-emotion relationships in this data is first of all considered to be related to the study setting and not so much a general phenomenon. A possible direction to investigate this relationship further could be to observe flow in social interaction scenarios, where such expressions might be more frequent [16, 32].

The second main finding is that variance-based metrics appear as useful flow intensity predictors. We consider two possible explanations. On the one hand, experiencing surprise might indicate explicit processing of novel situations, a process that is considered antithetical to implicit cognitive processing that supposedly occurs during flow

[33]. On the other hand, the relationship between flow and surprise might be a proxy for a larger class of automaticity-intervening emotional experiences. Negative emotions have repeatedly been suggested to be incompatible with flow experience [16, 33, 34]. Since no other emotions were expressed, further research is required to elaborate on the validity of these two propositions. For the negative relationship between flow intensities and head movement variation, similar reasoning is considered. Greater variation in head movement (generally, i.e. in either direction) might be indicating reduced flow intensity. As both relationships were identified for variance metrics, we propose altogether that flow might be usefully described as a state of increased behavioral consistency. This is aligned to a central concept in related literature, that flow is physiologically efficient [12, 33]. This means that irrelevant neuro-physiological processes are reduced during flow for optimal energetic resource distribution required to meet the demands of a challenging task. Evidence for this has been provided in the finding of reduced sympathetic activity during flow (reduced SD in electrodermal activity – see [7]) and in reduced variability of frontal brain regions [27]. Similarly, it is conceivable, that flow could be detected from reduced variability in behavioral metrics.

As our context is controlled and the sample still small, more research will be needed to confirm this proposition and to improve the understanding of how the observed constructs relate to flow (e.g. whether they are concomitants – related to flow through relationships to more abstract constructs like concentration or alertness). A strength of the presented approaches is that they provide fertile ground for intensified studies. The video-based flow observation thus gains testable propositions and economical approaches for natural settings. Cameras are already included in most computers and study realization might be even possible without providing additional hard- or software. Even extracting heart rate variability from video data could be a possible feature [35] that has been found useful for flow classification in related work [4]. This work should in the next steps be extended to capture flow experiences from additional contexts and should integrate additional data sources to assess the validity and robustness of flow detection. In summary, this work contributes to the NeuroIS community with a novel opportunity to intensify flow research during (adaptive) IS use.

References

1. Spurlin, S., Csikszentmihalyi, M.: Will work ever be fun again? In: Fullagar, C.J., Delle Fave, A. (eds.) Flow at Work: Measurement and Implications, pp. 176–187 (2017)
2. Riedl, R., Fischer, T., Léger, P.-M., Davis, F.D.: A decade of NeuroIS research: progress, challenges, and future directions. ACM SIGMIS Database DATABASE Adv. Inf. Syst. **51**, 13–54 (2020)
3. Knierim, M.T., Rissler, R., Hariharan, A., Nadj, M., Weinhardt, C.: Exploring flow psychophysiology in knowledge work. In: Davis, F.D., Riedl, R., vom Brocke, J., Léger, P.-M., Randolph, A.B. (eds.) Information Systems and Neuroscience. LNISO, vol. 29, pp. 239–249. Springer, Cham (2019). https://doi.org/10.1007/978-3-030-01087-4_29
4. Rissler, R., Nadj, M., Li, M.X., Knierim, M.T., Maedche, A.: Got flow? Using machine learning on physiological data to classify flow. In: Extended Abstracts of the 2018 CHI Conference on Human Factors in Computing Systems, pp. 1–6 (2018)

5. Moneta, G.B.: On the measurement and conceptualization of flow. In: Engeser, S. (ed.) Advances in Flow Research, pp. 23–50. Springer, New York (2012). https://doi.org/10.1007/978-1-4614-2359-1_2
6. Knierim, M.T., Rissler, R., Dorner, V., Maedche, A., Weinhardt, C.: The psychophysiology of flow: A systematic review of peripheral nervous system features. In: Davis, F.D., Riedl, R., vom Brocke, J., Léger, P.-M., Randolph, A.B. (eds.) Information Systems and Neuroscience. LNISO, vol. 25, pp. 109–120. Springer, Cham (2018). https://doi.org/10.1007/978-3-319-67431-5_13
7. Léger, P.M., Davis, F.D., Cronan, T.P., Perret, J.: Neurophysiological correlates of cognitive absorption in an enactive training context. Comput. Hum. Behav. **34**, 273–283 (2014)
8. Gaggioli, A., Cipresso, P., Serino, S., Riva, G.: Psychophysiological correlates of flow during daily activities. In: Wiederhold, B.K., Riva, G. (eds.) Annual Review of Cybertherapy and Telemedicine 2013, pp. 65–69. IOS Press (2013)
9. Rouast, P.V., Adam, M.T.P., Chiong, R., Lux, E.: Remote heart rate measurement using low-cost RGB face video: a technical literature review. Front. Comp. Sci. **1**, 1–15 (2016)
10. Mauri, M., Cipresso, P., Balgera, A., Villamira, M., Riva, G.: Why is Facebook so successful? Psychophysiological measures describe a core flow state while using Facebook. Cyberpsychol. Behav. Soc. Netw. **14**, 723–731 (2011)
11. Kivikangas, J.M., Puttonen, S.: Psychophysiology of flow experience: an explorative study (2006)
12. Peifer, C.: Psychophysiological correlates of flow-experience. In: Engeser, S. (ed.) Advances in Flow Research, pp. 139–164. Springer, New York (2012). https://doi.org/10.1007/978-1-4614-2359-1_8
13. Nakamura, J., Csikszentmihalyi, M.: Flow theory and research. In: Lopez, S., Snyder, C.R. (eds.) Oxford Handbook of Positive Psychology, pp. 195–206. Oxford University Press, New York (2009)
14. Peifer, C., Schulz, A., Schächinger, H., Baumann, N., Antoni, C.H.: The relation of flow-experience and physiological arousal under stress - can u shape it? J. Exp. Soc. Psychol. **53**, 62–69 (2014)
15. Labonté-Lemoyne, É., et al.: Are we in flow? Neurophysiological correlates of flow states in a collaborative game. In: Proceedings of the 2016 CHI Conference, pp. 1980–1988 (2016)
16. Engeser, S., Schiepe-Tiska, A.: Historical lines and an overview of current research on flow. Adv. Flow Res. **9781461423**, 1–22 (2012)
17. Kivikangas, J.M.: Psychophysiology of flow experience: an explorative study (2006)
18. de Manzano, Ö., Theorell, T., Harmat, L., Ullén, F.: The psychophysiology of flow during piano playing. Emotion **10**, 301–311 (2010)
19. Ullén, F., De Manzano, Ö., Theorell, T., Harmat, L.: The physiology of effortless attention: correlates of state flow and flow proneness. In: Bruya, B. (ed.) Effortless Attention: A New Perspective in the Cognitive Science of Attention and Action, pp. 205–217. MIT Press, Cambridge (2010)
20. Holodynski, M., Friedlmeier, W.: Development of Emotions and Emotion Regulation. Springer, Heidelberg (2006)
21. Höfling, T.T.A., Gerdes, A.B.M., Föhl, U., Alpers, G.W.: Read my face: automatic facial coding versus psychophysiological indicators of emotional valence and arousal. Front. Psychol. **11**, 1–15 (2020)
22. Calvo, R.A., Mello, S.D.: Affect detection : an interdisciplinary review of models, methods, and their applications. IEEE Trans. Affect. Comput. **1**, 18–37 (2010)
23. Dupré, D., Krumhuber, E.G., Küster, D., McKeown, G.J.: A performance comparison of eight commercially available automatic classifiers for facial affect recognition. PLoS ONE **15**, e0231968 (2020)

24. Skiendziel, T., Rösch, A.G., Schultheiss, O.C.: Assessing the convergent validity between the automated emotion recognition software Noldus FaceReader 7 and facial action coding system scoring. PLoS One **14**, 1–18 (2019)
25. Csikszentmihalyi, M., Hunter, J.: Happiness in everyday life: The uses of experience sampling. J. Happiness Stud. **4**, 185–199 (2003)
26. Engeser, S., Rheinberg, F.: Flow, performance and moderators of challenge-skill balance. Motiv. Emot. **32**, 158–172 (2008)
27. Knierim, M.T., Nadj, M., Hariharan, A., Weinhardt, C.: Flow neurophysiology in knowledge work: electroencephalographic observations from two cognitive tasks. In: 5th International Conference on Physiological Computing Systems (PhyCS), pp. 42–53 (2018)
28. Harris, D.J., Vine, S.J., Wilson, M.R.: Flow and quiet eye: the role of attentional control in flow experience. Cogn. Process. **18**(3), 343–347 (2017). https://doi.org/10.1007/s10339-017-0794-9
29. Quinn, R.W.: Flow in knowledge performance experience. Adm. Sci. Q. **50**, 610–641 (2005)
30. Keller, J.: The flow experience revisited: the influence of skills-demands-compatibility on experiential and physiological indicators. In: Harmat, L., Ørsted Andersen, F., Ullén, F., Wright, J., Sadlo, G. (eds.) Flow Experience, pp. 351–374. Springer, Cham (2016). https://doi.org/10.1007/978-3-319-28634-1_21
31. Bakdash, J.Z., Marusich, L.R.: Repeated measures correlation. Front. Psychol. **8**, 1–13 (2017)
32. Bakker, A.B.: Flow among music teachers and their students: the crossover of peak experiences. J. Vocat. Behav. **66**, 26–44 (2005)
33. Harris, D.J., Vine, S.J., Wilson, M.R.: Neurocognitive mechanisms of the flow state. Prog. Brain Res. **237**, 221–243 (2017)
34. Swann, C., Keegan, R.J., Piggott, D., Crust, L.: A systematic review of the experience, occurrence, and controllability of flow states in elite sport. Psychol. Sport Exerc. **13**, 807–819 (2012)
35. Rouast, P.V., Adam, M.T.P., Chiong, R., Cornforth, D.J., Lux, E.: Remote heart rate measurement using low-cost RGB face video: a technical literature review. Front. Comput. Sci. (2016). In Press

Understanding the Potential of Augmented Reality in Manufacturing Environments

Felix Kaufmann[(✉)], Laurens Rook, Iulia Lefter, and Frances Brazier

Faculty of Technology, Policy and Management, Delft
University of Technology, Delft, Netherlands
felixkaufmann@gmx.net, {L.Rook,I.Lefter,F.M.Brazier}@tudelft.nl

Abstract. Manufacturing companies are confronted with challenges due to increasing flexibility requirements and skill gaps. Augmented Reality applications offer an efficient way to overcome these tensions by enhancing the interaction between people and technology. The positive effects of Augmented Reality solutions are often described in individual models in the scientific literature. This research-in-progress aims to aggregate the empirical findings in the usage of Augmented Reality solutions in manufacturing environments. A meta-analysis is conducted to synthesise several small studies into one large study to achieve this. In particular, the meta-analysis will focus on the impact of Augmented Reality applications on cognitive load levels. Furthermore, the effect on processing time and error rates will be evaluated. Initial results of the meta-analysis will be expected and reported at this year's NeuroIS Retreat.

Keywords: Augmented reality · Meta-analysis · Manufacturing · Cognitive load

1 Introduction

Manufacturing companies are confronted with increasing variants and individualised products, with high-quality requirements and short product life cycles [1]. These companies find themselves in a field of tension between multiple requirements from the buyers' market and the labour market [2].

The heightened product diversity leads to interrupted learning curves, especially in maintenance applications, assembly, and machinery repair as part of manufacturing processes [3, 4]. The management of process complexity is further challenged by an ageing and heterogeneous workforce [5]. Despite these growing challenges, manufacturing systems must be reconfigurable and flexible to react quickly to changes in the buyers' market [6].

Highly experienced operators often meet the demand for flexibility with programming, maintenance, and diagnostic skills [7]. Human beings are still indispensable due to their cognitive abilities and flexibility. In particular, experienced operators can achieve flexible adaptation to changing situations and requirements. This ability to change can hardly be realised economically and technically by automated solutions [8].

© The Author(s), under exclusive license to Springer Nature Switzerland AG 2021
F. D. Davis et al. (Eds.): NeuroIS 2021, LNISO 52, pp. 128–138, 2021.
https://doi.org/10.1007/978-3-030-88900-5_15

Simultaneously, operators are exposed to alleviated cognitive and psychological load due to highly flexible employee deployment and continually changing working environments and methods [9]. The underlying information processes must be optimised to reduce both mental and psychological load [10]. However, many manufacturing companies still find themselves confronted with an impractical and inefficient presentation of information on the shop floor [11]. To overcome these challenges, Industry 4.0 solutions that support employees in an agile production environment are promising [12]. In particular, Augmented Reality applications offer a way to support the interaction between people and technology and combine the advantages of manual and automated processes [11].

Cognitive worker assistance systems, including Augmented Reality solutions, offer the potential to increase manufacturing systems' productivity and agility [13]. These devices enable efficient information distribution and support employees in the perception, reception, and processing of information [14]. In this context, the individual roles of employees, their qualifications, and personal characteristics are decisive. Taking them into account enables the provision of specific information adapted to the user and the environment [15]. In this way, an optimal distribution of information on the shop floor can be realised, strengthening manufacturing processes' competitiveness in high-wage geographical locations [16].

2 Research Gap

Manufacturing companies are undergoing major changes in today's world of globalization and digitization. Among others, companies – particularly in high-wage countries – are facing growing competition and disruptive market changes. To counter these challenges, Augmented Reality solutions are a promising technology. Among others, Danielsson et al. [17], Terhoeven et al. [18], Egger and Masood [19], Kohn and Harborth [20], and Vanneste et al. [21] illustrate the relevance and potential impact of Augmented Reality in industrial practice. Possible applications of Augmented Reality technologies are very diverse and mainly focus on applications in assembly, maintenance, and logistics processes [18, 19].

A common feature underlying all experiments is that they have not been investigated in practice-relevant, long-term field experiments. Furthermore, individual studies show ambiguous results, and a statistically powerful empirical assessment is still missing. For this reason, a meta-analysis is needed to determine the aggregated empirical influence of Augmented Reality by synthesising several small studies into one large study.

Danielsson et al. [17], Terhoeven et al. [18], Egger and Masood [19], Kohn and Harborth [20], and Vanneste et al. [21] highlight that an efficient implementation of Augmented Reality in manufacturing environments still requires additional research. In particular, a powerful empirical analysis of the effects of such technology is considered a knowledge gap.

3 Research Objective and Question

Individual studies in the scientific literature often describe the positive effects of Augmented Reality. Yet, little is known about the actual impact on employees' cognitive load levels or performance in manufacturing environments. Therefore, this research aims to explore the interrelationships in the usage of Augmented Reality solutions in manufacturing environments.

First, characteristics of Augmented Reality solutions in manufacturing environments will be identified to allow a quantification of the impact on variables relevant to manufacturing processes. Following, the target variables can be linked to Augmented Reality solutions' characteristics.

Given the motivation and the research objective of this study, the following central research question for this research-in-progress arises:

Can the use of Augmented Reality solutions benefit manufacturing activities and if so, how?

Based on the central research question, further sub-research questions can be derived to be able to answer the central research question:

1. *Which factors in manufacturing activities can be influenced using Augmented Reality solutions?*

Initial research shows that the focus of existing literature in manufacturing contexts lies in the influence of Augmented Reality solutions on the variables cognitive load, processing time, and error rate [19, 21]. This research-in-progress focuses on the influence of Augmented Reality solutions on those three variables.

2. *Can those factors be measured and if so, how?*

Cognitive load in the context of Augmented Reality solutions and manufacturing is mostly measured with the help of the NASA-TLX or NASA-RTLX test [22]. Consequently, this research-in-progress focuses on the assessment of Augmented Reality solutions with these tools. Processing time and error rates are measured during user tests and are comparable for similar test settings.

3. *Can a benefit be achieved and if so, by how much?*

The researchers expect that this research will provide a more powerful and significant evaluation of Augmented Reality's impact on cognitive load levels in the first place. Furthermore, improved processing times and error rates are expected as a result of reduced cognitive load.

4 Methodological Approach

The following section describes the methodological approach to answer and verify the central research question and the corresponding sub-questions. The methodological approach includes four sequential phases as explained in more detail in the following: (1) Meta-analysis, (2) derivation of hypotheses, (3) preparation empirical exploration, (4) execution empirical exploration.

4.1 Meta-analysis

First, a meta-analysis is carried out to analyse the state-of-the-art and the influence of Augmented Reality solutions on variables relevant to manufacturing processes. As identified by Egger and Masood [19] and Vanneste et al. [21], relevant variables include cognitive load, processing time, and error rates.

A meta-analysis is used if individual studies available show ambiguous results, and a statistically powerful assessment is still missing. By synthesising several small studies into one large study, a meta-analysis provides higher significance. As a result, a more powerful statistical influence of Augmented Reality on the evaluation criteria processing time, error rate, and cognitive load is expected. The meta-analysis follows six sequential phases: Formulation of the research question, data collection, evaluation of data, analysis and interpretation of data, sensitivity analysis, and presentation of results.

As part of the meta-analysis, a systematic literature search is conducted to collect and evaluate relevant data. The underlying literature search follows the framework by Vom Brocke et al. [22] (as shown in Fig. 1), which builds on five sequential steps: The definition of the review scope, the conceptualization of the topic, the literature search, the literature analysis and synthesis, and the research agenda. Each step includes individual systematic approaches that are in line with collecting and evaluating data as part of the meta-analysis.

The systematic literature review aims to build an extensive literature database covering empirical studies on Augmented Reality technologies in manufacturing environments. Here, the type of technology used to enable Augmented Reality is not specified in advance. Different types of technological enablers and use cases shall be compared, such that the impact and advantages of different technologies on the target variables can be distinguished. The publications contained in the database are evaluated based on an evaluation scheme (see Fig. 5). All identified publications are classified with the help of a homogeneity assessment to avoid the "apples and oranges problem" [24]. The remaining studies are evaluated with regard to minimal statistical requirements to allow extraction and synthesising. Thereupon, the meta-analysis will be carried out with a previously selected software. The software helps to run statistical calculations to allow an evaluation and interpretation of the data. As part of the subsequent sensitivity analysis, the results are verified by checking for statistical heterogeneity, publication bias, and other confounding factors.

4.2 Derive Hypotheses

Multiple hypotheses are derived based on the results of the meta-analysis. The hypotheses highlight the influence of Augmented Reality solutions on cognitive load levels. Additionally, the impact of different cognitive load levels on processing time and error rate are assessed. The results of the meta-analysis are expected to show a reduction of cognitive load and a decline of processing time as well as error rates through the usage of Augmented Reality during manufacturing activities.

4.3   Preparation and Execution Empirical Exploration

An empirical exploration will be prepared and executed based on the meta-analysis results and the derivation of hypotheses. User tests and surveys help to verify or adjust the formulated hypotheses. The analyses are based on use cases and user tests in the manufacturing department of a chemical and consumer goods company.

For this purpose, research questions are first formulated based on the developed hypotheses and then grouped according to main overarching topics. Based on the research questions, interview questions are formulated. These should encourage the users and experts to provide assessments, descriptions, and narratives on the topic.

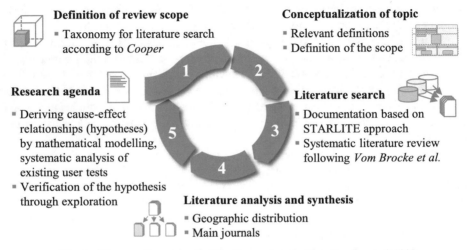

Definition of review scope
- Taxonomy for literature search according to *Cooper*

Conceptualization of topic
- Relevant definitions
- Definition of the scope

Research agenda
- Deriving cause-effect relationships (hypotheses) by mathematical modelling, systematic analysis of existing user tests
- Verification of the hypothesis through exploration

Literature search
- Documentation based on STARLITE approach
- Systematic literature review following *Vom Brocke et al.*

Literature analysis and synthesis
- Geographic distribution
- Main journals

Fig. 1. Framework systematic literature review by Vom Brocke et al. [22]

5   Initial Results Meta-analysis

This chapter presents the initial results of this research-in-progress. In particular, the first two steps of the meta-analysis, namely the formulation of a research question (Sect. 5.1) and data collection (Sect. 5.2) are described in more detail. Besides, Sect. 5.3 highlights the evaluation scheme that allows identifying primary studies relevant to the given research question. The evaluation of the data, the analysis and the interpretation of the data, the performance of a sensitivity analysis, and a summary of the meta-analysis are part of the research-in-progress.

5.1   Formulation of a Research Question

The first step in conducting a meta-analysis is to formulate a research question. As a result, only studies that support the research questions are taken into account in the further course of the meta-analysis. Turabian [25] distinguishes between three types of questions: Conceptual questions, practical questions, and applied questions.

Conceptual questions help readers to understand a certain problem better and to guide the thoughts [25]. Correspondingly, practical questions help develop an approach to change or improve a problematic or improvable situation [25]. Lastly, applied questions help the readers to first better understand a practical problem before solving it. An applied question helps to develop a step towards the solution of a practical problem [25].

This research-in-progress aims to understand the potential of Augmented Reality solutions in manufacturing environments with a meta-analysis. The underlying problem why Augmented Reality solutions are considered to support manufacturing activities is described in Sect. 1. This project thus does not address a conceptual question that helps the reader to understand a problem. In reality, the potential and influence of Augmented Reality solutions must first be researched to develop a concrete procedure to solve the underlying problems. For this reason, the present research question addresses an applied question.

Following Turabian [25], the applied research question of the meta-analysis is as follows:

What influence do Augmented Reality solutions have on workers' cognitive load, processing times, and error rates during manufacturing activities?

5.2 Collection of Data

As described in Sect. 4, a systematic literature review constitutes the data collection for the meta-analysis. Vom Brocke et al. [23] suggest Cooper's [26] taxonomy for a correct classification of the literature search.

Meta-analysis makes use of empirical studies and aims to achieve statistically more powerful assessments. To allow such assessment, the systematic literature review's focus lies on available research outcomes [26]. This project also aims to "integrate or synthesize past literature that is believed to relate to the same issue" [26]. At the same time, this project aims to identify central issues in Augmented Reality applications that have dominated past endeavors. The literature review attempts to represent the influence of Augmented Reality solutions neutrally. Following Booth [27], the exclusion criteria do not eliminate a particular point of view. Additionally, conclusions will be based on an exhaustive and selective review [26]. The organisation of the systematic literature review follows both a conceptual and methodological approach. Publications that relate to the same abstract ideas and employ similar methods are grouped [26]. Lastly, this review intends to address general scholars and practitioners. As a result, the review tries to pay "greater attention to the implication of the work being covered" [26] than on jargon and details. Figure 2 displays the described taxonomy by Cooper [26].

Next, a search string is created based on the classification of the literature search by Cooper [26]. The search string and different combinations of the keywords help to identify relevant publications in the first place. As shown in Fig. 3, the search string is constructed with three distinct segments: Technology, domain, and the target variable. The corresponding keywords result in 18 individual search strings.

Additionally, the STARLITE methodology is used as a documentation standard (see Fig. 4) [27]. As a result of an exhaustive and selective sampling strategy, this project considers all literature within predefined boundaries. The search for relevant literature

Characteristic	Categories			
Focus	**Research outcomes**	Research methods	Theories	Applications
Goal	**Integration**	Criticism		**Central Issues**
Perspective	**Neutral Representation**		Espousal of Position	
Coverage	Exhaustive	**Exhaustive and Selective**	Representative	Central
Organisation	Historical	**Conceptual**		**Methodological**
Audience	Specialized Scholars	**General Scholars**	**Practitioners**	General Public

Fig. 2. Completed taxonomy following Cooper [25]

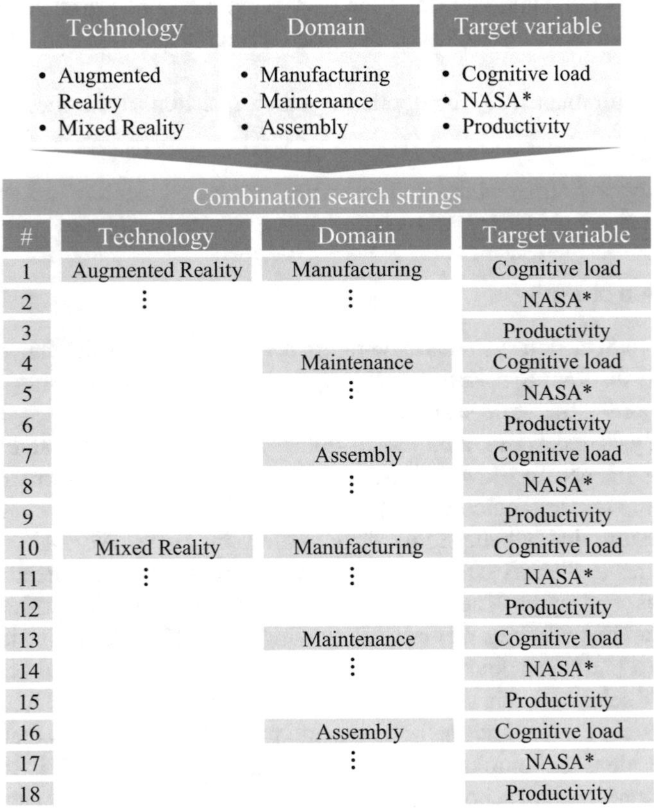

Technology	Domain	Target variable
• Augmented Reality	• Manufacturing	• Cognitive load
• Mixed Reality	• Maintenance	• NASA*
	• Assembly	• Productivity

Combination search strings			
#	Technology	Domain	Target variable
1	Augmented Reality	Manufacturing	Cognitive load
2	⋮	⋮	NASA*
3			Productivity
4		Maintenance	Cognitive load
5		⋮	NASA*
6			Productivity
7		Assembly	Cognitive load
8		⋮	NASA*
9			Productivity
10	Mixed Reality	Manufacturing	Cognitive load
11	⋮	⋮	NASA*
12			Productivity
13		Maintenance	Cognitive load
14		⋮	NASA*
15			Productivity
16		Assembly	Cognitive load
17		⋮	NASA*
18			Productivity

Fig. 3. Keywords and search string combinations

is limited to journal articles and books and is conducted with the help of an evaluation scheme following Vom Brocke et al. [23]. Augmented Reality applications have evolved significantly in recent years, and publications have increased considerably since 2014. Consequently, this thesis includes English and German articles between 2014 and 2021. Augmented Reality solutions are currently implemented in numerous different fields of application. As indicated in Sect. 1, this research project particularly focuses on the manufacturing industry. For this reason, literature without any empirical evaluation of

Augmented Reality solutions in manufacturing environments is excluded. Six different databases are chosen not to miss any relevant research outcomes.

(S)	Sampling Strategy	Exhaustive and selective
(T)	Type of Studies	Restriction to journal articles and books
(A)	Approaches	Keyword search in databases, forward search, backward search
(R)	Range of Years	Consideration of all sources published from 2014 to March 2021
(L)	Limits	Limitation to English and German sources
(I)	Inclusion & Exclusion	Focus on quantitative evaluation of Augmented Reality solutions in manufacturing environments
(T)	Terms used	Augmented Reality/ Mixed Reality, Manufacturing/ Maintenance/ Assembly, Cognitive load/ NASA*/ Productivity
(E)	Electronic Sources	IEEE, ISI Web of Knowledge, JSTOR, Science Direct, Scopus, ABI Informs

Fig. 4. Completed STARLITE approach following Booth [27]

5.3 Evaluation of Data

The evaluation of the data follows the described data collection phase and is highly dependent on the latter's results [28]. This phase identifies suitable primary studies to be included in the meta-analysis based on the data collection phase results. The collected primary studies are assessed with the help of a predefined and systematic evaluation scheme. Consequently, individual studies are eliminated, and the relevance and statistical independence of the meta-analysis are strengthened.

As part of the framework for systematic literature reviews (Fig. 1), Vom Brocke et al. [23] suggest a structured literature search process. The STARLITE methodology results form the evaluation scheme's basis and are included in the scheme's first two steps. Figure 5 displays the evaluation scheme used in preparation for the meta-analysis.

The first step of the evaluation scheme is the identification of primary studies. Primary studies are collected based on the STARLITE methodology and the corresponding keywords and search strings. Next, the duplicates are eliminated. Duplicates occur as the keyword search is conducted in multiple databases. Additionally, the results of the keyword search from off-topic journals are eliminated. Following, the eligibility of the remaining articles with regard to the research question is evaluated. Here, the depth of content increases gradually. First, the individual titles are assessed. Second, the abstracts of the remaining primary studies are evaluated. Finally, the full text is assessed. Last, further primary studies are identified through a forward and backward search. The chosen studies from the forward and backward search are evaluated according to the described procedure. As a result of the evaluation scheme, a relevant and predefined literature database is created.

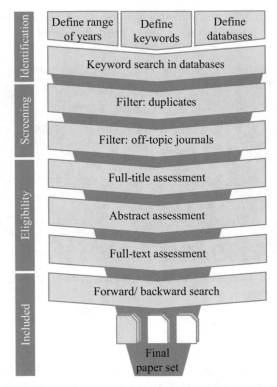

Fig. 5. Evaluation of data following Vom Brocke et al. [23]

6 Discussion and Conclusion

The present research-in-progress aims to understand the influence of Augmented Reality applications on manufacturing environments. A meta-analysis will be conducted to aggregate empirical user studies. The focus of this research-in-progress lies on the variables cognitive load, processing time, and error rates and its interrelationships.

This research-in-progress could in future work be used to explore the technostress produced by employees forced to work with Augmented Reality technologies. It would make perfect sense to assess the impact of technostress on cognitive load levels and productivity variables. In a next step, traditional electroencephalography (EEG) procedures for testing cortisol-inhibition linkages [29] could be adapted to empirical analysis.

References

1. Lušić, M., Fischer, C., Bönig, J., Hornfeck, R., Franke, J.: Worker information systems: state of the art and guideline for selection under consideration of company specific boundary conditions. Proc. CIRP **41**, 1113–1118 (2016). https://doi.org/10.1016/j.procir.2015.12.003
2. Teubner, S., Vernim, S., Dollinger, C., Reinhart, G.: Worker-centred production management – approaches to flexibility-and productivityen-hancing integration of production workers into an increasingly digitized and connected production system. ZWF Zeitschrift fuer Wirtschaftlichen Fabrikbetr. **113**(10), 647–651 (2018). https://doi.org/10.3139/104.111983

3. Gaimon, C., Singhal, V.: Flexibility and the choice of manufacturing facilities under short product life cycles. Eur. J. Oper. Res. **60**(2), 211–223 (1992). https://doi.org/10.1016/0377-2217(92)90094-P

4. Masoni, R., et al.: Supporting remote maintenance in industry 4.0 through augmented reality. Proc. Manuf. **11**, 1296–1302 (2017). https://doi.org/10.1016/j.promfg.2017.07.257

5. Hold, P., Erol, S., Reisinger, G., Sihn, W.: Planning and evaluation of digital assistance systems. Proc. Manuf. **9**, 143–150 (2017). https://doi.org/10.1016/j.promfg.2017.04.024

6. ElMaraghy, H., et al.: Product variety management. CIRP Ann. - Manuf. Technol. **62**(2), 629–652 (2013). https://doi.org/10.1016/j.cirp.2013.05.007

7. Sethi, A.K., Sethi, S.P.: Flexibility in manufacturing: a survey. Int. J. Flex. Manuf. Syst. **2**(4), 289–328 (1990). https://doi.org/10.1007/BF00186471

8. Stoessel, C., Wiesbeck, M., Stork, S., Zaeh, M.F., Schuboe, A.: Towards optimal worker assistance: investigating cognitive processes in manual assembly. In: Mitsuishi, M., Ueda, K., Kimura, F. (eds.) Manufacturing Systems and Technologies for the New Frontier, pp. 245–250. Springer, London (2008). https://doi.org/10.1007/978-1-84800-267-8_50

9. Vernim, S., Reinhart, G.: Usage frequency and user-friendliness of mobile devices in assembly. Procedia CIRP **57**, 510–515 (2016). https://doi.org/10.1016/j.procir.2016.11.088

10. Chen, B., Wan, J., Shu, L., Li, P., Mukherjee, M., Yin, B.: Smart factory of industry 4.0: key technologies, application case, and challenges. IEEE Access **6**, 6505–6519 (2017). https://doi.org/10.1109/ACCESS.2017.2783682

11. Burggräf, P., Dannapfel, M., Adlon, T., Riegauf, A., Schmied, J.: Automation configuration evaluation in adaptive assembly systems based on worker satisfaction and costs. In: Nunes, I.L. (ed.) AHFE 2019. AISC, vol. 959, pp. 12–23. Springer, Cham (2020). https://doi.org/10.1007/978-3-030-20040-4_2

12. Johansson, P.E.C., Malmsköld, L., Fast-Berglund, Å., Moestam, L.: Enhancing future assembly information systems - putting theory into practice. Proc. Manuf. **17**, 491–498 (2018). https://doi.org/10.1016/j.promfg.2018.10.088

13. Keller, T., Bayer, C., Bausch, P., Metternich, J.: Benefit evaluation of digital assistance systems for assembly workstations. Proc. CIRP **81**, 441–446 (2019). https://doi.org/10.1016/j.procir.2019.03.076

14. Syberfeldt, A., Danielsson, O., Gustavsson, P.: Augmented reality smart glasses in the smart factory: product evaluation guidelines and review of available products. IEEE Access **5**, 9118–9130 (2017). https://doi.org/10.1109/ACCESS.2017.2703952

15. Galaske, N., Anderl, R.: Approach for the development of an adaptive worker assistance system based on an individualized profile data model. In: Advances in Intelligent Systems and Computing, vol. 490, pp. 543–556 (2016). https://doi.org/10.1007/978-3-319-41697-7_47

16. Dachs, B., Kinkel, S., Jäger, A.: Bringing it all back home? Backshoring of manufacturing activities and the adoption of Industry 4.0 technologies. J. World Bus. **54**(6) (2019). https://doi.org/10.1016/j.jwb.2019.101017

17. Danielsson, O., Holm, M., Syberfeldt, A.: Augmented reality smart glasses for operators in production: survey of relevant categories for supporting operators. Proc. CIRP **93**, 1298–1303 (2020). https://doi.org/10.1016/j.procir.2020.04.099

18. Terhoeven, J., Schiefelbein, F.P., Wischniewski, S.: User expectations on smart glasses as work assistance in electronics manufacturing. Proc. CIRP **72**, 1028–1032 (2018). https://doi.org/10.1016/j.procir.2018.03.060

19. Egger, J., Masood, T.: Augmented reality in support of intelligent manufacturing – a systematic literature review. Comput. Ind. Eng. **140**, 106195 (2020). https://doi.org/10.1016/j.cie.2019.106195

20. Kohn, V., Harborth, D.: Augmented reality - a game changing technology for manufacturing processes? (2018). https://aisel.aisnet.org/ecis2018_rp/111/. Accessed 23 Feb 2021

21. Vanneste, P., Huang, Y., Park, J.Y., Cornillie, F., Decloedt, B., Van den Noortgate, W.: Cognitive support for assembly operations by means of augmented reality: an exploratory study. Int. J. Hum. Comput. Stud. **143** (2020). https://doi.org/10.1016/j.ijhcs.2020.102480

22. Jeffri, N.F.S., Awang Rambli, D.R.: A review of augmented reality systems and their effects on mental workload and task performance. Heliyon **7**(3) (2021). https://doi.org/10.1016/j.heliyon.2021.e06277

23. Vom Brocke, J., Simons, A., Niehaves, B., Riemer, K., Plattfaut, R., Cleven, A.: Reconstructing the giant: on the importance of rigour in documenting the literature search process (2009). www.uni.lihttp://aisel.aisnet.org/ecis2009/161/. Accessed 22 Feb 2021

24. Lipsey, M.W., Wilson, D.B.: Practical Meta-analysis. SAGE Publication, Inc. (2001). https://psycnet.apa.org/record/2000-16602-000. Accessed 23 Feb 2021

25. Turabian, K.L.: A Manual for Writers of Research Papers, Theses, and Dissertations: Chicago Style for Students and Researchers. University of Chicago Press (2013)

26. Cooper, H.M.: Organizing knowledge syntheses: a taxonomy of literature reviews. Knowl. Soc. **1**(1), 104–126 (1988). https://doi.org/10.1007/BF03177550

27. Booth, A.: "Brimful of STARLITE": toward standards for reporting literature searches. J. Med. Libr. Assoc. **94**, 421 (2006)

28. Forza, C., Di Nuzzo, F.: Meta-analysis applied to operations management: summarizing the results of empirical research. Int. J. Prod. Res. **36**(3), 837–861 (1998). https://doi.org/10.1080/002075498193714

29. Tops, M., Boksem, M.A.S.: Cortisol involvement in mechanisms of behavioral inhibition. Psychophysiology **48**(5), 723–732 (2011). https://doi.org/10.1111/j.1469-8986.2010.01131.x

On How Mind Wandering Facilitates Creative Incubation While Using Information Technology: A Research Agenda for Robust Triangulation

Frederike M. Oschinsky[1]([⊠]), Bjoern Niehaves[1], René Riedl[2,3], Michael Klesel[1,4], Selina C. Wriessnegger[5], and Gernot R. Mueller-Putz[5]

[1] University of Siegen, Siegen, Germany
{frederike.oschinsky,bjoern.niehaves,
michael.klesel}@uni-siegen.de
[2] University of Applied Sciences Upper Austria, Steyr, Austria
rene.riedl@fh-steyr.at
[3] Johannes Kepler University Linz, Linz, Austria
rene.riedl@jku.at
[4] University of Twente, Enschede, The Netherlands
m.klesel@utwente.nl
[5] Graz University of Technology, Graz, Austria
{s.wriessnegger,gernot.mueller}@tugraz.at

Abstract. Our minds tend to frequently drift away from present technology-related situations and tasks. Against this background, we seek to provide a better understanding of mind-wandering episodes while using information technology and its link to decisive variables of Information Systems research, such as performance, creativity and flow. Since the academic literature still lacks reliable and validated measurements that can fully account for all facets of mind-wandering episodes while using information technology, our work addresses this gap by presenting a way to triangulate data in the context of a digital insight problem-solving task. This new approach enables researchers to further investigate the effects of spontaneous thought in technology-related settings and is a promising building block for the development of neuroadaptive systems.

Keywords: Technology use · Mind wandering · Creative incubation · Insight · Triangulation · Experience sampling · Behavioral markers · Neuroimaging

1 Introduction

Information Systems (IS) research studies how to reason or interact with information technology (IT). Building on that, Neuro-Information-Systems (NeuroIS) seeks to understand the development, use, and impact of IT by including neurophysiological knowledge [1]. The emphasis is set on understanding how humans interact with IT, e.g., for designing neuroergonomic or neuroadaptive systems. Studies test for externally-focused

F. D. Davis et al. (Eds.): NeuroIS 2021, LNISO 52, pp. 139–147, 2021.
https://doi.org/10.1007/978-3-030-88900-5_16

concentration and internally-directed attention, as cognition is not limited to the processing of events in the environment. While intrinsically-generated thoughts such as mind wandering (MW) become increasingly relevant, however, measuring them comes along with methodological challenges. As NeuroIS research successfully coped with similar obstacles (e.g., studying technostress; see [2–5]), the triangulation of neurophysiological data and self-reports seems a particularly promising avenue.

MW is described as a shift in attention away from a primary task and toward dynamic, unconstrained, spontaneous thoughts [6, 7] – or as the mind's capacity to drift away aimlessly from external events and toward internally directed thoughts [8]. The emphasis on attentional engagement in IS research follows the implicit assumption that our thoughts are continuously focused [9–11]. However, a growing body of knowledge suggests the opposite – namely, that our minds regularly tend to proceed in a seemingly haphazard manner. Therefore, neglecting MW leaves important IT phenomena largely unexplored. It is complex in nature and can have both negative and positive effects. For example, MW can be a necessary and useful cognitive phenomenon that offers potential for technology-mediated creativity (e.g., in webinars). In contrast, it can go along with various deficits in performance (e.g., IT management), disturbed team dynamics (e.g., trust), or weakened IT security (e.g., data management). Building on the findings of current research on digital stress (e.g., on information overload or interruptions), and based on the increasing demand for healthy breaks and distraction-free phases at work, we focus on the potential of wandering thoughts. Research in complex technology-related contexts can benefit from both a clear conceptualization of MW and comprehensive triangulation that adequately captures its characteristics. Against this background, our study is novel because it addresses measurement of intrinsically-generated thought while using technology. Without reliable and valid measurement, it is hardly possible to understand whether to expect negative or positive consequences; moreover, it is difficult to design systems that either increase or reduce MW episodes. Against this background, the research question of this work-in-progress paper is: Which procedure is most suitable for measuring MW while using IT? In order to answer this question, we will briefly introduce the theoretical background as well as the neurophysiological correlates of the relevant concepts, propose a procedure for triangulation, and close with an outlook on our next steps.

2 Theoretical Background

Solving complex problems is often associated with creativity, as the solution seems new and useful. Looking at the creative process [12], insight problem-solving is often associated with incubation. Incubation stands for taking a step back from the problem, and for allowing the mind to wander. In this phase, unconscious thought processes take over, e.g., while going for a walk, taking a shower, or in Newton's case, while sitting under a tree. This stage is followed by illumination (i.e., "Eureka!"), as well as verification (or implementation), where we build, test, analyze, and evaluate the idea. Considering incubation is central when dealing with internally-directed attention, because it helps understand whether and why past studies have shown that letting the mind wander in this phase can lead to greater creativity [13]. The benefits of incubation appear to be greater when being engaged in an undemanding task, where MW is found to be more frequent, than in a demanding task or no task at all (ibid.). Therefore, task difficulty can be used as a manipulation in our experiment.

According to Christoff et al. [14], MW is "(…) a mental state, or a sequence of mental states, that arises relatively freely due to an absence of strong constraints on the contents of each state and on the transitions from one mental state to another" (p. 719). It often occurs during tasks that do not require sustained attention [15]. Literature refers to it as unguided, unintentional, task-unrelated, or stimulus-independent thought [16]. Because empirical evidence expresses concern to describe MW as unguided [17], unintentional [18, 19], or stimulus-independent [20–22], we follow the family-resemblances perspective by Seli et al., which treats it as a heterogeneous construct [16]. Against this background, it becomes all the more important to clearly measure and describe the specific aspects of MW when investigating it in technology-related settings. Given that MW is considered to represent a failure of attention and control [23–27], their potential to yield beneficial outcomes has been widely neglected. Only in the last decade have studies highlighted its advantages, which include more-effective brain processing, pattern recognition, and associative thinking as well as increased creativity [13, 15, 20, 28, 29]. Recent IS research shows that MW relates to enjoyment [30, 31], creativity [13, 32] as well as performance and knowledge retention [20, 33, 34].

Evidence shows that deep absorption undermines creativity, whereas distraction can enhance it [13, 35]. This speaks in favor of taking breaks, appreciating boredom, and doing simple, monotonous things when agonizing. In this context, the benefit of incubation seems greatest when being engaged in an undemanding task, compared to a demanding task or no task at all [15]. Because undemanding tasks evidently open the door for MW as attentional demand reduces MW [36], we expect that the success of incubation (i.e., insight problem-solving while using IT) relates to the opportunity for MW.

3 Methodology

MW is studied mostly by using thought sampling and questionnaires [15]. Facing the potential shortcomings of subjective self-reports (e.g., common methods, social desirability, subjectivity [37] (p. 688)), we depict triangulation as a more promising strategy, in which one applies different methods, types of data, and perspectives to the same phenomenon to achieve a higher validity of the results and to reduce systematic errors. In specific, we will conduct an experiment, in which we will triangulate neurophysiological data and self-reports. Because literature introduces a number of different methods of estimating MW, we briefly summarize the overview by Martinon et al. [34].

Experience Sampling. The gold standard measure estimates thoughts and feelings as they occur. However, the data relies on subjective inquiry. There are three groups: First, online experience sampling gathers self-reports of the participants' ongoing experience 'in the moment' while they are completing other activities. Either the probe caught method (open/closed) requires participants to be intermittently interrupted, often while performing a task, and describe the content of their experience. The self-caught method asks them to spontaneously report their mental state (e.g., MW) as soon as they notice it, e.g., by pressing a button. This accounts for meta-awareness. Second, retrospective experience sampling gathers data immediately after a task has been completed. The reports can be biased, as they rely on memory. Third, the assessment of disposition encompasses multiple dimensions of experience and includes personality traits.

Behavioral Markers. Behavioral indices provide evidence of the nature of an ongoing thought at a specific moment of time or in a particular task. They deliver additional insight into the processes underlying different aspects of experience and are a less subjective measure of the observable consequences associated with performing dull, monotonous tasks. There are numerous potential tasks, such as the Sustained Attention Response task (SART), the Oddball task, reading (comprehension) tasks, breath counting tasks, the Complex Working Memory task (CWM), or the Instructed Mind Wandering task (IMW). Task complexity can be varied. However, in isolation, behavioral markers struggle to provide evidence on underlying causal mechanisms, being only a superficial description of the nature of experience.

Neurophysiological Tools. Neurophysiological measures allow for a more detailed picture of whether participants' attention is directed externally or internally, by illustrating the level of engagement during different stages of ongoing thought [38]. They show that during MW, attention shifts from the processing of sensory input (suppression of external stimuli by perceptual decoupling) to internally-directed processes [39]. The measures include, but are not limited to, electroencephalography (EEG), eye-tracking, and functional magnetic resonance imaging (fMRI). First, EEG is a recognized brain imaging tool, which assesses MW non-invasively without interfering with a task [40]. The event-related potential (ERP) ("a waveform complex resulting from an external stimulus" [41]), and EEG oscillations ("the manifestation of the activity of populations of neurons in the brain" (ibid.)) can be assessed. During MW, perceptual input is reduced, pointing at P1-N1, P2, and P3 as discriminative ERP-features. Studies observe an increased activity of lower oscillation frequencies, namely theta and delta, as well as a decrease of higher frequencies, namely alpha and beta [40]. Second, fMRI measures brain activity by detecting changes associated with blood flow. It controls for individual variation, e.g., in the Default Mode Network (DMN), but it is highly intrusive (for details on the concept of intrusiveness in NeuroIS research, see [42]), more expensive than EEG, time-consuming, and does not allow for temporal conclusions on the milliseconds level as EEG. Third, eye tracking operates as a reliable "time-sensitive indicator of internal attention demands" by capturing specific eye behavior changes [39]. These psychophysiological changes are divided into three ocular mechanisms: visual disengagement, perceptual decoupling, and internal coupling (ibid.). Since eye-tacking is non-invasive, relatively inexpensive and has already been widely applied, recent studies increasingly integrate this tool [43] (p. 22). In the future, all of the three presented techniques offer great potential, for example, when it comes to developing machine learning estimators for MW detection, for non-invasive brain stimulations, or for building neuroadaptive systems that adapt to the mental state of technology users in real-time (e.g., [44–48]).

To the best of our knowledge, few studies have directly assessed the occurrence of MW during incubation [13]. Our work uses the incubation paradigm and seeks to enhance the meaningfulness and reliability of the involved measurement. Our proposed experiment will be based on the Unusual Uses Task (UUT) [13], a classic and widely used measure of divergent thinking [49]. It requires participants to generate as many unusual uses as possible for a common object, such as a food can, in a given amount of time. The originality of the responses is taken as an index of creative thinking [13]. The experimental procedure will replicate the work by Baird et al. [13]. Based on our past research [30–34], we propose to add neurophysiological measures, namely EEG and eye-tracking, to experience samplings [20, 30, 50–52]. The combination of self-reported information with the detailed measures of neural function promises to shed critical light on aspects of spontaneous thought while using IT.

Participants will be randomly assigned to work on two digital UUT problems (5 min each). They will tell their responses to the investigator who types them into a text box on a computer. After completing the baseline UUT, participants will be assigned to one of three groups (demanding task, undemanding task, rest) using a between-subjects design. The aim is to have approximately the same number of participants in the respective groups. Participants in the demanding-task condition will perform a 3-back task, whereas those in the undemanding-task condition will perform a simpler task (1-back). In the rest condition, participants will be asked to sit quietly. This step will be followed by incubation (12 min). Next, all participants will answer a MW questionnaire [based on e.g., 20, 30, 50–52], and then work on the same two UUT again (5 min). Finally, they will be thanked, debriefed and receive financial compensation. At each point, the cognitive processes of the participants will be recorded with an EEG and eye tracking device. The tools' high temporal resolution (milliseconds level) will make it possible to determine thought patterns and to work out the typical course of a MW episode. The self-reports will serve to validate the findings. In addition, the assessment of creativity by two raters controls the behavioral correlate. The following research agenda is inspired by Dimoka et al. [53] (Fig. 1).

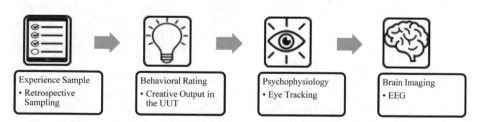

Fig. 1. Research agenda for robust triangulation

4 Outlook

Our work contributes to two crucial pillars of NeuroIS research [48], namely to designing information systems and developing neuroadaptive systems. First, we make a call for future research focusing on the relation between technology and creativity from various perspectives, e.g., on other phase of the creative process besides incubation. Future work can enhance our neuroscientific models of creativity while using IT, and further develop creativity-promoting tools. Moreover, we strongly believe that neuroadaptive systems offer significant potential, both from a theoretical and practical viewpoint. Although coming up with systems that adapt to the users' mental states in real time might sound utopic for mainstream IS and management researchers, efforts are already being made (not only in NeuroIS, but also in other fields that have been existing longer, such as affective computing, physiological computing, and brain-computer interfacing). Our work is a first step towards automatically observing and interpreting MW, which could help design human-computer interaction tasks and IT artifacts to increase the users' performance, productivity, and creativity. Note that the group of users explicitly also comprises programmers and software designer (because they are also users of computer systems). Creativity is a critical talent or skill in software development, and the potential of neuroscience for software engineering has been documented comprehensively in a recent review [54]. We see MW while using IT as a promising future research area based on the practical, methodological and theoretical values our project offers.

Acknowledgements. This research was supported by the Volkswagen Foundation (grant: 96982).

References

1. Riedl, R., Léger, P.-M.: Fundamentals of NeuroIS. Information Systems and the Brain. Springer, Heidelberg (2016). https://doi.org/10.1007/978-3-662-45091-8
2. Riedl, R.: On the biology of technostress. ACM SIGMIS Database **44**, 18–55 (2013)
3. Fischer, T., Riedl, R.: Technostress research: a nurturing ground for measurement pluralism? Commun. Assoc. Inf. Syst. **40**, 375–401 (2017)
4. Tams, S., Hill, K., de Guinea, A.O., Thatcher, J.B., Grover, V.: NeuroIS-alternative or complement to existing methods? Illustrating the holistic effects of neuroscience and self-reported data in the context of technostress research. J. Assoc. Inf. Syst. **15**, 723–753 (2014)
5. Fischer, T., Reuter, M., Riedl, R.: The digital stressors scale. Development and validation of a new survey instrument to measure digital stress perceptions in the workplace context. Front. Psychol. **12**(607598), 1–18 (2021)
6. Andrews-Hanna, J.R., Irving, Z.C., Fox, K.C.R., Spreng, R.N., Christoff, K.: The neuroscience of spontaneous thought. An evolving, interdisciplinary field. In: Fox, K.C.R., Christoff, K. (eds.) The Oxford Handbook of Spontaneous Thought. Mind-Wandering, Creativity, and Dreaming, pp. 143–163. Oxford University Press, New York (2018)
7. Smallwood, J., Schooler, J.W.: The restless mind. Psychol. Bull. **132**, 946–958 (2006)
8. Giambra, L.M.: Task-unrelated thought frequency as a function of age. A laboratory study. Psychol. Aging **4**, 136–143 (1989)
9. Addas, S., Pinsonneault, A.: E-mail interruptions and individual performance. Is there a silver lining? MIS Q. **42**, 381–405 (2018)
10. Agarwal, R., Karahanna, E.: Time flies when you're having fun. Cognitive absorption and beliefs about information technology usage. MIS Q. **24**, 665–694 (2000)
11. Devaraj, S., Kohli, R.: Performance impacts of information technology. Is actual usage the missing link? Manage. Sci. **49**, 273–289 (2003)
12. Amabile, T.M., Barsade, S.G., Mueller, J.S., Staw, B.M.: Affect and creativity at work. Adm. Sci. Q. **50**, 367–403 (2005)

13. Baird, B., Smallwood, J., Mrazek, M.D., Kam, J.W.Y., Franklin, M.S., Schooler, J.W.: Inspired by distraction. Mind wandering facilitates creative incubation. Psychol. Sci. **23**, 1117–1122 (2012)

14. Christoff, K., Irving, Z.C., Fox, K.C.R., Spreng, R.N., Andrews-Hanna, J.R.: Mind-wandering as spontaneous thought. A dynamic framework. Nat. Rev. Neurosci. **17**, 718–731 (2016)

15. Smallwood, J., Schooler, J.W.: The science of mind wandering. Empirically navigating the stream of consciousness. Ann. Rev. Psychol. **66**, 487–518 (2015)

16. Seli, P., et al.: Mind-wandering as a natural kind. A family-resemblances view. Trends Cogn. Sci. **22**, 479–490 (2018)

17. Agnoli, S., Vanucci, M., Pelagatti, C., Corazza, G.E.: Exploring the link between mind wandering, mindfulness, and creativity. A multidimensional approach. Creat. Res. J. **30**, 41–53 (2018)

18. Seli, P., Risko, E.F., Smilek, D., Schacter, D.L.: Mind-wandering with and without intention. Trends Cogn. Sci. **20**, 605–617 (2016)

19. Seli, P., Risko, E.F., Smilek, D.: On the necessity of distinguishing between unintentional and intentional mind wandering. Psychol. Sci. **27**, 685–691 (2016)

20. Wati, Y., Koh, C., Davis, F.: Can you increase your performance in a technology-driven society full of distractions? In: Proceedings of the 35th International Conference on Information Systems, Auckland, New Zealand, pp. 1–11 (2014)

21. Sullivan, Y.W., Davis, F., Koh, C.: Exploring mind wandering in a technological setting. In: Proceedings of the 36th International Conference on Information Systems, Fort Worth, United States of America, pp. 1–22 (2015)

22. Sullivan, Y.W., Davis, F.D.: Self-regulation, mind wandering, and cognitive absorption during technology use. In: Proceedings of the 53rd Hawaii International Conference on System Sciences, Honolulu, Hi, USA, pp. 4483–4492 (2020)

23. Baldwin, C.L., Roberts, D.M., Barragan, D., Lee, J.D., Lerner, N., Higgins, J.S.: Detecting and quantifying mind wandering during simulated driving. Front. Hum. Neurosci. **11**, 1–15 (2017)

24. Drescher, L.H., van den Bussche, E., Desender, K.: Absence without leave or leave without absence. Examining the interrelations among mind wandering, metacognition and cognitive control. PLoS ONE **13**, 1–18 (2018)

25. Mooneyham, B.W., Schooler, J.W.: The costs and benefits of mind-wandering. A review. Can. J. Exp. Psychol. **67**, 11–18 (2013)

26. Smallwood, J., Fishman, D.J., Schooler, J.W.: Counting the cost of an absent mind. Mind wandering as an underrecognized influence on educational performance. Psychonom. Bull. Rev. **14**, 230–236 (2007)

27. Zhang, Y., Kumada, T., Xu, J.: Relationship between workload and mind-wandering in simulated driving. PLoS ONE **12**, 1–12 (2017)

28. Smeekens, B.A., Kane, M.J.: Working memory capacity, mind wandering, and creative cognition. An individual-differences investigation into the benefits of controlled versus spontaneous thought. Psychol. Aesthet. Creat. Arts **10**, 389–415 (2016)

29. Fox, K.C.R., Beaty, R.E.: Mind-wandering as creative thinking. Neural, psychological, and theoretical considerations. Curr. Opin. Behav. Sci. **27**, 123–130 (2019)

30. Oschinsky, F.M., Klesel, M., Ressel, N., Niehaves, B.: Where are your thoughts? On the relationship between technology use and mind wandering. In: Proceedings of the 52nd Hawaii International Conference on System Sciences, Honolulu, Hi, USA, pp. 6709–6718 (2019)

31. Klesel, M., Oschinsky, F.M., Conrad, C., Niehaves, B.: Does the type of mind-wandering matter? Extending the inquiry about the role of mind-wandering in the IT use experience. Internet Res. (2021)

32. Baumgart, T.L., Klesel, M., Oschinsky, F.M., Niehaves, B.: Creativity loading – please wait! Investigating the relationship between interruption, mind wandering and creativity. In: Proceedings of the 53rd Hawaii International Conference on System Sciences. Honolulu, Hi, USA (2020)

33. Klesel, M., Oschinsky, F.M., Niehaves, B., Riedl, R., Müller-Putz, G.R.: Investigating the role of mind wandering in computer-supported collaborative work: a proposal for an EEG study. In: Davis, F.D., Riedl, R., vom Brocke, J., Léger, P.-M., Randolph, A., Fischer, T. (eds.) Information Systems and Neuroscience. LNISO, vol. 32, pp. 53–62. Springer, Cham (2020). https://doi.org/10.1007/978-3-030-28144-1_6

34. Klesel, M., Schlechtinger, M., Oschinsky, F.M., Conrad, C., Niehaves, B.: Detecting mind wandering episodes in virtual realities using eye tracking. In: Davis, F.D., Riedl, R., vom Brocke, J., Léger, P.-M., Randolph, A.B., Fischer, T. (eds.) NeuroIS 2020. LNISO, vol. 43, pp. 163–171. Springer, Cham (2020). https://doi.org/10.1007/978-3-030-60073-0_18

35. Dijksterhuis, A., Meurs, T.: Where creativity resides. The generative power of unconscious thought. Conscious. Cogn. **15**, 135–146 (2006)

36. Christoff, K., Gordon, A.M., Smallwood, J., Smith, R., Schooler, J.W.: Experience sampling during fMRI reveals default network and executive system contributions to mind wandering. Proc. Natl. Acad. Sci. U.S.A. **106**, 8719–8724 (2009)

37. Dimoka, A., Pavlou, P.A., Davis, F.D.: NeuroIS. The potential of cognitive neuroscience for information systems research. Inf. Syst. Res. **22**, 687–702 (2011)

38. Martinon, L.M., Smallwood, J., McGann, D., Hamilton, C., Riby, L.M.: The disentanglement of the neural and experiential complexity of self-generated thoughts. A users guide to combining experience sampling with neuroimaging data. NeuroImage 1–55 (2019)

39. Ceh, S.M., et al.: Neurophysiological indicators of internal attention: an electroencephalography-eye-tracking coregistration study. Brain Behav. e01790 (2020)

40. Arnau, S., Löffler, C., Rummel, J., Hagemann, D., Wascher, E., Schubert, A.-L.: The electrophysiological signature of mind wandering. bioRxiv, 819805 (2019)

41. Müller-Putz, G.R., Riedl, R., Wriessnegger, S.C.: Electroencephalography (EEG) as a research tool in the information systems discipline. Found. Meas. Appl. CAIS **37**, 911–948 (2015)

42. Riedl, R., Davis, F.D., Hevner, A.R.: Towards a NeuroIS research methodology. Intensifying the discussion on methods, tools, and measurement. JAIS **15**, 1–35 (2014)

43. Riedl, R., Fischer, T., Léger, P.M.: A decade of NeuroIS research. Status quo, challenges, and future directions. DATA BASE Adv. Inf. Syst. **51**, 13–54 (2020)

44. Loos, P., et al.: NeuroIS. Neuroscientific approaches in the investigation and development of information systems. Bus. Inf. Syst. Eng. **2**, 395–401 (2010)

45. Astor, P.J., Adam, M.T.P., Jerčić, P., Schaaff, K., Weinhardt, C.: Integrating biosignals into information systems. A NeuroIS tool for improving emotion regulation. J. Manage. Inf. Syst. **30**, 247–278 (2013)

46. Vom Brocke, J., Riedl, R., Léger, P.-M.: Application strategies for neuroscience in information systems design science research. J. Comput. Inf. Syst. **53**, 1–13 (2013)

47. Adam, M.T.P., Gimpel, H., Maedche, A., Riedl, R.: Design blueprint for stress-sensitive adaptive enterprise systems. Bus. Inf. Syst. Eng. **59**, 277–291 (2017)

48. Vom Brocke, J., Hevner, A., Léger, P.M., Walla, P., Riedl, R.: Advancing a NeuroIS research agenda with four areas of societal contributions. Eur. J. Inf. Syst. **29**, 9–24 (2020)

49. Guilford, J.P.: Creativity: yesterday, today and tomorrow. J. Creat. Behav. **1**, 3–14 (1967)

50. Mrazek, M.D., Phillips, D.T., Franklin, M.S., Broadway, J.M., Schooler, J.W.: Young and restless. Validation of the mind-wandering questionnaire (MWQ) reveals disruptive impact of mind-wandering for youth. Front. Psychol. **4**, 1–7 (2013)

51. Carriere, J.S.A., Seli, P., Smilek, D.: Wandering in both mind and body. Individual differences in mind wandering and inattention predict fidgeting. Can. J. Exp. Psychol./Rev. Can. Psychol. Exp. **67**, 19–31 (2013)
52. Mowlem, F.D., et al.: Validation of the mind excessively wandering scale and the relationship of mind wandering to impairment in adult ADHD. J. Atten. Disord. **23**, 624–634 (2016)
53. Dimoka, A., et al.: On the use of neurophysiological tools in IS research. Developing a research agenda for NeuroIS. MIS Q. **36**, 679–702 (2012)
54. Weber, B., Fischer, T., Riedl, R.: Brain and autonomic nervous system activity measurement in software engineering. A systematic literature review. J. Syst. Softw. (2021). (in press)

Consumers Prefer Abstract Design in Digital Signage: An Application of Fuzzy-Trace Theory in NeuroIS

Anika Nissen[1]($\boxtimes$), Gabriele Obermeier[2], Nadine R. Gier[3], Reinhard Schütte[1], and Andreas Auinger[2]

[1] Institute for Computer Science and Business Information Systems, University of Duisburg-Essen, Essen, Germany
{anika.nissen,reinhard.schuette}@icb.uni-due.de

[2] School of Business and Management, University of Applied Sciences Upper Austria, Steyr, Austria
{gabriele.obermeier,andreas.auinger}@fh-ooe.at

[3] Faculty of Business Administration and Economics, Heinrich Heine University Düsseldorf, Düsseldorf, Germany
nadine.gier@hhu.de

Abstract. Visual designs of digital signage (DS) content shape and influence consumers' decisions. Understanding the effect of DS design on consumer behavior requires a fundamental understanding of human reasoning and decision-making. This research explores the effect of different visual design cues of DS on a neural level and through the lens of Fuzzy-Trace Theory (FTT). The FTT suggests that humans have both a verbatim-based and a gist-based information processing. To explore the effect of FTT-based visual design, an experiment using functional near-infrared spectroscopy is conducted. DS are tested on three design levels: (1) verbatim: text, (2) verbatim: photographs, and (3) gist-based. Results show that only the gist-based design resulted in significantly higher self-reported results and activated brain areas in the medial prefrontal cortex, which are associated with emotional and rewarding processing. These results challenge the manifest differentiation only between image and text elements.

Keywords: Digital signage · Visual design · Fuzzy-trace theory · fNIRS · Neural measures

1 Introduction

In-store technologies provide the opportunity to enhance consumers' shopping experience [1, 2]. One of the most applied in-store technologies are digital signages (DS), which allow to dynamically advertise products at the physical Point-Of-Sale (POS), trying to create a pleasant store environment [3, 4]. Through their visual design, DS can add to the overall atmosphere, suggesting a modern image of the store [5–7]. This positive impact on store atmospherics seems to affect the consumers' decision-making by

© The Author(s), under exclusive license to Springer Nature Switzerland AG 2021
F. D. Davis et al. (Eds.): NeuroIS 2021, LNISO 52, pp. 148–161, 2021.
https://doi.org/10.1007/978-3-030-88900-5_17

facilitating purchase intentions [5, 8–10]. Although there is an increasing trend towards e-commerce, physical grocery stores are still the preferred place to purchase for European consumers [11]. Given that more than 90% of actual purchase decisions occur at the stationary POS [12, 13], triggering a reaction to promotions becomes particularly relevant for habituated or impulsive purchases of low-involvement products. Even though the effect of visual designs in POS advertising have been investigated in literature for some time (e.g., [14] and [15]), new technological advances in retailing and in neurophysiological measurement methods enable the analysis of the effect of different visual designs on consumer behavior.

Regarding visual design, research has shown that advertisements with more graphics (i.e., images and videos) elicit more positive overall effects than text-heavy advertisements [16]. However, the differentiation of visual designs into the manifest categories of text- and image-based design poses central problems in replicating results, since visual elements like images are diffuse and difficult to define [17, 18]. Even though images in visual designs are assumed to elicit higher emotional processing [19, 20], they do not necessarily affect consumers' behavioral outcomes, such as purchase intention [21]. Furthermore, an image could be represented through a real photograph or it could be an abstract visual design – both might affecting behavioral intentions differently. Therefore, a more particularized differentiation is needed because the manifest distinction of text and image may not adequately explain DS effects' variances.

One approach for such differentiation is to orient the design to the way humans process information to make decisions. Therefore, this paper utilizes Fuzzy-Trace Theory (FTT), which integrates considerations of cognitive psychology and neuroscience [22, 23], as a potential framework. In particular, FTT explains that human reasoning, judgment, and decision-making are based on a continuum between verbatim (i.e., detailed) and gist-based (e.g., abstract or meaning-based) representations of information and past events [24]. It has shown that gist-based processing is primarily used during trivial, daily decision-making [25, 26] or in the unconscious processing of visual stimuli in decision-making scenarios [27, 28]. Consequently, it might be assumed that visual cues on DS should be designed in a way that triggers gist-based rather than verbatim processing to facilitate behavioral intentions. Hence, this paper offers a first approach to receive insights into DS designs that trigger either verbatim or gist processing, assuming that FTT offers a promising approach to designing different visual stimuli according to the FTT processing representations. Therefore, this study aims to address the following research question: *how can the FTT-based differentiations of DS designs affect the neural processing of DS and impact behavioral intentions?*

To address this research question, this study focuses on the effect of three different visual DS designs (text vs. photograph vs. abstract imagery scene) on the consumers' perception and behavioral intentions as measured by store atmospherics, the attractiveness of the DS, and purchase intentions. The visual designs in this study intend to promote a more ecologically valuable and animal-friendly product choice by consumers in the low-involvement product category of eggs. Because both the neural processing and behavioral intentions are of interest, a neural study is conducted employing the neuroimaging method of functional near-infrared spectroscopy (fNIRS).

2   Related Literature

2.1   Digital Signage and Visual Design

DS at the physical POS offer a modern opportunity for retailers to inform consumers and create a pleasant store environment that try to add to the overall atmosphere and promote a specific brand or product [4]. Previous studies showed that DS can increase the positive perceptions of the retail environment and approach behaviors [29]. While research on the effectiveness of DS is still limited [2], recent studies have shown that the context and the content displayed on DS are important factors that can affect consumer behavior at the moment of decision [9, 30]. For instance, DS can increase traffic in the store and sales, especially when consumers find the message fits their current needs [31]. In addition, sensory-affective advertisements displayed on DS have been shown to elicit positive emotions and positively impact approach behavior, such as purchase intention, loyalty, and revisiting intention [10, 29, 32, 33]. However, the context of the DS also influences the required visual design. For instance, overly detailed, informative DS content seem to have a negative effect on consumers' positive emotions during their shopping journey in a grocery store [10], while it might positively affect purchase intentions in a restaurant environment [34].

Therefore, the effect of DS design seems to depend on the context of low-involvement (i.e., everyday products in a supermarket) and high-involvement decision-making situations (i.e., what the cafeteria offers on the menu today). Considering the effort needed to process information in general, consumers are assumend to either perform cognitive and thoughtful consideration of information or automatic, emotional processing of affective content [35] that subsequently elicit behavioral responses [36, 37].

Given that this study focuses on DS in physical retail stores, it offers a context of primarily low-involvement purchase decisions. Consequently, the cognitive resources of the consumers may be limited at the POS [35], potentially making less detailed visual designs more likely to influence the consumer's emotions and behavioral responses. Supporting this idea, the consumers' primary goal in physical grocery shopping is to navigate through the store finding the correct departments and selecting the desired products [31]. Therefore, consumers may not be able to process detailed, information-based content on DS that needs to be mentally encoded and understood [38, 39]. As stated, detailed content might not necessarily only include text-based elements but can also be represented through photographs. One theoretical approach that supports this differentiation offering an approach to guide DS visual design is FTT.

2.2   Fuzzy-Trace Theory and Visual Design

FTT proposes that during information processing, information is encoded in parallel on a continuum between two types of representations or memories: verbatim and gist [22]. Individuals rely on these memories in decision-making processes [22, 40]. Verbatim representations are assumed to represent exact numbers, words, or images, while gist memories are characterized by extracted "senses, pattern, and meaning" [22]. Transferring FTT as a framework to visual design, verbatim stimuli represent the exact replication of reality (e.g., a photograph) [41], while an abstract graphic stimulus would depict a less detailed, and, therefore, a gist-based representation that only transfers the main message of the content. According to the FTT, it is suggested that adults rely more on gist-based intuition, especially for known, habitualized, and routine processing [25].

At the physical POS, consumers are already exposed to a plethora of different stimuli that might exceed their cognitive capacities, potentially resulting in a feeling of overload [42]. In contrast to complex, verbatim-based visual designs, it could be assumed that abstract, gist-based designs more positively influence consumers' product choices because of the associated lower cognitive load. Hence, a categorization of visual stimuli based on the FTT framework could explain differences in DS effects; this would, however, lead to hypotheses that contradict the previous literature that apply the manifest image-text categorization of stimuli [10, 29, 33]. Following this manifest differentiation, photographs would stimulate consumer behavior primarily on an emotional level, even though this seems not to translate necessarily to consumers' intended behavior such as purchase intention [10, 21]. According to the FTT framework, photographs would be categorized as verbatim-based stimuli due to their visual complexity. These cues might increase consumers' cognitive load and elicit negative effects in a choice situation.

In contrast, the presumable preference for simple, gist-based processing requires less cognitive resources, influencing behavior more efficiently during decisions with limited cognitive capacities. This is supported by studies exploring FTT in the frame of graphic design that aim to promote health judgments, showing that gist-based designs were more effective in transferring and allowing the consumer to remember the message [43–45]. However, since this theory has not yet been applied to the visual design of DS in retail stores, this research offers an innovative and however, exploratory study design.

3   Method

This exploratory study focuses on the effect of three different visual DS designs according to the FTT processing representations (verbatim: text vs. verbatim: photograph vs. gist: abstract imagery scene) on consumers' perceptions of store atmospherics, the attractiveness of the DS ad, and purchase intentions. Additionally, the processing of the different visual stimuli is measured using the neuroimaging method of fNIRS. To reduce possible noise in the neural data, we investigated static DS as the first indicators. In addition, consumers' behavioral intentions were measured based on self-reported questionnaires.

Sample. The sample in this study consisted of $N = 28$ healthy participants (11 females, 17 males; 26 right-handed, 2 left-handed) with normal or corrected to normal sight. The average age was $M = 27.6$ years ($SD = 6.83$). Most participants were either full-time workers (43%) or students (43%). The majority hold a university degree (71%).

Stimuli. In order to simulate low-involvement purchase decisions, eggs were selected as promoted good by the DS stimuli [46]. Participants were shown images of DS placed on top of an egg shelf in a stationary retail store. Three different stimuli types were shown varying in their graphic and textual cues according to the FTT framework. Verbatim stimuli included either text or photographic cues, while gist-based stimuli were presented as abstract (visual) presentations [40, 41]. In order to investigate both verbatim stimuli types, text-based DS (1: text), and DS with real photographs (2: photos) were designed, as well as the icon-based, abstract imagery scene as gist-based DS (3: icons) (as shown in Fig. 1).

text: photos: icons:
text-based verbatim stimulus   *image-based verbatim stimulus*   *image-based gist-based stimulus*

Fig. 1. Used stimuli and reference to the FTT

Experimental Design. The data was collected in Germany, adhering to the strict governmental COVID-19 safety protocols. Prior to the start of the experiment, participants were informed about the overall goal of the study, experimental procedure, data storage, and the functionality of the fNIRS device in verbal and written form according to the Declaration of Helsinki, and participants gave their consent. During the event-related experimental paradigm, participants were repeatedly exposed to each of the three stimuli types for 4 s (s) followed by an evaluation of either the store atmospherics (adapted from [47]), the attractiveness of the DS ad [34], and purchase intentions (adapted from [34]) on a 5-point Likert scale (1 = totally disagree; 5 = totally agree). For each construct, three items were used. Each item was shown three times per DS, resulting in 27 trials per DS type. The order in which trials were shown was fully randomized (see Fig. 2). Afterwards, each participant received a compensation of €10.

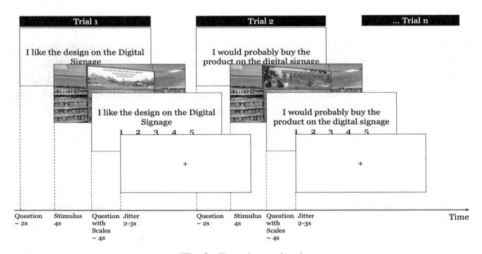

Fig. 2. Experimental task

Measurement Method. fNIRS allows the assessment of levels of oxygenated and deoxygenated hemoglobin (hbo and hbr, respectively) and has been proven to provide robust insights into the neural correlates of human decision-making [48–51]. In this paper, a mobile fNIRS device from NIRX is utilized, which holds 8 sources, 7 long-distance detectors (distance 30 mm), and 8 short-distance detectors (distance 8 mm). The fNIRS device sampled at wavelengths 760 nm and 850 nm with a sampling frequency of 7.81 Hz. Raw fNIRS data were pre-processed using the MATLAB Brain AnalyzIR toolbox [52]. The used fNIRS montage covered most cortical areas of the prefrontal cortex (PFC) (Fig. 3).

As depicted in Fig. 3, channels 2, 7, 9, 14, 16, and 21 represent approximately the right and left dorsolateral prefrontal cortex (dlPFC), which is generally related to performance monitoring and (adaption of) approach behavior [53, 54]. Channels 16, 17, 21, and 22 approximately describe regions of the ventrolateral prefrontal cortex (vlPFC), an area that receives emotional information and influences decision-making processes [55]. The area around channels 11 and 15 approximately indicates the orbitofrontal cortex (OFC), while the ventromedial prefrontal cortex (vmPFC) may be identified by channel 19 and 20. Both are typically involved in humans' reward system [56, 57]. The dorsomedial prefrontal cortex (dmPFC), represented by channel 5, 8, 10, 12, 13, 18, is especially responsible for processing emotional stimuli [58–60]. Both the left dlPFC and the left dmPFC are tightly connected to the amygdala, which influences emotion and memory [61, 62].

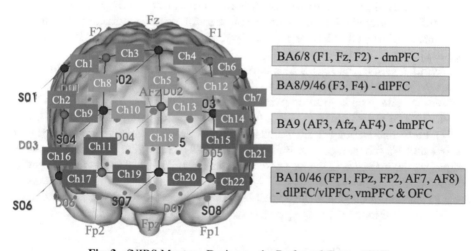

Fig. 3. fNIRS Montage Design on the Prefrontal Cortex (PFC)

Data Pre-Processing. First, the raw fNIRS signal was re-sampled to 4 Hz and used to calculate the optical density [63]. After that, short channel regression was used with the Linear Minimum Mean Square Estimations to filter out artefacts due to extracerebral blood flow, Mayer waves, and movements [64, 65]. After that, the hemoglobin values were calculated for the long-distance channels using the modified Beer-Lambert Law with a partial pathlength factor of .1 [66, 67]. On the within-subject level, a General Linear Model (GLM) with the AR-IRLS algorithm was set up [68], using canonical hemodynamic response function as the baseline. A mixed-effects model with DS as fixed effects and subjects as random effects was selected for the group analysis. As a threshold a false discovery rate of *p-values* $(= q) < .05$ [69] and a statistical *power* > .80 were used to identify significant fNIRS channels.

4 Results

Self-Reported Results. Self-reported results from the scales included in the experimental paradigm were analyzed using one-way repeated-measures ANOVAs. Because the assumption of sphericity was violated, Greenhouse-Geisser correction was applied. Overall, these tests revealed that the gist-based DS (icons) was rated significantly higher than at least one of the verbatim DS (text and photos) for the two outcome variables atmosphere (F_{atmos} (1.35, 36.38) = 6.01, $p = .012$, $\eta_p^2 = .182$ ($M_{icon} = 3.85$, $M_{photo} = 2.99$, $M_{text} = 2.99$)), and attractiveness ($F_{attract}$(1.26, 34.02) = 3.99, $p < .05$, $\eta_p^2 = .129$ ($M_{icon} = 3.79$, $M_{photo} = 3.42$, $M_{text} = 2.91$)), but not for purchase intentions (F_{PUI} (1.39, 37.65) = 3.07, $p > .075$, $\eta_p^2 = .102$ ($M_{icon} = 4.02$, $M_{photo} = 3.62$, $M_{text} = 3.42$)). The Tukey-Post Hoc tests showed that DS with gist-based content (icon) had a significantly ($p_{bonferroni} < .05$) higher impact on the perceived atmospherics than verbatim DS content (photos: $MD = .856$, text: $MD = .861$). Further, the gist-based DS scored significantly higher than the verbatim, text-based DS on attractiveness ($MD = .882$, $p_{bonferroni} < .05$), but only slightly higher for purchase intentions ($MD = .604$, $p_{bonferroni} = .055$). No significant difference between icon and photos with regard to attractiveness and purchase intention, as well as between the two verbatim DS (i.e., photos and text) could be identified.

Neural Results. The fNIRS results revealed significant effects for the gist-based versus verbatim design contrasts icon > text and icon > photos. Significant activation differences in the icon > text comparison were identified in the right dlPFC (ch2, $t_{hbr} = -4.979$) and vlPFC (ch17, $t_{hbo} = 3.657$), as well as in the right OFC (ch11, $t_{hbo} = 3.947$). Further activation was identified in the lower dmPFC (ch18, $t_{hbr} = -3.391$) and the left dmPFC (ch13, $t_{hbo} = 5.463$, $t_{hbr} = -6.044$). Activation in the latter may be unique for the gist-based DS since activation was also identified in the icon > photos contrast (ch13, $t_{hbo} = 4.054$, $t_{hbr} = -4.764$). Thus, when transferring these results to the FTT context, it may be assumed that especially the left dmPFC seems to modulate gist-based vs. verbatim stimuli, regardless of whether text or photos were used.

5 Discussion and Conclusion

The aim of this research was to investigate to what extent the FTT-based differentiation of visual cues can explain the effects of DS on a neural and behavioral level. Therefore, an experimental study was carried out in which data from a neurophysiological measurement with an fNIRS device and self-reported questionnaires were collected to identify the effect of different DS designs on consumer decision-making.

5.1 Self-reported Results

According to the self-reported results, participants preferred the gist-based icon condition in all included constructs. Regarding atmospherics, the results show significant differences between the gist (i.e., icon) and verbatim (i.e., photo and text) DS, which is the first indicator for supporting the applied FTT framework. The finding of no significant difference between the two verbatim DS (i.e., photos and text) for all three conditions may further reinforce the FTT approach to differentiate stimuli designs. With regard to the attractiveness of the advertisement shown on the DS and consumers' intention to purchase the promoted product, significant differences for the gist-based icon condition and the verbatim-based text condition indicate more positive consumer perceptions in favor of the gist-based DS. These results are in line with prior research with self-reported measurements in which the detailed and presumably more cognitive engaging DS design negatively affected consumers' positive attitudes during the shopping scenario [10]. The finding of no significant differences between the verbatim-based conditions photo and text indicate that the use of images in POS marketing may not positively affect the consumer per se. It is more likely that the choice of gist-based content helps to promote products effectively. From the self-reported results in this study, it can be concluded that the FTT framework provides more reliable differentiations about consumers' perception of visual DS designs at the POS than the manifest differentiation between text- and image-based design content.

5.2 Neural Results

When interpreting the neural results, three major conclusions can be drawn from our results and need to be confirmed in follow-up studies. Firstly, from the results, left dmPFC (ch13) brain activity is associated with the differences between gist-based and verbatim stimuli. This brain area has been related to the processing of emotional stimuli due to its tight connection to and controlling processes on the amygdala [58–60]. Consequently, the activation in this brain area presumes that gist-based DS content might lead to higher emotional processing in participants [62, 70]. In reference to the associated significantly higher evaluation of intentions to purchase, it may be considered that the activation of the emotional, gist-based processing influences consumer decisions. Further, elicited emotional processing can be interpreted as positive and pleasant because the identified activation is located in the left hemisphere, which is often attributed to more positive than negative emotional stimuli [71]. These results support the self-reported results, which revealed higher ratings of the icon design. Further, the comparison between the favored, gist-based design and the verbatim, text-based design revealed the additional activation differences. This results in the following implication:

Implication 1. Gist-based rather than verbatim-based visual digital signage design may lead to increased positive emotional attribution

The left dmPFC activation was accompanied by activation in the right dlPFC, vlPFC, and inferior dmPFC. The dlPFC is generally related to performance monitoring and behavioral adjustment and is further responsible for linking sensory information to taking action [53, 54]. Consequently, the activation in this area might be explained by a higher will to adapt one's behavior – more specifically, to alter a purchase decision towards the product promoted on the DS. Further, the right vlPFC receives emotional and motivational information, which influences human decision-making [55]. Due to the vlPFC also being functionally connected to the amygdala, it can be linked to the processing of emotional stimuli similar to the dmPFC prior discussed [61, 62]. This leads to the following implication:

Implication 2.   Gist-based rather than textual, verbatim-based visual digital signage design may lead to increased behavioral intentions of the consumers

The activation in the right OFC/vmPFC, which was only identified for the icon versus text-based design, may be related to participants' perceiving the icon design as more pleasant and aesthetically pleasing [72, 73]. In shopping contexts, this brain area has also been related to increased purchase intentions due to its role in reward processing [56, 57]. This finding is in line with the self-reported results that only show significant differences in purchase intentions between the gist- and verbatim-text contrast. Consequently, the increased activation in this area is likely related to the higher intention to purchase the advertised eggs when the gist-based icon design is used. The few prior studies, which investigated FTT in a neuroimaging experiment, identified that gist-based processing is primarily associated with activation in the amygdala and, when remaining within the PFC, in the right inferior area of the PFC [74, 75]. The latter is reflected in the identified activation of the right vlPFC and OFC/vmPFC and is suggested by our results. Thus, the following final implication is derived:

Implication 3.   Gist-based rather than textual, verbatim-based visual digital signage design may lead to increased aesthetic attribution

5.3   Limitations, Future Research and Conclusion

The self-reported and neural results of this exploratory study legitimate reasons to apply the FTT in research on the visual design of DS and their effects on consumer perceptions and behavior. However, given that this is an exploratory study, it comes with several limitations.

First, a real-world scenario of DS in virtual physical retail store was used. The study, however, was conducted in a laboratory environment and requires validation in the field. In line with this, monetary aspects were not considered in the current study, even though the price is a relevant factor when choosing between product alternatives of the same category [76]. Moreover, the consumers' choice might be further investigated by considering time constraints [77]. In addition, related research has identified that DS might be prone to the probable effect of display blindness [78, 79], as shown in a study about DS in a coffee shop [80]. Consequently, future research could repeat this study in a field experiment to include the factors mentioned above to validate these preliminary results further.

Secondly, static images were used in the DS to control for bias in the neural data. However, and especially as a factor that distinguishes DS from traditional in-store advertisements, animated content was not considered. Since animations add more visual complexity to a design, the effects identified in this study might differ. Finally, although most Europeans still prefer in-store shopping to ecommerce, there is an undeniable trend towards online shopping [81]. Future research may therefore tackle an application of FTT to the visual design of ecommerce websites, since its basic assumptions might hold true for online environments, too. That is, websites with abstract, designed images as website elements might be preferred to websites incorporating real photographs as design elements. Whether this assumption holds true, however, is still to be investigated.

Despite its limitations, this study has provided preliminary insights into a different categorization of the visual design of DS in food retailers. Results have shown that gist-based design on DS elicits more positive emotional and aesthetic attribution and might lead to higher behavioral intentions than verbatim-based designs.

Acknowledgements. Part of this work was funded by the European Union's Horizon 2020 research and innovation program under the Marie Skłodowska-Curie grant agreement No. 765395.

References

1. Grewal, D., Hulland, J., Kopalle, P.K., Karahanna, E.: The future of technology and marketing: a multidisciplinary perspective. J. Acad. Mark. Sci. **48**(1), 1–8 (2019). https://doi.org/10.1007/s11747-019-00711-4
2. Willems, K., Smolders, A., Brengman, M., Luyten, K., Schöning, J.: The path-to-purchase is paved with digital opportunities: an inventory of shopper-oriented retail technologies. Technol. Forecast. Soc. Change **124**, 228–242 (2017). https://doi.org/10.1016/j.techfore.2016.10.066
3. Sanden, S., Van De, Willems, K., Brengman, M.: How do consumers process digital display ads in-store? The effect of location, content, and goal relevance. J. Retail. Consum. Serv. **56**, 102177 (2020). https://doi.org/10.1016/j.jretconser.2020.102177
4. Pantano, E., Vannucci, V.: Who is innovating? An exploratory research of digital technologies diffusion in retail industry. J. Retail. Consum. Serv. **49**, 297–304 (2019). https://doi.org/10.1016/j.jretconser.2019.01.019
5. Newman, A., Dennis, C., Wright, L.T., King, T.: Shoppers' experiences of digital signage-a cross-national qualitative study. Int. J. Digit. Content Technol. Appl. **4**, 50–57 (2010). https://doi.org/10.4156/jdcta.vol4.issue7.5
6. Ballantine, P.W., Jack, R., Parsons, A.G.: Atmospheric cues and their effect on the hedonic retail experience. Int. J. Retail Distrib. Manag. **38**, 641–653 (2010). https://doi.org/10.1108/09590551011057453
7. Roggeveen, A.L., Grewal, D., Schweiger, E.B.: The DAST framework for retail atmospherics: the impact of in- and out-of-store retail journey touchpoints on the customer experience. J. Retail. **96**, 128–137 (2020). https://doi.org/10.1016/j.jretai.2019.11.002
8. Bauer, C., Garaus, M., Strauss, C., Wagner, U.: Research directions for digital signage systems in retail. Procedia Comput. Sci. **141**, 503–506 (2018). https://doi.org/10.1016/j.procs.2018.10.135
9. Jäger, A.-K., Weber, A.: Increasing sustainable consumption: message framing and in-store technology. Int. J. Retail Distrib. Manag. **48**, 803–824 (2020). https://doi.org/10.1108/IJRDM-02-2019-0044
10. Garaus, M., Wagner, U., Manzinger, S.: Happy grocery shopper: the creation of positive emotions through affective digital signage content. Technol. Forecast. Soc. Change **124**, 295–305 (2017). https://doi.org/10.1016/j.techfore.2016.09.031

11. Wunsch, N.-G.: Forecasted online grocery market size in selected European nations from 2018 to 2023. https://www.statista.com/statistics/960484/online-grocery-market-sizes-europe/. Accessed 01 April 2021
12. Beharrell, B., Denison, T.J.: Involvement in a routine food shopping context. Br. Food J. **97**, 24–29 (1995). https://doi.org/10.1108/00070709510085648
13. Behe, B.K., Huddleston, P.T., Childs, K.L., Chen, J., Muraro, I.S.: Seeing through the forest: the gaze path to purchase. PLoS ONE **15**, e0240179 (2020). https://doi.org/10.1371/journal.pone.0240179
14. Scott, L.M.: Images in advertising: the need for a theory of visual rhetoric. J. Consum. Res. **21**, 252 (1994). https://doi.org/10.1086/209396
15. Lavidge, R.J., Steiner, G.A.: A model for predictive measurements of advertising effectiveness. J. Mark. **25**, 59 (1961). https://doi.org/10.2307/1248516
16. Percy, L., Rossiter, J.R.: Effects of picture size and color on brand attitude responses in print advertising. ACR North Am. Adv. Consum. Res. **10**, 17–20 (1983)
17. Bock, M.A.: Theorising visual framing: contingency, materiality and ideology. Vis. Stud. **35**, 1–12 (2020). https://doi.org/10.1080/1472586X.2020.1715244
18. Rodriguez, L., Dimitrova, D.V.: The levels of visual framing. J. Vis. Lit. **30**, 48–65 (2011). https://doi.org/10.1080/23796529.2011.11674684
19. Powell, T.E., Boomgaarden, H.G., De Swert, K., de Vreese, C.H.: Framing fast and slow: a dual processing account of multimodal framing effects. Media Psychol. **22**, 572–600 (2019). https://doi.org/10.1080/15213269.2018.1476891
20. Entman, R.M.: Framing U.S. coverage of international news: contrasts in narratives of the KAL and iran air incidents. J. Commun. **41**, 6–27 (1991). https://doi.org/10.1111/j.1460-2466.1991.tb02328.x
21. Kim, M., Lennon, S.: The effects of visual and verbal information on attitudes and purchase intentions in internet shopping. Psychol. Mark. **25**, 146–178 (2008). https://doi.org/10.1002/mar
22. Reyna, V.F., Brainerd, C.J.: Fuzzy-trace theory: an interim synthesis. Learn. Individ. Differ. **7**, 1–75 (1995). https://doi.org/10.1016/1041-6080(95)90031-4
23. Reyna, V.F., Brainerd, C.J.: Fuzzy-trace theory and framing effects in choice: gist extraction, truncation, and conversion. J. Behav. Decis. Mak. **4**, 249–262 (1991). https://doi.org/10.1002/bdm.3960040403
24. Corbin, J.C., Reyna, V.F., Weldon, R.B., Brainerd, C.J.: How reasoning, judgment, and decision making are colored by gist-based intuition: a fuzzy-trace theory approach. J. Appl. Res. Mem. Cogn. **4**, 344–355 (2015). https://doi.org/10.1016/j.jarmac.2015.09.001
25. Reyna, V.F.: A new intuitionism: meaning, memory, and development in fuzzy-trace theory. Judgm. Decis. Mak. **7**, 332–359 (2012)
26. Setton, R., Wilhelms, E., Weldon, B., Chick, C., Reyna, V.: An overview of judgment and decision making research through the lens of fuzzy trace theory. Adv. Psychol. Sci. **22**, 1837 (2014). https://doi.org/10.3724/SP.J.1042.2014.01837
27. Dijksterhuis, A., Smith, P.K., van Baaren, R.B., Wigboldus, D.H.J.: The unconscious consumer: effects of environment on consumer behavior. J. Consum. Psychol. **15**, 193–202 (2005). https://doi.org/10.1207/s15327663jcp1503_3
28. Barrett, L.F., Bar, M.: See it with feeling: affective predictions during object perception. Philos. Trans. R. Soc. B Biol. Sci. **364**, 1325–1334 (2009). https://doi.org/10.1098/rstb.2008.0312
29. Dennis, C., Newman, A., Michon, R., Josko Brakus, J., Tiu Wright, L.: The mediating effects of perception and emotion: digital signage in mall atmospherics. J. Retail. Consum. Serv. **17**, 205–215 (2010). https://doi.org/10.1016/j.jretconser.2010.03.009

30. Roggeveen, A.L., Nordfält, J., Grewal, D.: Do digital displays enhance sales? Role of retail format and message content. J. Retail. **92**, 122–131 (2016). https://doi.org/10.1016/j.jretai.2015.08.001

31. Burke, R.R.: Behavioral effects of digital signage. J. Advert. Res. **49**(2), 180–185 (2009). https://doi.org/10.2501/S0021849909090254

32. Dennis, C., Joško Brakus, J., Alamanos, E.: The wallpaper matters: digital signage as customer-experience provider at the Harrods (London, UK) department store. J. Mark. Manag. **29**, 338–355 (2013). https://doi.org/10.1080/0267257X.2013.766628

33. Dennis, C., Joško Brakus, J., Gupta, S., Alamanos, E.: The effect of digital signage on shoppers' behavior: the role of the evoked experience. J. Bus. Res. **67**, 2250–2257 (2014). https://doi.org/10.1016/j.jbusres.2014.06.013

34. Lee, H., Cho, C.-H.: An empirical investigation on the antecedents of consumers' cognitions of and attitudes towards digital signage advertising. Int. J. Advert. **38**, 97–115 (2019). https://doi.org/10.1080/02650487.2017.1401509

35. Petty, R.E., Cacioppo, J.T.: The elaboration likelihood model of persuasion. In: Communication and Persuasion, pp. 1–24. Springer, New York, NY (1986). https://doi.org/10.1007/978-1-4612-4964-1_1

36. Bakker, I., van der Voordt, T., Vink, P., de Boon, J.: Pleasure, arousal, dominance: mehrabian and russell revisited. Curr. Psychol. **33**(3), 405–421 (2014). https://doi.org/10.1007/s12144-014-9219-4

37. Mehrabian, A., Russell, J.A.: An Approach to Environmental Psychology. The MIT Press, Cambridge, MA, US (1974)

38. Lang, A.: The limited capacity model of mediated message processing. J. Commun. **50**, 46–70 (2000). https://doi.org/10.1111/j.1460-2466.2000.tb02833.x

39. Lang, A.: Using the limited capacity model of motivated mediated message processing to design effective cancer communication messages. J. Commun. **56**, 57–80 (2006). https://doi.org/10.1111/j.1460-2466.2006.00283.x

40. Broniatowski, D.A., Reyna, V.F.: A formal model of fuzzy-trace theory: variations on framing effects and the allais paradox. Decision **5**, 205–252 (2018). https://doi.org/10.1037/dec0000083

41. Clark, H.H., Clark, E.V.: JCL volume 4 issue 2 Cover and Back matter. J. Child Lang. **4**, b1–b3 (1977). https://doi.org/10.1017/S0305000900001562

42. Jacoby, J., Speller, D.E., Kohn, C.A.: Brand choice behavior as a function of information load. J. Mark. Res. **11**, 63–69 (1974). https://doi.org/10.1177/002224377401100106

43. Cho, Y.H., You, M., Choi, H.: Gist-based design of graphics to reduce caffeine consumption among adolescents. Health Educ. J. **77**, 778–790 (2018). https://doi.org/10.1177/0017896918765024

44. Wilhelms, E.A., Reyna, V.F.: Effective ways to communicate risk and benefit. AMA J. Ethics **15**, 34–41 (2013). https://doi.org/10.1001/virtualmentor.2013.15.1.stas1-1301

45. Blalock, S.J., Reyna, V.F.: Using fuzzy-trace theory to understand and improve health judgments, decisions, and behaviors: a literature review. Heal. Psychol. **35**, 781–792 (2016). https://doi.org/10.1037/hea0000384

46. Frank, P., Brock, C.: Bridging the intention-behavior gap among organic grocery customers: the crucial role of point-of-sale information. Psychol. Mark. **35**, 586–602 (2018). https://doi.org/10.1002/mar.21108

47. Grewal, D., Baker, J., Levy, M., Voss, G.B.: The effects of wait expectations and store atmosphere evaluations on patronage intentions in service-intensive retail stores. J. Retail. **79**, 259–268 (2003). https://doi.org/10.1016/j.jretai.2003.09.006

48. Gefen, D., Ayaz, H., Onaral, B.: Applying functional near infrared (fNIR) spectroscopy to enhance MIS research. AIS Trans. Hum.-Comput. Interact. **6**, 55–73 (2014)

49. Nissen, A., Krampe, C., Kenning, P., Schütte, R.: Utilizing mobile fNIRS to investigate neural correlates of the TAM in eCommerce. In: International Conference on Information Systems (ICIS), pp. 1–9. Munich (2019)
50. Krampe, C., Gier, N.R., Kenning, P.: The application of mobile fnirs in marketing research—detecting the "first-choice-brand" effect. Front. Hum. Neurosci. **12**, 433 (2018). https://doi.org/10.3389/fnhum.2018.00433
51. Gier, N.R., Kurz, J., Kenning, P.: Online reviews as marketing placebo? First insights from neuro-is utilising fNIRS. In: Twenty-eighth European Conference on Information Systems (ECIS2020), pp. 1–11. Marrakech (2020)
52. Santosa, H., Zhai, X., Fishburn, F., Huppert, T.: The NIRS brain AnalyzIR toolbox. Algorithms **11**, 73 (2018). https://doi.org/10.3390/a11050073
53. Taren, A.A., Venkatraman, V., Huettel, S.A.: A parallel functional topography between medial and lateral prefrontal cortex: evidence and implications for cognitive control. J. Neurosci. **31**, 5026–5031 (2011)
54. Heekeren, H.R., Marrett, S., Ruff, D.A., Bandettini, P.A., Ungerleider, L.G.: Involvement of human left dorsolateral prefrontal cortex in perceptual decision making is independent of response modality. Proc. Natl. Acad. Sci. **103**, 10023–10028 (2006). https://doi.org/10.1073/pnas.0603949103
55. Sakagami, M., Pan, X.: Functional role of the ventrolateral prefrontal cortex in decision making. Curr. Opin. Neurobiol. **17**, 228–233 (2007). https://doi.org/10.1016/j.conb.2007.02.008
56. Plassmann, H., O'Doherty, J., Rangel, A.: Orbitofrontal cortex encodes willingness to pay in everyday economic transactions. J. Neurosci. **27**, 9984–9988 (2007). https://doi.org/10.1523/JNEUROSCI.2131-07.2007
57. Kühn, S., Gallinat, J.: The neural correlates of subjective pleasantness. Neuroimage **61**, 289–294 (2012). https://doi.org/10.1016/j.neuroimage.2012.02.065
58. Britton, J.C., Phan, K.L., Taylor, S.F., Welsh, R.C., Berridge, K.C., Liberzon, I.: Neural correlates of social and nonsocial emotions: an fMRI study. Neuroimage **31**, 397–409 (2006). https://doi.org/10.1016/j.neuroimage.2005.11.027
59. Etkin, A., Egner, T., Kalisch, R.: Emotional processing in anterior cingulate and medial prefrontal cortex. Trends Cogn. Sci. **15**, 85–93 (2011). https://doi.org/10.1016/j.tics.2010.11.004
60. Dolcos, F., Iordan, A.D., Dolcos, S.: Neural correlates of emotion–cognition interactions: a review of evidence from brain imaging investigations. J. Cogn. Psychol. **23**, 669–694 (2011). https://doi.org/10.1080/20445911.2011.594433
61. Wager, T.D., Davidson, M.L., Hughes, B.L., Lindquist, M.A., Ochsner, K.N.: Prefrontal-subcortical pathways mediating successful emotion regulation. Neuron **59**, 1037–1050 (2008). https://doi.org/10.1016/j.neuron.2008.09.006
62. Ellard, K.K., Barlow, D.H., Whitfield-Gabrieli, S., DE Gabrieli, J., Deckersbach, T.: Neural correlates of emotion acceptance vs worry or suppression in generalized anxiety disorder. Soc. Cogn. Affect. Neurosci. **12**, 1009–1021 (2017). https://doi.org/10.1093/scan/nsx025
63. Huppert, T.J.: Commentary on the statistical properties of noise and its implication on general linear models in functional near-infrared spectroscopy. Neurophotonics **3**, 010401 (2016). https://doi.org/10.1117/1.NPh.3.1.010401
64. Saager, R.B., Berger, A.J.: Direct characterization and removal of interfering absorption trends in two-layer turbid media. J. Opt. Soc. Am. A **22**, 1874 (2005). https://doi.org/10.1364/JOSAA.22.001874
65. Scholkmann, F., et al.: A review on continuous wave functional near-infrared spectroscopy and imaging instrumentation and methodology. Neuroimage **85**, 6–27 (2014). https://doi.org/10.1016/j.neuroimage.2013.05.004

66. Delpy, D.T., Cope, M., van der Zee, P., Arridge, S., Wyatt, S.W.S.: Estimation of optical pathlength through tissue from direct time of flight measurement. Phys. Med. Biol. **33**, 1433 (1988)
67. Kocsis, L., Herman, P., Eke, A.: The modified beer-lambert law revisited. Phys. Med. Biol. **51**, N91 (2006)
68. Barker, J.W., Aarabi, A., Huppert, T.J.: Autoregressive model based algorithm for correcting motion and serially correlated errors in fNIRS. Biomed. Opt. Express. **4**, 1366 (2013). https://doi.org/10.1364/BOE.4.001366
69. Benjamini, Y., Hochberg, Y.: Controlling the false discovery rate: a practical and powerful approach to multiple testing. J. R. Stat. Soc. Ser. B **57**, 289–300 (1995). https://doi.org/10.1111/j.2517-6161.1995.tb02031.x
70. Schienle, A., Wabnegger, A., Schoengassner, F., Scharmüller, W.: Neuronal correlates of three attentional strategies during affective picture processing: an fMRI study. Cogn. Affect. Behav. Neurosci. **14**(4), 1320–1326 (2014). https://doi.org/10.3758/s13415-014-0274-y
71. Killgore, W.D.S., Yurgelun-Todd, D.A.: The right-hemisphere and valence hypotheses: could they both be right (and sometimes left)? Soc. Cogn. Affect. Neurosci. **2**, 240–250 (2007). https://doi.org/10.1093/scan/nsm020
72. Brown, S., Gao, X., Tisdelle, L., Eickhoff, S.B., Liotti, M.: Naturalizing aesthetics: brain areas for aesthetic appraisal across sensory modalities. Neuroimage **58**, 250–258 (2011). https://doi.org/10.1016/j.neuroimage.2011.06.012
73. Cela-Conde, C.J., et al.: Dynamics of brain networks in the aesthetic appreciation. Proc. Natl. Acad. Sci. **110**, 10454–10461 (2013). https://doi.org/10.1073/pnas.1302855110
74. Reyna, V.F., Helm, R.K., Weldon, R.B., Shah, P.D., Turpin, A.G., Govindgari, S.: Brain activation covaries with reported criminal behaviors when making risky choices: a fuzzy-trace theory approach. J. Exp. Psychol. Gen. **147**, 1094–1109 (2018). https://doi.org/10.1037/xge0000434
75. Turney, D.: Teaching computers the meaning of words. https://www.smh.com.au/technology/teaching-computers-the-meaning-of-words-20131002-hv1v2.html. Accessed 31 March 2021
76. Garrido-Morgado, Á., González-Benito, Ó., Martos-Partal, M., Campo, K.: Which products are more responsive to in-store displays: utilitarian or hedonic? J. Retail. **67**(3), 477–491 (2020). https://doi.org/10.1016/j.jretai.2020.10.005
77. Eisenbeiss, M., Wilken, R., Skiera, B., Cornelissen, M.: What makes deal-of-the-day promotions really effective? The interplay of discount and time constraint with product type. Int. J. Res. Mark. **32**, 387–397 (2015). https://doi.org/10.1016/j.ijresmar.2015.05.007
78. Müller, J., et al.: Display blindness: the effect of expectations on attention towards digital signage. In: Tokuda, H., Beigl, M., Friday, A., Brush, A.J.B., Tobe, Y. (eds.) Pervasive 2009. LNCS, vol. 5538, pp. 1–8. Springer, Heidelberg (2009). https://doi.org/10.1007/978-3-642-01516-8_1
79. Dalton, N.S., Collins, E., Marshall, P.: Display Blindness?, pp. 3889–3898 (2015). https://doi.org/10.1145/2702123.2702150
80. Willems, K., Brengman, M., van de Sanden, S.: In-store proximity marketing: experimenting with digital point-of-sales communication. Int. J. Retail Distrib. Manag. **45**, 910–927 (2017). https://doi.org/10.1108/IJRDM-10-2016-0177
81. Bhatti, A., et al.: E-Commerce trends during COVID-19 Pandemic. Int. J. Futur. Gener. Commun. Netw. **13**, 1449–1452 (2020)

Topographic Analysis of Cognitive Load in Tacit Coordination Games Based on Electrophysiological Measurements

Dor Mizrahi[✉], Ilan Laufer, and Inon Zuckerman

Department of Industrial Engineering and Management, Ariel University, Ariel, Israel
dor.mizrahi1@msmail.ariel.ac.il, {ilanl,inonzu}@ariel.ac.il

Abstract. Tacit coordination games are coordination games in which communication between the players is not possible. Various studies have shown that people succeed in these games beyond what is predicted by classical game theory. This success is attributed to the identification of focal points (also known as Schelling points). Focal points are pronounced solutions based on salient features of the game that somehow attracts the players' attention. Experiments with tacit coordination games show that some players manage to "see" the focal points and reason about the selections made by the co-player, while others fail to do so, and might turn to guessing. According to the Cognitive Hierarchy Theory (CHT), the task of *coordinating*, that is, reasoning about what the other player would choose is performed on cognitive levels greater than or equal to 1. In contrast, the task of just *picking* an answer, without an explicit need to coordinate is done at cognitive level 0. With that in mind, our study has two main purposes. First, to examine whether the same task that is defined each time at a different cognitive level (*picking* or *coordination*) causes a different psychological cognitive load in the participating players. Second, to examine the distribution of cognitive load across the scalp during coordination tasks.

Keywords: Tacit coordination games · EEG · Theta/Beta ratio · Cognitive Hierarchy Theory (CHT)

1 Introduction

A tacit coordination problem is one in which two individuals are rewarded for making the same choice from the same set of alternatives when communication is not possible [1–3]. Until now, there is still no generally accepted explanation of how players manage to converge on the same solution [4]. In these games the more pronounced solutions, which are based on salient features attracting the player's attention, are referred to as focal points (also known as Schelling Points) [1]. In order to identify the patterns of players' actions, a number of studies were conducted that quantified and classified the players' behaviors at the individual level [5–7]. In addition, other works have examined the impact of a variety of variables, such as social orientation, cultural background [8] and loss-aversion [9] on players' behavior patterns and how the focal point is chosen.

© The Author(s), under exclusive license to Springer Nature Switzerland AG 2021
F. D. Davis et al. (Eds.): NeuroIS 2021, LNISO 52, pp. 162–171, 2021.
https://doi.org/10.1007/978-3-030-88900-5_18

Alongside the empirical measures and modeling that have been performed in various behavioral experiments, there are also economic-behavioral theories that attempt to describe players' behavior in tacit coordination games. One approach, known as the Cognitive Hierarchy theory (CHT) [10–12], is based on the *level-k thinking* model [4, 13–15]. This model assumes that players at level k assume that all the rest of the players are drawn from a distribution of lower k-levels (e.g. [13, 14, 16]). For example, players in which k = 0 (sometimes referred to as L_0 players) will choose randomly between the available actions, while L_1 players assume that all other players are L_0 reasoners and will act according to this assumption. In general, L_K players will respond based on the assumption that the rest of the players are L_{k-1} level.

That is, according to the CHT, the greater the value of k, the greater is the depth of reasoning, which entails increased cognitive load as a function of the thinking steps [17, 18]. It has also been suggested that available cognitive resources may affect the strategic behavior of players, and therefore the cost of reasoning may play an important role in making level k adjustments depending on the strategic environment, e.g. whether the player is in an advantageous position or not [18]. Therefore, it is important to find a suitable biomarker that could possibly serve as an objective correlate of level k or reasoning depth. The information added by this marker could potentially aid in constructing behavioral models that can more adequately anticipate the player's decision making in the context of coordination scenarios. In this study we aim to use the theta to beta ratio (TBR, [19, 20]) as a correlate of cognitive load by employing signal processing of EEG data.

Cognitive load refers to the amount of working memory resources required to perform a particular task [21] and there are two basic approaches for estimating it from EEG data. The first approach for assessing cognitive load relies on power spectrum analysis of continuous EEG that reveals the distribution of signal power over frequency. The EEG signal is divided into different frequency bands (i.e., delta, theta, alpha, and beta) to detect the bands that are sensitive to variations in load as a function of task demands. The second approach involves measuring the neural signal complexity that has been associated with both memory ability [22] and cognitive load [23]. Common methods in this category include fractal dimension (e.g. [24]), multi-scale entropy (e.g. [25]), and detrended fluctuation analysis [26, 27].

This study has two main objectives. First, we aim to test whether tacit coordination games (at cognitive hierarchy level $>= 1$) are cognitive tasks that produce a higher cognitive load on the subjects in relation to a parallel picking task (at cognitive hierarchy level $= 0$). In addition, we would like to examine how the cognitive load is distributed across the scalp and if there is any compatibility with the research literature (e.g. [28–31]) that holds that most of the load is concentrated in the frontal and pre-frontal locations.

2 Materials and Methods

In this study players were presented with a tacit coordination task in which they had to select a word from a given set of four words (in Hebrew). This task consists of 12 different instances each with a different set of words. The experiment consists of two experiment sessions, in the first session, the task presented to the players as a picking task. That is, to get the full score (100 points) for each game board they must select one of the options within the time frame of the task otherwise, they will get nothing. In the second experimental session, the players faced the same game boards. In each of the games, the players were told that they need to coordinate with an unknown randomly selected co-player by choosing the same word from the given set of words. Participants were further informed that they will receive an amount of 100 points each in case of successful coordination and that otherwise, they will get nothing. For example, game board #1 displayed in Fig. 1 (B) contains the set {"Water", "Beer", "Wine", "Whisky"}. All the words belong to the same semantic category, however, there is at least one word that stands out from the rest of the set because it is different in some prominent feature, e.g., in the current example, a non-alcoholic beverage ("water") which stands out among other alcoholic beverages. The more salient is the outlier, the easier it is to converge on the same focal point [1, 32].

{A}   {B}

Fig. 1. (A) Stand by screen (B) Game board #1 ["Water", "Beer", "Wine", "Whisky"]

Figure 2 portrays the outline of the experiment. The list of four words were embedded within a sequence of standby screens each presented for $U(2,2.5)$ sec. The slide presenting the list of words was presented for a maximal duration of 8 s and the next slide appeared following a button press. The order of the 12 games was randomized in each session. The participants were 10 students from the university that were enrolled in one of the courses on campus (right-handed, mean age = ~ 26, SD = 4).

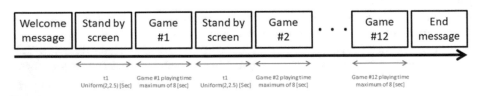

Fig. 2. Experimental paradigm with timeline

The EEG Data acquisition process during the game sessions was recorded by a 16-channel g.USBAMP bio signal amplifier (g.tec, Austria) at a sampling frequency of 512 Hz. 16 active electrodes were used for collecting EEG signals from the scalp based on the international 10–20 system. Recording was done by the OpenVibe [33] recording software. Impedance of all electrodes was kept below the threshold of 5K [ohm] during all recording sessions.

Theta/Beta Ratio (TBR) - In this study we estimated the cognitive load in each epoch using the Theta/Beta ratio (TBR). The TBR is known from the literature to reflect cognitive load in various cognitive tasks and to covary with activity in the executive control and default mode networks [19, 20]. It was shown that the smaller the TBR, the cognitive load is higher [19, 20]. Xie and Salvendy [34, 35] have differentiated between several main indices meant to quantify mental workload. These measures include instantaneous load (dynamic changes in load during task performance), peak load, average load, overall load, and accumulated load. In the current study we have created a hybrid index as follows. For each epoch we have first calculated the accumulated cognitive load [19], by calculating the energy ratio between theta and beta bands for each participant on each single epoch. Then, we have averaged the ratio across all epochs of an individual player to obtain the average cognitive load.

3 Data Processing and Analysis

3.1 Preprocessing and Feature Extraction

The data processing stage started by pre-processing the EEG signals recorded to maximize the signal-to-noise ratio. The preprocessing pipeline consisted of band-pass filtering of [1, 32] Hz and notch filter of 50 Hz for an artifact removal following by iCA. Then, the data was re-referenced to the average reference and down sampled to 64 Hz following a baseline correction. Data was analyzed on a 1-s epoch window from the onset of each game, as presented in Fig. 3.

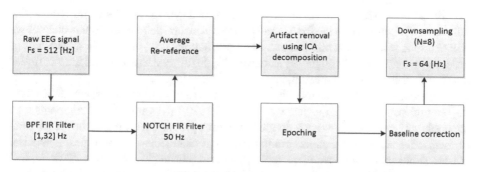

Fig. 3. Preprocess pipeline

Next, we calculated the intensity of the cognitive load in each task using the TBR index. To calculate the energy in the Theta and Beta bands, for each epoch, we have used the Discrete Wavelet Transform (DWT) [36, 37]. The DWT is based on a multiscale feature representation. Every scale represents a unique thickness of the EEG signal [38]. Each filtering step contains two digital filters, a high pass filter, $g(n)$, and a low pass filter $h(n)$. After each filter, a downsampler with factor 2 is used to adjust time resolution. In our case, we used a 3-level DWT, with the input signal having a sampling rate of 64 Hz. This specific DWT scheme resulted in the coefficients of the four EEG main frequency bands (i.e., Delta, Theta, Alpha, Beta). To calculate the cognitive load, which is expressed by the TBR, the DWT was applied on all the epochs to calculate the TBR. That is the ratio of the average power (see Fig. 4 and Eq. 1) in each one of the Theta and Beta bands (Theta/Beta) was calculated in each of the epochs.

$$P_x = \frac{1}{T} \sum_{t=1}^{T} x^2(t)$$ (1)

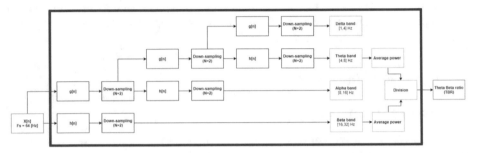

Fig. 4. Theta-Beta ratio calculation based on 3 level DWT scheme

3.2   The Distribution of TBR Across the Scalp in Coordination Tasks

The literature indicates that electrophysiological changes during the cognitive activity will be most significant in the frontal and prefrontal electrode regions. In this chapter we will examine whether this claim is true for coordination tasks, or whether these types of tasks do not require much cognitive resources. To do this, we have divided the electrodes into three clusters: frontal, central, and posterior (see Fig. 5).

A one-way ANOVA was used to examine the effect of electrode scalp distribution on TBR intensity. Overall, the data set contained 1920 TBR values (10 participants x 16 channels x 12 coordination tasks) divided into three groups according to the distribution of electrodes across the scalp:1) frontal cluster(6 channels - 720 samples; 2) central cluster (4 channels - 480 samples); and 3) posterior cluster (6 channels - 720 samples) (Fig. 5). Results showed that the main effect of topographic location was significant $F(2, 1917) = 3.65$, $p < 0.05$. Post hoc analyses using Tukey's honestly significant difference criterion [39] indicated that the average mean TBR value of the frontal electrodes was significantly lower than the rear electrode values ($p < 0.005$). As for the comparison results of the frontal-central and central-posterior, we received no statistical significance with p-value of 0.10 and 0.97, respectively.

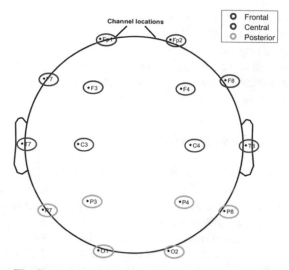

Fig. 5. Division of electrodes into topographic groups

The distribution of TBR values in relation to the position of the electrodes can be estimated based on the data of the different percentages shown in Table 1.

Table 1. Distribution of TBR values in relation to electrode position

Electrodes cluster	Min value	25th percentile	50th percentile	75th percentile	Max value
Frontal	0.1798	1.5009	2.9274	5.4127	82.4868
Central	0.2166	2.0823	3.6765	6.3265	108.385
Posterior	0.3281	2.2678	4.0459	6.8264	97.2420

3.3 Distribution of Cognitive Load in Relation to the Cognitive Hierarchy Level

In the second part of this study, we would like to compare the cognitive load of the different participants in the experiment depending on the cognitive hierarchy level of the same task. When it comes to the *picking* tasks, the player acts randomly because he has no incentive in choosing one solution over another. Therefore, the cognitive hierarchy level of the picking task is 0. In contrast, in a *coordination* task, the player must predict the behavior of the opposing player, so the cognitive level of this task equal to 1. In this section, we would like to test whether the change in the cognitive level in a specific task, on the same set of game boards, will be reflected in electrophysiological changes in the various participants by estimating the contingent load. Following the results of Sect. 3.2 we have only analyzed the results of the frontal electrode cluster.

The 720 TBR values (10 participants x 6 frontal channels x 12 tasks) extracted from the coordination task (M = 4.835, SD = 7.859) were compared to the 720 TBR values extracted from the picking tasks (M = 5.903, SD = 5.489) (t(1438) = 2.9892, p < 0.01). A higher cognitive load was observed in coordination tasks relative to the picking tasks. This result indicates that the reasoning process applied in the coordination task requires more cognitive resources for reasoning than in the picking task.

4 Conclusions and Future Work

To the best of our knowledge, this is the first time in which findings support the notion that the Theta/Beta ratio might serve as a marker of cognitive load in the context of picking and coordination tasks. In this study we have first corroborated that complex cognitive tasks depend on prefrontal [28] and frontal [19] cortex activation (e.g. [28–31]). Then, we have shown that within this prefrontal/frontal electrode cluster, TBR was found to be higher in the coordination condition relative to the picking condition. The correlation of TBR with task type might suggest the existence of more steps of k-level reasoning involved in the case of coordination as compared to picking, and thus may support CHT.

The results obtained show that there is a significant increase in the cognitive load in selection tasks compared to coordination tasks. This finding may correspond with the results of Tversky and Kahneman regarding human bounded rationality [40, 41]. The significant increase in cognitive load, which is reflected in the decrease in TBR values, implies that players use *system-2* for coordination and not *system-1*. That is, they adopt thought-analytical processes, which cause an increase in cognitive load, and do not use intuitive processes that rely on fast heuristics [1, 11] and are, therefore, more automatic.

The decrease in the TBR index found in the current study might be explained by a dissociation between beta and theta rhythms that was found in a previous study [42]. While beta 2 (20–30 Hz) in that study was found to increase as a function of working memory load, theta increases were only found among high performance participants, whereas theta decreases among low performance participants. In our study the transition from picking tasks (CHT = 0) to coordination tasks (CHT > 0) was accompanied with a relatively small increase in theta power of only 1.97% (20.23% to 20.63%). However, in case of the beta frequency band (which was largely overlapping with Beta 2) a much higher relative increase was evident, of 24.28% (7.66% to 9.52%).

These results might be in line with the dissociation found in the latter study [42], since in our study beta 2 has increased and theta has been decreased with the transition from picking to coordination tasks. Thus, enhanced task difficulty in the coordination condition might have diminished an increase in theta due to low performance participants, resulting in only a relatively small increase in theta and a much larger increase in beta 2 power. Moreover, an increase in beta power as a function of task complexity is consistent with the literature indicating that increases in beta power are associated with elevated mental workload levels during mental tasks [43] and concentration [44].

Overall, in view of the above it might be possible that the increased task difficulty in the coordination condition entails a more complex reasoning process that involves the reliance on system 2 and increased depth of reasoning suggested by CHT. that might be reflected by the reduction in TBR. Nevertheless, the connection between system 2 and level k reasoning, as well as the connection of TBR to increased load in coordination games should be further examined by future studies, which should also consider the individual coordination ability of the participants.

The results obtained open several future research options. For example, previous studies (e.g. [45, 46]) have shown that it is possible to produce an autonomous agent that predicts the player's behavior in coordination games, following the results of this study it is possible to incorporate electrophysiological indices in order to improve the prediction of the agent's results. Moreover it will be interesting to investigate the interaction between electrophysiological metrics, such as TBR, and variables that have been found to be significant for human behavior in coordination games, such as loss-aversion [9], social orientation [46, 47], culture background [48, 49] and fairness [50, 51].

References

1. Schelling, T.C.: The strategy of conflict. Cambridge (1960)
2. Mehta, J., Starmer, C., Sugden, R.: The nature of salience: an experimental investigation of pure coordination games. Am. Econ. Rev. **84**, 658–673 (1994)
3. Dong, L., Montero, M., Possajennikov, A.: Communication, leadership and coordination failure. Theor. Decis. **84**(4), 557–584 (2017). https://doi.org/10.1007/s11238-017-9617-9
4. Bardsley, N., Mehta, J., Starmer, C., Sugden, R.: Explaining focal points : cognitive hierarchy theory versus team reasoning. Econ. J. **120**, 40–79 (2009)
5. Mizrahi, D., Laufer, I., Zuckerman, I.: Individual strategic profiles in tacit coordination games. J. Exp. Theor. Artif. Intell. **33**(1), 63–78 (2021)
6. Mizrahi, D., Laufer, I., Zuckerman, I.: Modeling individual tacit coordination abilities. In: Liang, P., Goel, V., Shan, C. (eds.) Brain Informatics. BI 2019. Lecture Notes in Computer Science, vol. 11976, pp. 29–38. Springer, Cham (2019). https://doi.org/10.1007/978-3-030-37078-7_4
7. Mizrahi, D., Laufer, I., Zuckerman, I.: The effect of individual coordination ability on cognitive-load in tacit coordination games. In: Davis, F., Riedl, R., Brocke, J. vom, Léger, P.-M., Randolph, A., Fischer, T. (eds.) NeuroIS Retreat 2020. Vienna, Austria (2020). https://doi.org/10.1007/978-3-030-60073-0
8. Mizrahi, D., Laufer, I., Zuckerman, I., Zhang, T.: The effect of culture and social orientation on player's performances in tacit coordination games. In: Wang, S., et al. (eds.) Brain Informatics. BI 2018. Lecture Notes in Computer Science, vol. 11309, pp. 437–447. Springer, Cham (2018). https://doi.org/10.1007/978-3-030-05587-5_41
9. Mizrahi, D., Laufer, I., Zuckerman, I.: The effect of loss-aversion on strategic behaviour of players in divergent interest tacit coordination games. In: Mahmud, M., Vassanelli, S., Kaiser, M.S., Zhong, N. (eds.) BI 2020. LNCS (LNAI), vol. 12241, pp. 41–49. Springer, Cham (2020). https://doi.org/10.1007/978-3-030-59277-6_4
10. Stahl, D.O., Wilson, P.W.: On players' models of other players: theory and experimental evidence. Games Econ. Behav. **10**, 218–254 (1995)
11. Bacharach, M., Stahl, D.O.: Variable-frame level-n theory. Games Econ. Behav. **32**, 220–246 (2000)
12. Camerer, C., Ho, T.-H., Chong, K.: A cognitive hierarchy model of games. Q. J. Econ. **119**, 861–898 (2004)
13. Costa-Gomes, M.A., Crawford, V.P., Iriberri, N.: Comparing models of strategic thinking in Van Huyck, Battalio, and Beil's coordination games. J. Eur. Econ. Assoc. **7**, 365–376 (2009)

14. Faillo, M., Smerilli, A., Sugden, R.: The roles of level-k and team reasoning in solving coordination games (2013)
15. Gold, N., Colman, A.M.: Team reasoning and the rational choice of payoff-dominant outcomes in games. Topoi **39**(2), 305–316 (2018). https://doi.org/10.1007/s11245-018-9575-z
16. Strzalecki, T.: Depth of reasoning and higher order beliefs. J. Econ. Behav. Organ. **108**, 108–122 (2014)
17. Jin, Y.: Does level-k behavior imply level-k thinking?. Exp. Econ. **24**(1), 330–353 (2021)
18. Zhao, W.: Cost of reasoning and strategic sophistication. Games **11**, 40 (2020)
19. Bagyaraj, S., Ravindran, G., Shenbaga Devi, S.: Analysis of spectral features of EEG during four different cognitive tasks. Int. J. Eng. Technol. **6**, 725–734 (2014)
20. van Son, D., de Rover, M., De Blasio, F.M., van der Does, W., Barry, R.J., Putman, P.: Electroencephalography theta/beta ratio covaries with mind wandering and functional connectivity in the executive control network. Ann. N. Y. Acad. Sci. **1452**, 52–64 (2019)
21. Antonenko, P., Paas, F., Grabner, R., van Gog, T.: Using electroencephalography to measure cognitive load. Educ. Psychol. Rev. **22**, 425–438 (2010)
22. Sheehan, T.C., Sreekumar, V., Inati, S.K., Zaghloul, K.A.: Signal complexity of human intracranial EEG tracks successful associative-memory formation across individuals. J. Neurosci. **38**, 1744–1755 (2018)
23. Friedman, N., Fekete, T., Gal, K., Shriki, O.: EEG-based prediction of cognitive load in intelligence tests. Front. Hum. Neurosci. **13** (2019)
24. Stokić, M., Milovanović, D., Ljubisavljević, M.R., Nenadović, V., Čukić, M.: Memory load effect in auditory–verbal short-term memory task: EEG fractal and spectral analysis. Exp. Brain Res. **233**(10), 3023–3038 (2015). https://doi.org/10.1007/s00221-015-4372-z
25. Escudero, J., Abásolo, D., Hornero, R., Espino, P., López, M.: Analysis of electroencephalograms in Alzheimer's disease patients with multiscale entropy. Physiol. Meas. **27** (2006)
26. Peng, C.K., Havlin, S., Stanley, H.E., Goldberger, A.L.: Quantification of scaling exponents and crossover phenomena in nonstationary heartbeat time series. Chaos **5**, 82–87 (1995). https://doi.org/10.1063/1.166141
27. Rubin, D., Fekete, T., Mujica-Parodi, L.R.: Optimizing complexity measures for fMRI data: algorithm, artifact, and sensitivity. PLoS One **8** (2013). https://doi.org/10.1371/journal.pone. 0063448
28. Gartner, M., Grimm, S., Bajbouj, M.: Frontal midline theta oscillations during mental arithmetic: Effects of stress. Front. Behav. Neurosci. **9**, 1–8 (2015)
29. De Vico Fallani, F., et al.: Defecting or not defecting: How to "read" human behavior during cooperative games by EEG measurements. PLoS One **5** (2010)
30. Boudewyn, M., Roberts, B.M., Mizrak, E., Ranganath, C., Carter, C.S.: Prefrontal transcranial direct current stimulation (tDCS) enhances behavioral and EEG markers of proactive control. Cogn. Neurosci. **10**, 57–65 (2019)
31. Moliadze, V., et al.: After-effects of 10 Hz tACS over the prefrontal cortex on phonological word decisions. Brain Stimul. **12**, 1464–1474 (2019)
32. Mehta, J., Starmer, C., Sugden, R.: Focal points in pure coordination games: an experimental investigation. Theory Decis. **36**, 163–185 (1994)
33. Renard, Y., et al.: Openvibe: an open-source software platform to design, test, and use brain–computer interfaces in real and virtual environments. Presence Teleoperators Virtual Environ. **19**, 35–53 (2010)
34. Xie, B., Salvendy, G.: Review and reappraisal of modelling and predicting mental workload in single-and multi-task environments. Work Stress **14**, 74–99 (2010)
35. Xie, B., Salvendy, G.: Prediction of mental workload in single and multiple tasks environments. Int. J. Cogn. Ergon. **4**, 213–242 (2000). https://doi.org/10.1207/S15327566IJC E0403

36. Shensa, M.J.: The discrete wavelet transform: wedding the a trous and mallat algorithms. IEEE Trans. Signal Process. **40**, 2464–2482 (1992)
37. Jensen, A., la Cour-Harbo, A.: Ripples in Mathematics: The Discrete Wavelet Transform. Springer Science & Business Media, Heidelberg (2001)
38. Hazarika, N., Chen, J.Z., Tsoi, A.C., Sergejew, A.: Classification of EEG signals using the wavelet transform. Signal Process. **59**, 61–72 (1997)
39. Tukey, J.W.: Comparing individual means in the analysis of variance. Biometrics **5**, 99–114 (1949)
40. Tversky, A., Kahneman, D.: Prospect theory: an analysis of decision under risk. Econometrica **47**, 263–291 (1979)
41. Kahneman, D.: Thinking, Fast and Slow. Macmillan, New York (2011)
42. Pavlov, Y.G., Kotchoubey, B.: EEG correlates of working memory performance in females. BMC Neurosci. **18**, 1–14 (2017)
43. Coelli, S., Sclocco, R., Barbieri, R., Reni, G., Zucca, C., Bianchi, A.M.: EEG-based index for engagement level monitoring during sustained attention. In: 37th Annual International Conference of the IEEE Engineering in Medicine and Biology Society (EMBC) (2015)
44. Kakkos, I., et al.: Mental workload drives different reorganizations of functional cortical connectivity between 2D and 3D simulated flight experiments. IEEE Trans. Neural Syst. Rehabil. Eng. **27**, 1704–1713 (2019)
45. Zuckerman, I., Kraus, S., Rosenschein, J.S.: Using focal point learning to improve human-machine tacit coordination. Auton. Agents Multi-Agents Syst. **22**, 289–316 (2011)
46. Mizrahi, D., Zuckerman, I., Laufer, I.: Using a stochastic agent model to optimize performance in divergent interest tacit coordination games. Sensors **20**, 7026 (2020)
47. Cheng, K.L., Zuckerman, I., Nau, D., Golbeck, J.: The life game: cognitive strategies for repeated stochastic games. In: 2011 IEEE Third International Conference on Privacy, Security, Risk and Trust and 2011 IEEE Third International Conference on Social Computing, pp. 95–102 (2011)
48. Cox, T.H., Lobel, S.A., Mcleod, P.L.: Effects of ethnic group cultural differences on cooperative and competitive behavior on a group task. Acad. Manag. J. **34**, 827–847 (1991)
49. Mizrahi, D., Laufer, I., Zuckerman, I.: Collectivism-individualism: strategic behavior in tacit coordination games. PLoS One **15** (2020)
50. De Herdt, T.: Cooperation and fairness: the flood-Dresher experiment revisited. Rev. Soc. Econ. **61**, 183–210 (2003). https://doi.org/10.1080/0034676032000098219
51. Fehr, E., Schmidt, K.M.: A theory of fairness, competition, and cooperation. Q. J. Econ. **114**, 817–868 (1999)

Active Learning Techniques for Preparing NeuroIS Researchers

Arjan Raven[1] and Adriane B. Randolph[2]([⊠])

[1] Department of Management Information Systems, Temple University, Philadelphia, PA, USA
[2] Department of Information Systems and Security, Kennesaw State University, Kennesaw, GA, USA
arandol3@kennesaw.edu

Abstract. The field of neuroIS is rapidly evolving, and there is a need to create a research and work force at various levels of the academy ranging from undergraduate students to professors. Motivation is not an issue with neuroIS as students are typically excited to learn, but how do we teach them the skills they need to succeed? Active learning is a pedagogical technique that has a natural fit with neuroIS. It focuses on the higher levels of learning that are essential in the field. This paper is an introduction to active learning for the benefit of the neuroIS community. It discusses examples of what can be done as well as challenges that need to be overcome.

Keywords: Active learning · NeuroIS tools · Pedagogy

1 Introduction and Motivation

Although neuroIS tools are being used to examine "neuro-education" to better understand student learners and their cognitive processes [1–3], there is also a need to examine the education of neuroIS researchers. This becomes even more important as the need for people with neuroIS skills is increasing and as the required skill base is expanding. The neuroIS community may benefit from educational pedagogy particularly well-suited to teach the necessary skills through active learning [4].

NeuroIS researchers and workers need to understand the different technologies such as functional magnetic resonance imaging (fMRI), electroencephalography (EEG), skin conductance response (SCR), functional near-infrared (fNIR) imaging, and eye tracking. They need to have experience with some of the equipment, but they also need to understand the intricacies of brain-based interface design as they consider the development of neuroadaptive interfaces [5]. Perhaps the most important skill they need is to be able to learn about new developments in the field and to quickly adapt new technologies. Any equipment and software they train on today may be replaced within a few years.

Riedl and Leger have created a foundational textbook that can be used as a corner stone for courses in neuroIS [6]. With the publication of the textbook, a reference syllabus was created that creates a basic structure for a course built around the book. As the authors point out in their discussion of the syllabus, neuroIS is a younger field that is

F. D. Davis et al. (Eds.): NeuroIS 2021, LNISO 52, pp. 172–177, 2021.
https://doi.org/10.1007/978-3-030-88900-5_19

rapidly developing [7]. It can be difficult to create course material and assignments and to keep all the material up-to-date from semester to semester. In the five years since the publication of the book the fundamental science has not changed, but equipment has continued to develop and a plethora of applications of neuroIS has emerged, including even brain-controlled toys. Recently published articles and newspaper stories can be used to teach the latest developments.

The combination of the syllabus, textbook, and recent materials provide the basic structure for a neuroIS course, but they do not provide guidelines for how they can be used in the courses. The aim of this paper is not to update the reference syllabus but to discuss what educational techniques can be used in courses on neuroIS. Teaching neuroIS is essential, but it is also challenging. Perhaps the biggest challenge is to create lab time for every student so that they can individually get hands-on experience. A pedagogical development that has been gaining interest over the years and now seems to have real momentum is active learning [8]. Here we explore how an active learning framework may be especially relevant for neuroIS.

2   Active Learning as a Pedagogical Technique

It would be ideal if neuroIS classes could be taught in an active lab where every student has access to all the tools and can work alongside experienced researchers [5]. However, the equipment is expensive and the number of neuroIS scientists is limited. At the same time, students must learn how to work directly with brain-based interfaces and how to develop software. The closest simulation of real-life neuroIS situations is to create an active learning classroom environment for the students.

There are many ways in which students can perform activities that will give them a deeper and more practical understanding of the material. Bonwell and Eison [8] (p. iii) define strategies to promote active learning as "…instructional activities involving students in doing things and thinking about what they are doing…They must read, write, discuss, or be engaged in solving problems. Most important, to be actively involved, students must engage in such higher-order tasks as analysis, synthesis, and evaluation."

These levels of tasks and learning are classically examined in education. Bloom's Taxonomy [9, 10] identifies three lower levels of learning (Remember, Apply, Understand) as well as three higher levels (Analyze, Evaluate, Create). Active learning is a great way to reach the higher levels of learning. While the taxonomy addresses the learning goals, active learning can be seen as a guide to the methods to attain these goals.

Building on Bonwell and Eison's work, Prince [11] (p. 223) defines active learning as "any instructional method that engages students in the learning process. In short, active learning requires students to do meaningful learning activities and think about what they are doing." He notes that meta-studies on active learning at times show conflicting results and notes a need to more clearly distinguish between the different types of active learning. He identifies collaborative learning, cooperative learning, and problem-based learning as important subsets. His investigation of the different types of active learning identifies extensive support for the success of active learning. In a 2014 study, Freeman et al. [12] conducted a meta-analysis of 225 studies in science, engineering, and mathematics, and found that student performance improved by almost half a standard deviation in active learning vs traditional lecturing.

Active learning is applied in many different fields of study that are relevant for neuroIS courses. It is, for instance, considered to be beneficial for medical students in psychiatry [13]. In Information Systems (IS) education, active learning has been used for business intelligence education. A recent empirical study discusses collaborative active learning resulted in better performance, more reflection, and better retention of the material [4]. Two studies discuss how IS core courses at different universities are redesigned to include more active learning. Both articles found mixed results from the changes [14, 15].

3   Application of the Active Learning Approach at Kennesaw State University

There is an evolving set of learning objectives in neuroIS following as the field is evolving [5, 16]. It is a highly specialized field where students need to get the hands-on experience of working with the tools. There is a large tacit knowledge component to it. For example, students need to be able to look at the data and recognize a pattern that indicates that an external signal is interfering and bleeding into the main signal, as with EEG. They also need to know what the current hardware and software are and be able to quickly learn new technologies.

To have a stronger grasp of the underlying concepts and to encourage higher-level learning, students should be encouraged to see the technology in action by offering demonstrations or field trips and engagement with the material. Expect that many students will want to try the technology first-hand despite stated risks for an fMRI scan or inconvenience of gel in one's hair as with most EEG systems. In classes about technology, it can be a problem to motivate the students, but in neuroIS, we have the luxury of students who tend to be highly-motivated and are excited about the work.

An introductory course on neuroIS was taught three times by the second author, at Kennesaw State University (KSU). It was offered as an upper-level elective targeted for undergraduate MIS majors. The last time this was taught was in the fall of 2015, and at that point the Riedl book [6] was not yet available. Instead, the course relied on academic articles and media reports. The course design overlaps with that of the reference syllabus [7], but did not include discussions of measurement of the peripheral nervous system, oculometry, facial muscular movement, or hormones. It also did not include a discussion on how to establish a neuroIS lab, but the BrainLab at KSU was used as an example of what a lab could look like and had been established for eight years at the time. Table 1 presents examples of activities from a neuroIS class taught by the second author. The activities are mapped to their associated learning objective. They range from passive to increasingly more active [17].

There are limitations to how active the assignments in a class can be. Lab space is often physically limited, and there especially is a dearth of equipment unless a space has received deep funding support. Students have to take turns working with the neurophysiological equipment. Presently, there are additional concerns regarding safety during a pandemic although protocols have been recommended by vendors of neurophysiological equipment such as by Cortech Solutions in

Table 1. NeuroIS class activity examples

Learning Objective	Activity
Identify parts of the brain utilized for cognitive processing and control	**Passive:** Listen to guest lecture from a professor in cognitive neuroscience. Even though the material lecture was very interesting, the students did not work with the material they were learning
Compare different control-abilities of end-users	**More Active:** Read, review, and discuss academic papers on individual differences and neural control. In reviewing the material the students were asked to critically review what they were reading. In the next step they discussed their observations and thus were exposed to different viewpoints
Discuss applications for using brain-imaging techniques to assess human mental states	**Yet More Active:** Visit local hospital to see an fMRI in action. The second author organized a site visit in which the students were able to see fMRI as it is used in a clinical setting. They interacted with the medical staff, and were able to compare this real-world setting with what they had read about
Design and demonstrate useful integration of information systems with novel input from the brain	**Yet More Active:** Conduct case analysis, design, or experiment in teams and present to the class with panel of expert visitors. For this project component of the class the students had a choice of activity. For instance, a student group designed an interface for a system that allows a locked-in patient to change settings in the environmental controls in their home. The project had the students apply Bloom's lower-level skills as they studied existing systems as well as the higher-level skills while they created the new interface
Identify current technologies that incorporate neural or psychophysiological recordings	**Yet More Active:** Participate in technology demonstrations in-class. In this kind of setting students are either subjects or conduct the demonstrations. In both roles they get first-hand experience of doing neuroIS research work in a lab setting. The activities integrate what they have more passive learned through reading and observation with the physical experience of using the equipment

the United States (https://cortechsolutions.com/special-considerations-for-human-neu roscience-research-in-the-midst-of-the-covid-19-pandemic/).

The problems become even more challenging in on-line classes where equipment may have to be sent from student to student. Then, the student has to be their own subject. To achieve this, a trade-off often must be made between using multiple, less-expensive devices that may be considered less robust versus one or a few more research-grade devices [18]. This tradeoff may be made inconsequential depending on level of focus for the outcomes of the class. In addition to measurement equipment, there is a need for software tools that can be used to design a brain-computer or neuroadaptive interface.

4   Conclusion

Active learning is a very effective way to help students achieve the higher levels of learning in Bloom's taxonomy. The need for neuroIS classes is increasing, which means that there will also be more innovation in teaching methods. As a field, we can share more of those methods through case studies, best practice descriptions, and research on the pedagogy of neuroIS. Furthermore, we can help each other identify affordable measurement equipment and software development tools that still meet our learning needs. As more neuroIS courses are outlined and offered, they are being done so by researchers from established labs as well as newcomers. This paper helps codify a method of how new researchers in the field may obtain the knowledge beyond what are the details of that knowledge.

References

1. Charland, P., et al.: Measuring implicit cognitive and emotional engagement to better understand learners' performance in problem solving. Zeitschrift für Psychologie **224**(4), 294 (2017)
2. Charland, P., et al.: Assessing the multiple dimensions of engagement to characterize learning: a neurophysiological perspective. J. Vis. Exp. JoVE 101 (2015)
3. Randolph, A., Mekbib, S., Calvert, J., Cortes, K., Terrell, C.: Application of neurois tools to understand cognitive behaviors of student learners in biochemistry. In: Davis, F.D., Riedl, R., vom Brocke, J., Léger, P.-M., Randolph, A., Fischer, T. (eds.) Information Systems and Neuroscience. LNISO, vol. 32, pp. 239–243. Springer, Cham (2020). https://doi.org/10.1007/978-3-030-28144-1_26
4. Romanow, D., Napier, N.P., Cline, M.K.: Using active learning, group formation, and discussion to increase student learning: a business intelligence skills analysis. J. Inf. Syst. Educ. **31**(3), 218–231 (2020)
5. vom Brocke, J., et al.: Advancing a neurois research agenda with four areas of societal contributions. Eur. J. Inf. Syst. **29**(1), 9–24 (2020)
6. Riedl, R., Léger, P.-M.: Fundamentals of NeuroIS. Studies in Neuroscience, Psychology and Behavioral Economics. Springer, Berlin, Heidelberg (2016)
7. Riedl, R., Léger, P.-M.: Neuro-Information-Systems (NeuroIS). A.f.I. Systems (2016). EDUglopedia.org
8. Bonwell, C.C., Eison, J.A.: Active Learning: Creating Excitement in the Classroom. 1991 ASHE-ERIC Higher Education Reports. ERIC (1991)

9. Anderson, L.W., Bloom, B.S.: A taxonomy for learning, teaching, and assessing: a revision of Bloom's taxonomy of educational objectives. Longman (2001)
10. Bloom, B.S.: Taxonomy of Educational Objectives. vol. 1: Cognitive Domain, no. 20, p. 24. McKay, New York (1956)
11. Prince, M.: Does active learning work? A review of the research. J. Eng. Educ. **93**(3), 223–231 (2004)
12. Freeman, S., et al.: Active learning increases student performance in science, engineering, and mathematics. Proc. Natl. Acad. Sci. **111**(23), 8410–8415 (2014)
13. Sandrone, S., et al.: Active learning in psychiatry education: current practices and future perspectives. Front. Psychiatr. **11**, 211 (2020)
14. Lavin, A., Martin, M.-C., Sclarow, S.: Radically redesigning introductory MIS large-scale lectures: creating enhanced learning environments. In: Proceedings of the 2018 AIS SIGED international Conference on Information Systems Education and Research, Philadelphia (2019)
15. Riordan, R.J., Hine, M.J., Smith, T.C.: An integrated learning approach to teaching an undergraduate information systems course. J. Inf. Syst. Educ. **28**(1), 59–70 (2017)
16. Riedl, R., et al.: A decade of NeuroIS research: progress, challenges, and future directions. ACM SIGMIS Database DATABASE Adv. Inf. Syst. **51**(3), 13–54 (2020)
17. Randolph, A.B.: Learning what is top-of-mind: a course on neuro-information systems. In: 11th Pre-ICIS Workshop on HCI Research in MIS. 2012: Orlando, FL (2012)
18. Riedl, R., Minas, R.K., Dennis, A.R., Müller-Putz, G.R.: Consumer-grade EEG instruments: insights on the measurement quality based on a literature review and implications for NeuroIS research. In: Davis, F.D., Riedl, R., vom Brocke, J., Léger, P.-M., Randolph, A.B., Fischer, T. (eds.) NeuroIS 2020. LNISO, vol. 43, pp. 350–361. Springer, Cham (2020). https://doi.org/10.1007/978-3-030-60073-0_41

Examining the Impact of Social Video Game Tournaments on Gamers' Mental Well-Being

Tanesha Jones, Adriane B. Randolph[✉], and Sweta Sneha

Department of Information Systems and Security, Kennesaw State University, Kennesaw, GA, USA
tjone148@students.kennesaw.edu, {arandol3,ssneha}@kennesaw.edu

Abstract. We examine the impact that gaming on a social tournament platform while playing multiplayer games has on the mental well-being of college students. In this early-stage study, we used the Scale of Positive and Negative Experiences and the Player Experience and Need Satisfaction Scale to measure well-being, gaming motivation, and enjoyment. We complement these survey tools with facial expression analysis of students during gameplay for a more holistic understanding of their emotional states and the impact of social gaming.

Keywords: Mental well-being · Motivation · Video game tournaments · Gaming · Game skill level · Facial expression analysis

1 Introduction

There has been a recent push to research positive mental health and gaming [1–4] in contrast to a focus on the negative effects [5, 6]. It is believed that players who experience a high degree of relatedness during video game play (or "gaming") will likely experience higher well-being [4]. For the purpose of this study, mental well-being is defined as feeling good about oneself and life, while mental health encompasses emotional, psychological, and social well-being. We will be looking at well-being and the emotional and social well-being aspects of mental health [7, 8].

Nonvoluntary isolation, as with the COVID-19 pandemic, may cause individuals to suffer negative mental well-being effects. It is our belief that playing games and connecting virtually can produce similar positive mental well-being effects as in-person social experiences. Gaming is a non-physical activity that can be done alone or socially and is absent of any need for physical contact through touching.

Touch is a sense that contributes to well-being and can overshadow or mask other contributing factors affecting a person's well-being. Touch and touch therapy are relatively new methods being researched and used to combat poor mental health and well-being [9]. This makes gaming a great opportunity to isolate other factors that may also affect well-being as the study will look at an activity where touch and physical presence of others have been removed. We will use self-reported measures through surveys for primary data collection.

F. D. Davis et al. (Eds.): NeuroIS 2021, LNISO 52, pp. 178–183, 2021.
https://doi.org/10.1007/978-3-030-88900-5_20

However, neuroIS researchers have found that self-reported data has its limitations. With added neurophysiological measures, we may find a wealth of knowledge untapped and unknown even to the participants, themselves, to create a more holistic picture [10]. It has been found that these methods can complement more common study methods by using neurophysiological measures to explain additional variances beyond what psychological measures can offer.

In this study, we attempt to use supplementary facial expression analysis of emotion to complement insights gleaned from traditional surveys on mental well-being, gaming motivation, and enjoyment. The overall goal is to determine if participants experience changes in mental well-being from playing in social tournaments. The results may serve as fuel for further studies to find if these virtual gaming experiences create similar positive effects as in-person gaming experiences. The following presents results from data collected in five social tournaments for this work-in-progress.

2 Methodology

This work examines the mental well-being of college students at a large university in the southeastern United States. For our study, we used the validated Scale of Positive and Negative Experiences (SPANE) [11–13] to measure well-being. We asked participants how they had been feeling in the past four weeks and how often they experienced each of six positive and six negative feelings according to the SPANE tool. Participants were surveyed pre- and post- competing in the tournaments for their well-being and gaming motivation.

We used Player Experience and Need Satisfaction Scale (PENS) for post tournament experience [14]. This survey was administered post tournament and as a final questionnaire near the end of the study to record participants' overall experience of the tournaments. There are two groups: 1) a cohort that plays all or most of the tournaments for a historical look at their experience and emotions over time, and 2) rotations of participants who play in one tournament, for a total of 125 targeted participants.

The participants play multiple rounds of Rocket League, League of Legends, or Overwatch in individual tournaments, playing one game throughout the tournament session. Separate analysis will be done on each group. Within each group there are subgroups of participants who label themselves as Elite Gamer, Aspirational Gamer, Casual Gamer, or Beginner. These are self-assigned gaming skill levels that serve as control variables.

The participants will be randomly sampled and categorized post-hoc according to characteristics captured in a demographics portion of the survey. They will share their username/gamer name for longitudinal tracking in the surveys but are anonymized in final reporting. Players will voluntarily record video of their facial expressions during gameplay and share the recordings with the research team for post-hoc analysis of emotion using iMotions AFFDEX, a recording and analytic software based on the Facial Action Coding System (FACS) [15]. A metric for overall positive/negative emotion experienced during gameplay will be analyzed against the survey measures of well-being and supplemented with their facial recordings to create a more in-depth profile of the gamer's experience and mental health as participants may not be fully attuned with their emotions and self-assess incorrectly. Having the visual data of their facial emotions will supplement the survey responses [13].

The model we will be using as illustrated in Fig. 1 is inspired by the research of Deci and Ryan's Self Determination Theory (SDT) [16] which resulted in the creation of the Gaming Motivation Scale (GAMS) [18, 19] and another model by Sterling [17] that looks at the relationship between online gamers' psychological needs and gaming behavior and motivation. Self Determination Theory looks at human motivation and personality being constructed of three components: autonomy, competence, and relatedness [16]. We base our study on gaming motivation as used in Sterling's study that was measured using GAMS. Gaming behavior was measured using frequency and duration among online gamers, and psychological needs were measured using the Basic Needs Satisfaction in General Scale (BNSG-S) [17].

For this study, we will be looking at gaming motivation as the independent variable and mental well-being as the dependent variable. We are hypothesizing that within gaming skill levels like Elite Gamer, Aspirational Gamer, Casual Gamer, and Beginner, that the player's gaming motivation will have a direct correlation to the well-being level of the player.

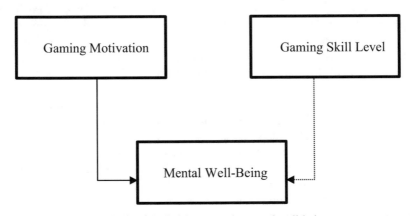

Fig. 1. Model of video gamer's mental well-being.

3   Preliminary Results

During the first pilot tournament, we found that participants were overwhelmed by the instructions on the tournament website. We reduced the number of words and added text formatting like bolding and italics for emphasis. Participants also were not finishing the post survey, which is believed to be due to the last question which implied a mandatory submission of their video recording without optional language to convey they could submit the survey without the video. The participants, not having done the recordings, thus exited out of the survey rather than submitting without the recording. We changed the survey to ask if they had done the optional recording and if not then they were given a submit button, otherwise they were given the link to submit the video.

We found that the next tournament did not have any participants that signed up, so we increased recruitment efforts by adding flyers in the school weekly news email and emphasized the gaming platform was offering prize raffle incentives. We also decided to remove the Game Experience Questionnaire (GEQ) section from the survey. We found it to be too tedious and made the survey much longer which we feared would reduce the number of participants who would complete the survey. The PENS was found to be validated more heavily and more in line with the type of question we are trying to answer. The tournament after this had nine completed post surveys which showed that our changes were effective.

After four more tournaments we had enough data to do an initial analysis with a sample size of 39. SPANE B is the mental well-being score summed from SPANE P (positive score) minus SPANE N (negative score). The SPANE B score ranges from -24 (lowest unhappiness score) to 24 (highest happiness score) [5, 18]. Before participating in the tournaments, we saw a mean SPANE B of 5.63 and post tournament a mean SPANE B of 5.51. This signifies on average that participants had a low level positive mental well-being score and also their score decreased after the tournament. Many factors could be attributed to this such as losing their match or fatigue from playing causing them to answer erroneously. We hope the supplementary facial expression analysis will offer more insights to determine if more negative emotions were experienced. We were also unable to break out the scores by gaming skill level to see if experienced gamers scored higher or lower than beginner gamers.

Using iMotions' AFFDEX, we analyzed a sample video of a student engaged in gameplay and were able to gain insights as to the person's emotions and facial expressions during gameplay. The software accurately identified, at an 80% threshold, attention, engagement, negative emotions, lip suck and press, open mouth, chin raise, dimpler, and jaw drop. With these cues it identified moments of anger, contempt, joy, sadness, engagement, and disgust throughout the video as illustrated in Fig. 2.

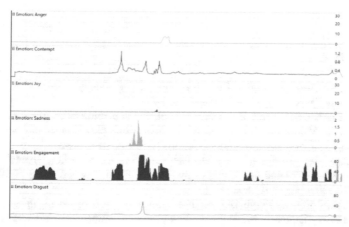

Fig. 2. Example of video gamer's analysis of emotion from facial expressions.

iMotions identified other facial expressions outside the threshold like lip stretch and lid tighten, brow rise, eye widen, smile, and smirk. We hope to pair this analysis with the survey to confirm or enhance the participants' self-assessments of their emotions and mental well-being. This facial expression analysis will help ensure we have accurate measurements when looking to see if playing the social tournaments had an impact on their mental well-being.

4   Conclusion

We have found ways to improve our data collection process by changing the language of our outreach and participation directions, emphasizing incentives and optional steps, creating recruitment videos to better explain directions, implementing a project website for participants to reference, and modifying the flow of the survey. Based on the tournament participation following the first two events, we saw an increase in participation signup and completed surveys. We hope that at the end of data collection we will have achieved our goal of 125 participants. Based on an initial sample of 39 participants, we saw a decrease in mental well-being scores but were unable to break the sample into gaming skill level groups and look at the participants' motivations. As we collect more data, these individual groups will hopefully allow us to better understand trends and correlations between gaming motivation and skill level against the well-being scores. We will also have facial expression videos of participants to gain a deeper insight of involuntary and unconscious information to provide a more holistic picture of well-being and gameplay.

Acknowledgements. We would like to thank United in Gaming for customized access to their gaming platform and enthusiastic support of this effort. Additionally, we thank the members of the Kennesaw State University Esports community for their participation under the guidance of Andrew Harvill and Norman Reid.

References

1. Daugette, A.: The Psychological Benefits of Video Games. Sekg, 13 December 2019. https://www.sekg.net/psychology-benefits-video-games/
2. Granic, I., Lobel, A., Engels, R.C.M.E.: The benefits of playing video games (2014)
3. Johannes, N., Vuorre, M., Przybylski, A.K.: Video game play is positively correlated with well-being, 13 November 2020. https://doi.org/10.31234/osf.io/qrjza
4. Jones, C.M., Scholes, L., Johnson, D., Katsikitis, M., Carras, M.C.: Gaming well: links between videogames and flourishing mental health. Front. Psychol. **5**, 260 (2014). https://doi.org/10.3389/fpsyg.2014.00260
5. Ryan, R.M., Rigby, C., Przybylski, A.: The motivational pull of video games: a self- determination theory approach. Motiv. Emot. **30**(4), 347–363 (2006). https://doi.org/10.1007/s11031-006-9051-8
6. Smyth, J.M.: Beyond self-selection in video game play: an experimental examination of the consequences of massively multiplayer online role-playing game play. Cyberpsychol. Behav. **10**(5), 717–721 (2007). https://doi.org/10.1089/cpb.2007.9963

7. National Center for Chronic Disease Prevention and Health Promotion, Division of Population Health: Well-Being Concepts I HRQOL I CDC. Centers for Disease Control and Prevention (2018). https://www.cdc.gov/hrqol/wellbeing.htm
8. Laidlaw, A., McLellan, J., Ozakinci, G.: Understanding undergraduate student perceptions of mental health, mental well-being and help-seeking behaviour. Stud. High. Educ. **41**(12), 2156–2168 (2015). https://doi.org/10.1080/03075079.2015.1026890
9. Field, T.: Touch. The MIT Press, Cambridge, MA (2014)
10. Tams, S., Hill, K., de Guinea, A.O., Thatcher, J., Grover, V.: NeuroIS – alternative or complement to existing methods? Illustrating the holistic effects of neuroscience and self-reported data in the context of technostress research. J. Assoc. Inf. Syst. **15**(10) 723–752 (2014)
11. Diener, E., et al.: New measures of well-being: flourishing and positive and negative feelings. Soc. Indic. Res. **39**, 247–326 (2009)
12. Scale of Positive and Negative Experience (SPANE) – NovoPsych Psychometrics (2018). NovoPsych. https://novopsych.com.au/assessments/scale-of-positive-and-negative-experience-spane/
13. Scale of Positive and Negative Experience (SPANE) (n.d.). http://labs.psychology.illinois.edu/~ediener/SPANE.html. Accessed 21 Jan 2021
14. Johnson, D., Gardner, M.J., Perry, R.: Validation of two game experience scales: the player experience of Need SATISFACTION (PENS) and game experience questionnaire (GEQ). Int. J. Hum Comput. Stud. **118**, 38–46 (2018). https://doi.org/10.1016/j.ijhcs.2018.05.003
15. Farnsworth, B.: Facial action coding SYSTEM (FACS) - a Visual guidebook, 18 August 2019. https://imotions.com/blog/facial-action-coding-system/. Accessed 04 March 2021
16. Deci, E.L., Ryan, R.M.: Intrinsic Motivation and Self-Determination in Human Behavior. Plenum, New York, NY (1985)
17. Sterling, R.: Influence of Psychological Needs and Gaming Motivation on Well-Being of Adult Gamers. Scholarworks (2017). https://scholarworks.waldenu.edu/cgi/viewcontent.cgi?article=4829&context=dissertations
18. Ryan, R.M., Deci, E.L.: Self-determination theory and the facilitation of intrinsic motivation, social development, and well-being. Am. Psychol. **55**(1), 68–78 (2000). https://doi.org/10.1037/0003-066X.55.1.68
19. Lafrenière, M.A.K., Verner-Filion, J., Vallerand, R.J.: Development and validation of the gaming motivation scale (GAMS). Personal. Individ. Differ. **53**(7), 827–831 (2012)

Continuing Doctoral Student Training for NeuroIS and EEG During a Pandemic: A Distance Hands-On Learning Syllabus

Théophile Demazure[1](✉), Alexander Karran[1], and Pierre-Majorique Léger[1,2]

[1] Tech3Lab, HEC Montréal, Montréal, QC, Canada
{theophile.demazure,alexander.karran,
pierre-majorique.leger}@hec.at
[2] Department of Information Technologies, HEC Montréal, Montréal, QC, Canada

Abstract. There is a need to train newcomers to NeuroIS on neuroscientific tools and methodologies. Due to the pandemic, existing syllabi required adaptation to enable remote training. In this paper, we present a syllabus aimed at providing hands-on distance learning and EEG training. The proposed syllabus was pretested during a Ph.D. course on Neuroscience and IT. We report in this manuscript our lessons learned and recommendations for conducting remote neuroscience training.

Keywords: NeuroIS · EEG · Syllabus · IT · Neuroscience · Remote · Distance learning

1 Introduction

Researchers in the field of NeuroIS have identified the need to train new information systems scholars in certain aspects of cognitive neuroscience, both theory and methods, to better understand the theoretical underpinnings of how the human brain responds, adapts, and utilizes information systems and technology when proposing new systems or new ways to interact with technology [1]. The field of NeuroIS attempts to bridge the human response to information systems (IS) through the systemic investigation of IS topics that intersect with neuroscience and the neurophysiological response [2].

Riedl and Léger [3] proposed a syllabus to teach NeuroIS to provide a broad overview of NeuroIS theory and cover the fundamental themes of the what and why of NeuroIS, how to conduct a NeuroIS study, and how to select appropriate measures for specific research questions. This proposed syllabus was a first step towards guiding newcomers to the field through the process of creating a research question, choosing a psychological construct to explain behavior, measuring the neurophysiological response, and finally defining the neurophysiological inference of the chosen construct. However, the proposed syllabus places a focus on developing fundamental knowledge rather than the development of practical skills and vom Brocke, Hevner, Léger, Walla and Riedl [4] have recently called to update reference syllabi continuously.

© The Author(s), under exclusive license to Springer Nature Switzerland AG 2021
F. D. Davis et al. (Eds.): NeuroIS 2021, LNISO 52, pp. 184–191, 2021.
https://doi.org/10.1007/978-3-030-88900-5_21

This manuscript proposes a blended approach to teaching NeuroIS based on the "technological pedagogical content knowledge framework" (TPACK) [5] to provide a pedagogical foundation and the recently proposed iterative process framework (IPF) for performing NeuroIS experiments at a distance [6] to provide practical experience. The iterative approach created in the manuscript was integrated into these workshops to facilitate the learning related to EEG, research methods, and technical skills. Indeed, EEG is one of the main research tools in the community and used in 27 papers published on 73 surveyed between 2008 and 2018 [7]. EEG is a popular tool, developing an introductory and foundational understanding can be deemed helpful to newcomers.

The motivation for the development of the syllabus was based not just upon a gap in the NeuroIS training literature but also upon necessity. In these socially distant times, attending school, college, university, or indeed a research laboratory has become challenging to say the least. However, research and education have received significant blows to productivity and output. In our case, to continue to train Ph.D. students on NeuroIS tools and methodologies without access to a laboratory, we needed a training program to ensure doctoral students have the skills and the hands-on practice required to perform research activities for their thesis dissertations. Thus, the series of workshops was provided in parallel with a Ph.D. seminar on NeuroIS following a similar structure than past syllabus [3] in which students discussed seminal manuscripts in NeuroIS (e.g., [1, 8–11]) with some sessions focusing on fundamental EEG resources such as Müller-Putz et al. introduction to Electroencephalography [12] or Newman's book [13]. The workshops were conducted aside, focusing on the practical skills.

In the early development phase of both the proposed syllabus and the IPF, we utilized a series of virtual workshops, in which participants were supplied with their own portable BCI grade EEG headsets, participants had access to a broadband internet connection, webcam, and conferencing software. From this, we derived a series of recommendations to aid in the successful completion of NeuroIS research at a distance. Based upon its success, we applied the IPF to a pedagogical problem, namely for the teaching of NeuroIS. Ordinarily, laboratory practical sessions are seen as a secondary activity necessary to the acquisition of a skill. However, in our case, we took the approach of "learning through participation" in which the learner undertakes NeuroIS practical activities with a community of other learners and a number of subject matter experts to create an engaging environment in which no questions go unanswered and practical and theoretical aspects of NeuroIS research are addressed and assessed formatively in real-time and summatively in an iterative fashion with each successive workshop session.

Utilizing this approach, we integrate the primary purposes 1) integrating theory and practice, 2) enhancing the learning process, 3) personal development, 4) community engagement. Thus, our purpose was to develop practical research skills for NeuroIS, linking theory taught in the online Ph.D. seminar to practice in a workshop environment, creating an engaging learning and collaborative process.

Furthermore, our approach integrates elements of the TPACK framework such as content knowledge, pedagogical knowledge, and technological knowledge through the aggregation of practical skills, underpinning knowledge, and applying these within NeuroIS research both remotely and in a laboratory setting. This approach fits with the third pathway to TPACK as it is referred to in the literature, in which technology is systematically integrated into the learning process, in our case compelled by the need to teach at a distance during a global pandemic.

This article presents a syllabus of activities to perform with learners developed to increase the knowledge and skills of learners new to the field of NeuroIS and prepare them for a career within the research or industrial communities. We have pretested this syllabus in a Ph.D. course on Neuroscience and IT, and we report our lessons learned and recommendations for conducting remote neuroscience training in this manuscript.

2   Proposed Syllabus of Workshop's Activities

For the purpose of this syllabus, all teaching and practical activities were performed online using a combination of video conferencing software (i.e., Zoom) and BCI grade EEG sensor hardware. More specifically, in the case of the EEG sensor hardware, each learner was provided with an educational version of the Unicorn Black EEG headset from g.tech (g.Tec GmbH, Austria) delivered to their homes to ensure safe and sanitary conditions. Additionally, all learners ($n = 15$) had access to broadband internet, a web camera, and a means to communicate, such as a headset and microphone. Finally, all learners were encouraged to enable their webcams at all times to create a collaborative and engaging environment.

The decision to utilize the g.tech Unicorn Black for the purposes of teaching NeuroIS was based upon a previous assessment which found the simple design of the Unicorn cap to offer excellent mobility, practicality in terms of its size and lack of complexity, and high degree of self-applicability. The Unicorn offers 8 hybrids dry - wet electrodes placed at Fz, C3, Cz, C4, Pz, PO7, Oz, PO8. The system uses a rightmost sensor placed at the mastoid for a reference and a leftmost sensor also placed at the mastoid as ground. The hybrid spiky electrodes offer direct contact with the skin and can be easily adjusted. Additionally, these electrodes are of a robust construction that adds to the headset's practicality aspect, reducing wear when learners continuously adjust the headset during early workshop sessions as they gain confidence and skill with the application. The EEG signal is sampled at 250 Hz per channel with a 24-bit resolution, the amplifier is attached to the back of the head and offers a wireless connection via Bluetooth. Additionally, G.tec provides a python API for data acquisition that facilitates integration with a stimuli presentation software.

To provide learners with all the materials necessary for learning and completing the course syllabus, collaborative Wiki pages were created. Materials were composed of written support for the live video workshop, code snippets, pre-recorded neurophysiological data, live links to supporting material such as python libraries, and recordings of each workshop which was cumulatively added after each workshop session. Each workshop session's duration was developed to be between 1 h and 1 h and 30 min. Shown in Fig. 1. is a representation of the hardware and software infrastructure required to complete the syllabus and scheduled activities.

For the software and experiment development environment, we use the Python environment. The Python environment has many advantages; its containerized design allows the instructor to build a closed environment that is similar between learners, which eases

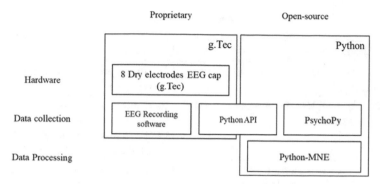

Fig. 1. Hardware and software infrastructure required for proposed syllabus activities

the process of code sharing between both instructor and learners. Another advantage to Python is the availability of many libraries for data manipulation and analysis, in our case, we opted for MNE-python [14] for EEG analysis and Psychopy [15] for stimuli presentation. Moreover, installing and running the Python environment does not require expensive computing hardware, and everything is open-source and continually supported by various communities. Finally, utilizing Python to perform all the required steps to complete a NeuroIS style project facilitated the searching of information by the learners in between weekly workshops, which aided in building and reinforcing expertise and confidence.

The information shown in Table 1 outlines the syllabus of learning activities performed during 8 weeks NeuroIS training activities, each weekly workshop builds upon the activities of the previous ones and includes both formative and summative elements.

Table 1. 8 weeks NeuroIS practical training on EEG.

Week	Objectives	Activities
1	Present pedagogical approach and tools	EEG key concepts (15 min), tools presentation (30 min), Software installation (15 min)
2	Introduction to Programming environment and Experimentation, hands-on experience with Python, and creating an experimental paradigm	Creating and running a Python environment (15 min), basic knowledge about the set of libraries (15 min), running a basic stimuli presentation (30 min)
3	EEG in detail - self-application and safety	EEG headset installation (20 min), sensor placement (20 min), impedance and data quality (20 min)

(continued)

Table 1. (*continued*)

Week	Objectives	Activities
4	Introduction to data acquisition, a first small study	End to end process of a NeuroIS experiment (1 h)
5	Data analysis EEG band power and ERP using Python-MNE	Relating cognitive construct to neurophysiological measure (20 min), EEG data processing pipeline (20 min), data visualization (20 min)
6	Interactive Workshop - construct experimental protocol using PsychoPy	Group project: Scientific method, research question, and hypothesis design, protocol creation Creating a stimulus in PsychoPy (1 h)
7	Data collection	Study moderation and participation of projects (2 to 3 h per participants/moderators)
8	Project presentation/submission	Summative assessment

3 Discussion and Lessons Learned

The proposed workshop syllabus was conducted in parallel with a Ph.D. course on neuroscience and information systems using Riedl and Leger (2015) as a reference syllabus (https://www.hec.ca/cours/detail/?cours=TECH80747A). The theoretical and critical knowledge relating to neuroscience methods and application was provided during the main class during which seminal resources on EEG and NeuroIS methods were discussed [12, 13, 16]. This allowed the workshop to focus on the technical aspects of EEG research while at the same time contributing to theoretical development through reinforcement and adding new knowledge and skills concerning stimuli creation, data processing, and analysis which are further reinforced during each successive workshop. Learners have received by mail the g.tech Unicorn Black one week before the semester.

Fifteen students participated in this workshop. Students were paired in teams of two or three, leading to seven student groups. Each team had to propose a research idea for mid-semester. Consumer-grade EEG instruments have important limitations, such as insufficient evidence on addressing complex tasks such as in NeuroIS, low-density and fixed electrode placements, low sampling rates, and reduced data quality (see [17] for a review on consumer-grade EEG). In light of these limitations, the professor and teaching assistants assessed the feasibility and provided feedback based on a one-pager. Upon validation, the teams had to make a five pages proposition for the 6th workshop and develop a stimulus in PsychoPy based on the framework provided (i.e., time-locked, visual, within-subject). For example, students were advised to increase the number of trials compared to the literature to increase the dry EEG system's signal-to-noise ratio and focus on neurophysiological indices that produce robust brain responses capturable with EEG not made for research purpose. The sequential data collection (participant -

moderator) allowed every student to collect data from each other, providing an average sample size of 15 subjects per study.

At the moment of preparing this manuscript, the students have completed the data collection of each of their proposed experiments. Such that, seven groups projects collected data from 10 to 12 participants; every student participated in each project, except their own project. While the data analysis is not completed, visual inspection of the data suggests that signal quality appears adequate and should allow students to complete the analysis as expected. Based on a first pass visual artifact rejection, approximately 40% of trials were rejected, as expected given the non-optimal experimental environment, justifying the need for the increased number of trials.

Our approach to distance learning supported by video conferences and the collaborative platform proved to be effective. Learners were engaged during practical activities encouraged through group activities to collaborate and provide support to others during workshops and during at home activities. Every learner was provided access to the software and hardware, allowing them to follow instructions and perform the experimental procedures independently. Furthermore, learners were instructed to troubleshoot as a peer group, using screen sharing facilitates and feedback from peers. For example, learners were trained to apply the EEG headset and to visually check impedance, data quality and artifacts together and provided small "at-home" group tasks to supplement workshop activities.

To avoid the steep learning curve associated with computer programming for inexperienced Ph.D. students, we developed a simple and generalizable PsychoPy framework that supported a time-locked visual presentation experiment paradigm followed by a behavioral response. The objective was twofold, to facilitate the process of creating further visual stimuli through the adjustment of pre-existing code for the student and enable swift debugging for the teaching assistant during live demonstrations.

To aid with logistical considerations related to experimental the need for subjects - moderators, we adopted a sequential approach where each learner becomes a participant and then a moderator until all learners have experienced the different roles. This avoids participant recruitment (especially given the pandemic circumstances) and provides experience to the student as both participant and moderator of an experimental protocol. Using the sequential approach allowed participants to remain blind to their colleagues' hypothesis, they took part in projects as a participant and then implement their own experimental protocol as moderator. Moderation is essential with trained subjects to ensure self-application, data quality, and control movement.

The approach we selected was to provide a large corpus of content at the onset of training via the wiki, and this, combined with recorded workshop sessions, allowed workshops to run on schedule and avoid videoconference fatigue. During the workshops, the teaching assistant demonstrated the practical activities, process and methods to the student for them to reproduce. This created a rich offline environment that allowed students to study and practice at their own pace between formal classes and the workshops further reinforcing the acquisition of knowledge and skills. Furthermore, pre-set "at-home" activities performed with peers "virtually" between classes and workshops created a positive reinforcement loop further adding breadth and depth particularly in the case of fairly technical workshops such as stimuli construction where precise knowledge was transferred that could have been complicated to internalize for students.

One risk we identified associated with the distance learning approach in our case was to convey the importance related to the fragility of the equipment. We provided low-cost BCI grade EEG headsets to meet learners' needs, which do not accurately represent laboratory-based equipment which is both orders of magnitude more expensive and sensitive to damage. From this, we can advise that a clear and concise protocol should be transmitted to the participants relating the intrinsic differences between mobile and laboratory NeuroIS technologies. Of course, the experimental environment couldn't reach laboratory methodological quality and standards [12], but the objective to provide a hands-on learning experience to conduct a complete NeuroIS study was undoubtedly reached.

4 Conclusion

We believe our multi-method approach to teaching NeuroIS at a distance provides a flexible but structured foundation from which to acquire knowledge and skills and train doctoral students to ideate and create novel experiments for the field while at the same time managing expectations and supporting a pragmatic approach to science in a time of pandemic and social distancing. There is still much work to do to improve training for NeuroIS utilizing different methods once the current global crisis is over. However, it is our belief that integrating the syllabus and training methods proposed in this manuscript with more traditional face-to-face training would accelerate knowledge and skills acquisition under normal circumstances.

References

1. Riedl, R., Léger, P.-M.: Fundamentals of NeuroIS Studies in Neuroscience. Psychology and Behavioral Economics, Springer, Heidelberg (2016). https://doi.org/10.1007/978-3-662-450 91-8
2. Loos, P., et al.: NeuroIS: neuroscientific approaches in the investigation and development of information systems. Bus. Inf. Syst. Eng. **2**, 395–401 (2010)
3. Riedl, R., Léger, P.-M.: Neuro-information-systems (NeuroIS). Association for Information Systems, Reference Syllabi (2015)
4. vom Brocke, J., Hevner, A., Léger, P.M., Walla, P., Riedl, R.: Advancing a neurois research agenda with four areas of societal contributions. Eur. J. Inf. Syst. **29**, 9–24 (2020)
5. Koehler, M.J., Mishra, P., Kereluik, K., Shin, T.S., Graham, C.R.: The technological pedagogical content knowledge framework. In: Spector, J., Merrill, M., Elen, J., Bishop, M. (eds.) Handbook of research on educational communications and technology, pp. 101–111. Springer, New York (2014). https://doi.org/10.1007/978-1-4614-3185-5_9
6. Demazure, T., Karran, A., Boasen, J., Léger, P.-M., Sénécal, S.: Distributed remote EEG data collection for NeuroIS research: a methodological framework. In: International Conference on Human-Computer Interaction. Springer (Forthcoming)
7. Riedl, R., Fischer, T., Léger, P.-M., Davis, F.D.: A decade of NeuroIS research: progress, challenges, and future directions. ACM SIGMIS Database: DATABASE Adv. Inf. Syst. **51**, 13–54 (2020)

8. Dimoka, A., Pavlou, P.A., Davis, F.D.: Research commentary—NeuroIS: the potential of cognitive neuroscience for information systems research. Inf. Syst. Res. **22**, 687–702 (2011)
9. Ortiz de Guinea, A., Webster, J.: An investigation of information systems use patterns: technological events as triggers, the effect of time, and consequences for performance. Mis Q. **37** (2013)
10. Riedl, R., Hubert, M., Kenning, P.: Are there neural gender differences in online trust? An fMRI study on the perceived trustworthiness of eBay offers. MIS Q. 397–428 (2010)
11. Vance, A., Jenkins, J.L., Anderson, B.B., Bjornn, D.K., Kirwan, C.B.: Tuning out security warnings: a longitudinal examination of habituation through fMRI, eye tracking, and field experiments. MIS Q. **42**, 355–380 (2018)
12. Müller-Putz, G.R., Riedl, R., Wriessnegger, S.C.: Electroencephalography (EEG) as a research tool in the information systems discipline: foundations, measurement, and applications. CAIS **37**, 46 (2015)
13. Newman, A.: Research Methods for Cognitive Neuroscience. Sage (2019)
14. Gramfort, A., et al.: MEG and EEG data analysis with MNE-Python. Front. Neurosci. **7**, 267 (2013)
15. Peirce, J.W.: PsychoPy—psychophysics software in Python. J. Neurosci. Methods **162**, 8–13 (2007)
16. Riedl, R., Davis, F.D., Hevner, A.R.: Towards a NeuroIS research methodology: intensifying the discussion on methods, tools, and measurement. J. Assoc. Inf. Syst. **15**, I (2014)
17. Riedl, R., Minas, R.K., Dennis, A.R., Müller-Putz, G.R.: Consumer-grade EEG instruments: insights on the measurement quality based on a literature review and implications for NeuroIS research. In: Davis, F.D., Riedl, R., vom Brocke, J., Léger, P.M., Randolph, A.B., Fischer, T. (eds.) NeuroIS 2020. LNISO, vol. 43, pp. 350–361. Springer, Heidelberg (2020). https://doi.org/10.1007/978-3-030-60073-0_41

Design Mode, Color, and Button Shape: A Pilot Study on the Neural Effects of Website Perception

Anika Nissen[1][(✉)] and René Riedl[2,3]

[1] University Duisburg-Essen, Essen, Germany
anika.nissen@uni-due.de
[2] University of Applied Sciences Upper Austria, Steyr, Austria
rene.riedl@fh-steyr.at
[3] Johannes Kepler University Linz, Linz, Austria

Abstract. The investigation of website aesthetics has a long history and has already been addressed in NeuroIS research. The extant literature predominantly studied website complexity, symmetry, and colors. However, other design factors have not yet been examined so far. We studied two new factors (design mode: light vs. dark, button shape: rounded vs. sharp angled) along with color (blue vs. red). Specifically, we examined the impact of these three factors on several outcomes. Results from a repeated-measures MANOVA indicate: (i) design mode (light vs. dark) significantly affects users' pleasure, arousal, trust, attitude, and use intention, (ii) color (blue vs. red) significantly influences pleasure, arousal, and use intentions, while (iii) button shape (rounded vs. sharp) does not significantly influence any of the dependent measures. Based on these results, follow up functional near-infrared spectroscopy studies are developed which aim to further complement our self-report findings.

Keywords: Website aesthetics · Button shape · Color · Dark website design · Attitude · fNIRS · Brain

1 Introduction

Website aesthetics is a major determinant of user perception and use intention [1–4]. Aesthetics impact perceived trustworthiness, usability, and user experience (UX) [5, 6], and is therefore a critical factor in web design (e.g., ecommerce websites). Within the field of website aesthetics, several determinants of aesthetic and beautiful websites have been identified, such as symmetry, complexity, balance, colors, and shapes [1, 4, 5, 7]. Several aesthetic factors have also already been investigated based on neurophysiological measurement, such as symmetry [8] and color [9, 10]. However, other design factors have received much less attention or have not been studied at all, including shapes of different user interface (UI) elements, in particular the shape of buttons [11]. Research indicates that different shape forms (curved, sharp) influence perceptions of aesthetics, emotion, and purchase intentions [12, 13]. The stimuli used in these studies comprise lines, abstract

© The Author(s), under exclusive license to Springer Nature Switzerland AG 2021
F. D. Davis et al. (Eds.): NeuroIS 2021, LNISO 52, pp. 192–203, 2021.
https://doi.org/10.1007/978-3-030-88900-5_22

figures, products, and interior designs [14–17]. More importantly, it has been consistently found that sharp objects and designs often lead to increased activation in arousal and fear-related brain areas, particularly in the amygdala, while curved designs elicit activations in reward-related brain areas, particularly in the anterior cingulate cortex (ACC) [12, 13, 15]. At this point, it needs to be noted that all neurophysiological studies on shapes were conducted with functional magnetic resonance imaging (fMRI) [12, 15]. Although fMRI offers high spatial resolution, it is limited in its external validity due to participants' being required to stay in position. Especially for measurements which involve looking at or using a website, more natural measurements must complement fMRI research since restraining the movements of users may have an impact on the intensity of emotional experience [18]. Thus, an investigation of button shapes on websites not only offers room for upcoming studies with self-reported and behavioral measures, but also with mobile neuroimaging methods such as functional near-infrared spectroscopy.

In addition, a recent trend in UI design in general, and in website design in particular, is the use of dark mode. Although this design trend has been around for several years already, scientific research on the effects of different design modes (dark vs. light) is scarce [19, 20]. First results indicate that while no significant impact on user performance seems to exist [19], trust ratings are higher for the light mode [20]. However, practice more often adapts this trend in major operating systems and apps (i.e. Windows 10, macOS X), as well as on websites. While the use of dark mode is often justified with saving resources of OLED displays, an investigation on how this impacts users' experiences is still lacking [20].

Consequently, both button shape and design mode may have a significant impact on users' perceptions and attitude towards a website. Given that, to the best of our knowledge, only little scientific research on the effects of round versus sharp button design exists, along with the fact that only few studies on the effects of dark design mode are available, this study aims to instigate such research. Because color in website design may strongly affect website perception, and hence may even override the effects elicited by button shapes [21], we also consider color in our study. Against this background, based on the context of ecommerce websites, this pilot study investigates three design factors with an online survey: sharp versus curved buttons, dark versus light design mode, and red versus blue color scheme of the website. The results of this survey will then be used to select stimuli and experimental designs for a follow-up neuroimaging study.

The remainder of this paper is structured as follows: First, we discuss related work on human evaluations of curved versus sharp designs, as well as the use of color and design mode in website design. Next, we develop our working hypotheses. Afterwards, the experimental design and method is presented, followed by the results. We close this paper with a brief discussion and an outlook on the follow-up study.

2 Related Work and Hypotheses

2.1 Human Preference for Curved Shapes

Humans tend to prefer round or curved shapes and objects over sharp, rectangular, or angular shapes and objects [12, 14, 16, 17]. This has been consistently found across several research contexts such as car design, consumer product designs, abstract shapes, or interior architectural design [12, 13, 15, 17, 21]. Since this preference can also be observed in new born babies and great apes, evolution has likely shaped this preference [22]. Also, curved designs tend to be regarded as more natural and harmonious compared to sharp angled design [23]. Following the processing fluency theory of aesthetic pleasure, curved design might be more fluently processable by humans and is therefore regarded as more beautiful [24, 25]. This finding is substantiated by an fMRI study in which participants judged curved designs as more beautiful, which was associated with activations in a neural network related to aesthetic and reward processing [15]. Next to aesthetic evaluations, curved designs also tend to be associated with quiet or calm sounds, green color, and relieved emotion while sharp designs are more attributed to loud and dynamic sound, red color, and excited emotion [26]. For consumer products, sharp or angular design seems to elicit negative emotions associated with threat [12, 16, 27–30], while curved designs lead not only to more positive aesthetic evaluations, but also to increased purchase intentions [13]. Consequently, we propose that websites with round buttons will elicit a more positive emotional experience (measured with pleasure), which further results in approach behavior (measured with use intention):

H 1. Websites with round buttons will be rated higher in pleasure and use intention, if compared to websites with sharp buttons.

2.2   Color and Design Mode

Research on the impact of color on human perception and emotion has a long history [31–33], and has also been frequently researched in the context of website design [10, 34, 35]. In essence, blue colors on websites have a calming, positive emotional effect, while red colors tend have a more negative emotional effect and are related to arousal [2, 35–44]. Furthermore, blue websites are perceived as more trustworthy than red or yellow websites [40]. The described effects were consistently found across a variety of studies, all of which were including different shades of blue and red, and identified the effects in behavioral and neural measurements [10]. Consequently, we propose that color has a significant effect on pleasure and arousal as measures for the emotional experience, and perceived trustworthiness and attitude as indicators for behavioral intention:

H2. Blue websites will be rated higher in pleasure, perceived trustworthiness, and positive attitude towards the website, but lower in arousal, if compared to red websites.

Related to colors in website design is design mode, which can either be dark or light. Design mode is an uprising UX design element, and a dark design is currently a trend in UI design as signified by its integration into major operating systems such as Windows, macOS, iOS, and Android [45]. Interestingly, there is a paucity of academic literature on the effects of dark mode on UX. Generally, dark mode could delay visual fatigue and increase visual comfort, especially in environments with poor lighting conditions [46, 47]. However, although users perceive their performance to be increased on dark over light designs [47], studies have found that there are no significant differences in reading performance between the designs [19, 48]. With respect to perceived trustworthiness, first results suggest that websites designed in dark mode can lead to lower perceived trustworthiness than light mode [20]. Consequently, we propose:

H3. Websites designed in dark mode will lead to lower ratings in perceived trustworthiness, if compared to websites in light mode.

3   Method

To test our three hypotheses and to identify potential stimulus material for the neuroimaging follow-up study, different versions of an ecommerce website (e-learning courses) were created and tested in this pilot study through an online survey. In this survey, participants viewed the different website designs and evaluated their perceived pleasure, arousal, trust, attitude towards the website, and use intention for each website version.

3.1   Sample

The online survey was distributed through the platform clickworker to 137 participants from Germany, Austria, and Switzerland. The data of 20 participants were excluded due to missing data or because the rating of the attention check question was wrong (this question was randomly hidden between the other scales and asked participants to select '4' on the Likert scale). Consequently, a dataset of $N = 117$ participants remained for further analysis. The average age of the sample was $M_{age} = 40.7$ years ($SD_{age} = 13.1$ years) with 54.7% being male and 45.3% being female. Regarding work status, 76.1% are currently employed, 12% are students, 6.8% searching for employment, 3.4% retired, .9% are in internship, and .9% are still pupils. Moreover, we asked the participants to indicate disposition to trust towards websites and their familiarity with booking e-learning courses on the web with validated scales taken from [49, 50]. Both ratings are based on a 5-point Likert scale resulting in $M_{dispotrust} = 3.01$ ($SD_{dispotrust} = .771$) for disposition to trust, and in $M_{familiarity} = 3.62$ ($SD_{familiarity} = .773$) for familiarity.

3.2   Stimuli and Study Design

We used a $2 \times 2 \times 2$ factorial within-subjects study design with color (red, blue), button shape (round, sharp), and design mode (light, dark) as independent variables, and pleasure, arousal, trust, attitude, and use intention as dependent variables. The scales for pleasure were adapted from [51], arousal scales were adapted from [52], trust scales were taken and adapted from [49, 53], attitude towards the website scales were taken and adapted from [54], and finally, use intention scales were adapted from [55]. Since the study was conducted in Germany, examples of the used stimuli can be seen in Fig. 1 (in German). The time participants took was not manipulated by the survey so that each participant could proceed at his/her own speed.

All of the website versions and the scales were shown in randomized order. Each website version was shown with the scales of the dependent variables on one page of the questionnaire. When all scales were evaluated for the shown website version, participants

a) Light design, red color and round buttons   b) Dark design, blue color and sharp buttons

Fig. 1. Examples for stimuli in the pilot study

had to click on a button to proceed with the next website. After having evaluated three different website versions, a questionnaire page with scales of the control variables disposition to trust and familiarity were shown without the websites. After that, further three website variations were presented with the scales of the dependent variables each, which was followed by a questionnaire page with demographic questions. Finally, the remaining two website versions were shown. This procedure was chosen to avoid fatigue in participants due to them having to rate the same items eight times without breaks.

3.3   Results

The data was analyzed with a repeated measures multivariate analysis of variance (RM-MANOVA) with the independent variables color, button shape, and design mode as repeated measures. Because we only included two levels per measure, sphericity is assumed. Although multivariate normality was not provided for half of the website design variations (measured with Mardia's skewness and kurtosis, threshold $p < .05$) [56], RM-MANOVA still seems to provide more robust results than non-parametric alternatives [57]. Yet, the following results may be treated with care.

Across all included dependent variables, significant differences could be identified for color ($F(5, 112) = 6.159$, $p < .001$, $\eta^2_p = .216$) and design mode ($F(5, 112) = 4.161$, $p = .002$, $\eta^2_p = .157$), but not for button shape ($F(5, 112) = 1.707$, $p > .05$, $\eta^2_p = .031$). Furthermore, only a slightly significant interaction effect between color and design mode could be detected ($F(5, 112) = 2.405$, $p = .033$, $\eta^2_p = .097$). No interaction effects between all three independent variables (color x button shape x design mode), nor between color and button shape, or between design mode and button shape were found (all with $p > .05$).

Regarding button shape, no significant differences were identified in any of the dependent variables (all with $p > .05$). Thus, based on the data of this pilot study, H1 has to be rejected. However, a medium strong effect could be identified for color for which the most significant difference was found for arousal ratings ($F(1, 116) = 12.873$, $p < .001$, $\eta^2_p = .100$), followed by slightly significant different ratings of pleasure ($F(1, 116) = 4.139$, $p = .044$, $\eta^2_p = .034$), and use intention ($F(1, 116) = 4.264$, $p = .041$, $\eta^2_p = .035$); all of which were rated more positive for the blue website versions. However, no significant differences could be identified for trust ($F(1, 116) = 2.389$, $p > .05$, η^2_p

= .02) and attitude ($F(1, 116) = 1.465, p > .05, \eta^2_p = .012$) between the blue and red website versions. Hence, H2 is only partly supported. Finally, the strongest effect was found for design mode, thus, all included dependent variables revealed significant differences between light and dark design, with all ratings being in favor of the light design mode ($F(1, 116) = 26.03, p < .001, \eta^2_p = .183$ for pleasure, $F(1, 116) = 13.269, p < .001, \eta^2_p = .103$ for arousal, $F(1, 116) = 20.913, p < .001, \eta^2_p = .153$ for trust, $F(1, 116) = 21.304, p < .001, \eta^2_p = .155$ for attitude, and $F(1, 116) = 22.794, p < .001, \eta^2_p = .164$ for use intention). Consequently, H3 is supported by these results.

4 Discussion and Follow-Up Neuroimaging Study

4.1 Discussion of Self-reported Results

Given the consistency of findings in the related literature on sharp versus round object designs over several research contexts, the rejection of H1 was unexpected. Paired with the significant differences in arousal for color, and the significant differences between light and dark designed websites for all included outcome variables, we conclude that color and design mode tend to have stronger effects than button shape. In particular, this pilot study shows that an in-depth investigation of the effects of design mode on website users' perceptions is crucial.

Our investigated outcomes can have a significant effect on actual website use and thus, business success. What follows is that further investigation of design mode has high relevance for both academia and practice. Specifically, based on the findings of this pilot study, we will study potential differences in the neural processing of light and dark designed websites to gain further insights into how both design modes are perceived and processed in the brain. Also, because the extant brain imaging literature on sharp versus round designs has revealed significant differences in neural activity related to emotion and threat processing [12, 58], we surmise that even though no significant differences in self-reports could be observed in this pilot study, an identification of significant neurophysiological differences is likely. Furthermore, the increase in self-reported arousal for the red website is supported by neural investigations of color on websites which revealed increased activations for websites designed in reddish color [9, 10]. In an attempt to further validate these results, color manipulation is also planned to be included in the follow-up brain imaging study. At this point, it needs to be mentioned that the impact of color on user perceptions and processing might depend on the intensity and shade of the used colors. In our pre-study, only one level of intensity and shade was considered which was held constant across the red and the blue website version. Follow-up research on color could therefore focus on different shades and intensity levels to detect possible effects of these features on user perceptions and behavioral intentions.

Finally, as already pointed out, design mode had a significant effect on users' perceptions in this pilot study. In fact, the impact of design mode might have been so overwhelming that effects due to button shape for the outcome variables were not consciously recognized by the participants.

4.2 Follow-Up Neuroimaging Study

The upcoming neuroimaging study will thus primarily focus on the effects of button shape and colors by means of the mobile neuroimaging method functional near-infrared spectroscopy (fNIRS) in an attempt to overcome the limited external validity of fMRI measurements. fNIRS offers a lightweight method which is portable and thus, it can be applied to realistic use contexts [59, 60]. Although fNIRS has shown to be a promising method to investigate NeuroIS constructs [10, 60–63], its application in NeuroIS research is still scarce [64].

Further, the neural effects of color and shapes have been primarily measured with fMRI [12, 15, 65, 66] which also assesses the hemodynamic response function (hrf) as measure to identify neural activations. While fNIRS also relies on the hrf, it is applicable in more realistic scenarios than fMRI and does not require participants to lay down. Still, since it does measure hrf, too, results from fNIRS heavily correlate with fMRI results which offers potential to compare fMRI and fNIRS results between studies [67–70]. Next to fNIRS, electroencephalography (EEG) also offers a mobile method that allows for field experiments and is frequently applied in NeuroIS research [64, 71]. Although EEG has higher temporal resolution [72], fNIRS is more robust against task un-related activations on cortical surfaces due to respiration or movements [73], and also offers higher spatial resolution than EEG [74]. While both EEG and fNIRS are limited to measuring neural activations on cortical structures, we will further focus on areas of the prefrontal cortex (PFC) as regions of interest.

Although fMRI studies on color and shapes have found activations primarily in the amygdala, they have also found significant activations in the PFC [15, 66]. Explanations for this can be found in several studies which investigate the links between the PFC and other brain regions. That is, across several contexts, the dorsolateral and ventrolateral PFC were found to exhibit top-down control on activations in the visual cortex due to their role in attention control [75–77]. Furthermore, the dorsomedial and ventromedial PFC were found to exhibit control on amygdala activations for means of emotion regulation [78–82]. These findings make the PFC an interesting region to study in relation to the processing of color and button shapes, which is also well assessable by means of fNIRS. This approach is further supported by related research focusing on neural activation changes in the PFC depending on color use in the context of ecommerce websites which were assessed with fNIRS [9, 10].

Beyond the planned upcoming study, future research could investigate design elements such as color, button shapes, and design mode with combined EEG-fNIRS measurements to cover both the electrical potential, as well as hrf related to website design variations. Since applications of fNIRS in this research domain are still scarce, combined measurements are even rarer [83]. Therefore, this offers a fruitful approach for other follow-up studies.

As a first step, however, fNIRS offers a still novel, yet appropriate approach to validate whether prior results apply to website design, too, that go beyond the conscious recognition of design elements.

References

1. Moshagen, M., Thielsch, M.T.: Facets of visual aesthetics. Int. J. Hum. Comput. Stud. **68**, 689–709 (2010). https://doi.org/10.1016/j.ijhcs.2010.05.006
2. Cyr, D., Head, M., Larios, H.: Colour appeal in website design within and across cultures: a multi-method evaluation. Int. J. Hum. Comput. Stud. **68**, 1–21 (2010). https://doi.org/10.1016/j.ijhcs.2009.08.005
3. Lavie, T., Tractinsky, N.: Assessing dimensions of perceived visual aesthetics of web sites. Int. J. Hum. Comput. Stud. **60**, 269–298 (2004). https://doi.org/10.1016/j.ijhcs.2003.09.002
4. Ngo, D.C.L., Teo, L.S., Byrne, J.G.: Modelling interface aesthetics. Inf. Sci. (Ny) **152**, 25–46 (2003). https://doi.org/10.1016/S0020-0255(02)00404-8
5. Tuch, A.N., Bargas-Avila, J.A., Opwis, K.: Symmetry and aesthetics in website design: it's a man's business. Comput. Hum. Behav. **26**, 1831–1837 (2010). https://doi.org/10.1016/j.chb.2010.07.016
6. Lee, S., Koubek, R.J.: Understanding user preferences based on usability and aesthetics before and after actual use. Interact. Comput. **22**, 530–543 (2010)
7. Bauerly, M., Liu, Y.: Computational modelling and experimental investigation of effects of compositional elements on interface and design aesthetics. Int. J. Hum. Comput. Stud. **64**, 670–682 (2006). https://doi.org/10.1016/j.ijhcs.2006.01.002
8. Vasseur, A., Léger, P.-M., Sénécal, S.: The impact of symmetric web-design: a pilot study. In: Davis, F.D., Riedl, R., vom Brocke, J., Léger, P.-M., Randolph, A., Fischer, T. (eds.) Information Systems and Neuroscience: NeuroIS Retreat 2019, pp. 173–180. Springer International Publishing, Cham (2020). https://doi.org/10.1007/978-3-030-28144-1_19
9. Nissen, A.: Why we love blue hues on websites: a fNIRS investigation of color and its impact on the neural processing of ecommerce websites. In: Davis, F.D., Riedl, R., vom Brocke, J., Léger, P.-M., Randolph, A.B., Fischer, T. (eds.) NeuroIS 2020. LNISO, vol. 43, pp. 1–15. Springer, Cham (2020). https://doi.org/10.1007/978-3-030-60073-0_1
10. Nissen, A.: Psychological and physiological effects of color use on ecommerce websites: a neural study using fNIRS. In: International Conference on Information Systems (ICIS), Hyderabad, India (2020)
11. Tao, D., Yuan, J., Liu, S., Qu, X.: Effects of button design characteristics on performance and perceptions of touchscreen use. Int. J. Ind. Ergon. **64**, 59–68 (2018). https://doi.org/10.1016/j.ergon.2017.12.001
12. Bar, M., Neta, M.: Visual elements of subjective preference modulate amygdala activation. Neuropsychologia **45**, 2191–2200 (2007). https://doi.org/10.1016/j.neuropsychologia.2007.03.008
13. Westerman, S.J., Gardner, P.H., Sutherland, E.J., et al.: Product design: preference for rounded versus angular design elements. Psychol. Mark. **29**, 595–605 (2012). https://doi.org/10.1002/mar.20546
14. Guthrie, G., Wiener, M.: Subliminal perception or perception of partial cue with pictorial stimuli. J Pers Soc Psychol **3**, 619–628 (1966). https://doi.org/10.1037/h0023197
15. Vartanian, O., Navarrete, G., Chatterjee, A., et al.: Impact of contour on aesthetic judgments and approach-avoidance decisions in architecture. Proc. Natl. Acad. Sci. **110**, 10446–10453 (2013). https://doi.org/10.1073/pnas.1301227110
16. Bar, M., Neta, M.: Humans prefer curved visual objects. Psychol. Sci. **17**, 645–648 (2006). https://doi.org/10.1111/j.1467-9280.2006.01759.x
17. Silvia, P.J., Barona, C.M.: Do people prefer curved objects? Angularity, expertise, and aesthetic preference. Empir. Stud. Arts **27**, 25–42 (2009). https://doi.org/10.2190/em.27.1.b
18. Kemp, A.H., Krygier, J., Harmon-Jones, E.: Neuroscientific perspectives of emotion. In: Calvo, R.A., D'Mello, S., Gratch, J., Kappas, A. (eds.) The Oxford Handbook of Affective Computing, pp. 38–53. Oxford University Press, New York (2015)

19. Pedersen, L.A., Einarsson, S.S., Rikheim, F.A., Sandnes, F.E.: User interfaces in dark mode during daytime – improved productivity or just cool-looking? In: Antona, M., Stephanidis, C. (eds.) HCII 2020. LNCS, vol. 12188, pp. 178–187. Springer, Cham (2020). https://doi.org/10.1007/978-3-030-49282-3_13

20. Nazeriha, S., Jonsson, A.: Does "Dark Mode" affect users' trust towards E-commerce websites ? KTH (2020)

21. Leder, H., Tinio, P.P.L., Bar, M.: Emotional valence modulates the preference for curved objects. Perception **40**, 649–655 (2011). https://doi.org/10.1068/p6845

22. Munar, E., Gomez-Puerto, G., Call, J., Nadal, M.: Common visual preference for curved contours in humans and great apes. PLoS ONE **10**, 1–15 (2015). https://doi.org/10.1371/journal.pone.0141106

23. Gómez-Puerto, G., Munar, E., Nadal, M.: Preference for curvature: a historical and conceptual framework. Front. Hum. Neurosci. **9**, 1–8 (2016). https://doi.org/10.3389/fnhum.2015.00712

24. Reber, R., Schwarz, N., Winkielman, P.: Processing fluency and aesthetic pleasure: is beauty in the perceiver's processing experience? Pers. Soc. Psychol. Rev. **8**, 364–382 (2004). https://doi.org/10.1207/s15327957pspr0804_3

25. Bertamini, M., Palumbo, L., Redies, C.: An advantage for smooth compared with angular contours in the speed of processing shape. J. Exp. Psychol. Hum. Percept. Perform. **45**, 1304–1318 (2019). https://doi.org/10.1037/xhp0000669

26. Blazhenkova, O., Kumar, M.M.: Angular versus curved shapes: correspondences and emotional processing. Perception **47**, 67–89 (2018). https://doi.org/10.1177/0301006617731048

27. Aronoff, J., Woike, B.A., Hyman, L.M.: Which are the stimuli in facial displays of anger and happiness? Configurational bases of emotion recognition. J. Pers. Soc Psychol **62**, 1050–1066 (1992). https://doi.org/10.1037/0022-3514.62.6.1050

28. Aronoff, J., Barclay, A.M., Stevenson, L.A.: The recognition of threatening facial stimuli. J. Pers. Soc. Psychol. **54**, 647–655 (1988). https://doi.org/10.1037/0022-3514.54.4.647

29. Lundholm, H.: The affective tone of lines: experimental researches. Psychol. Rev. **28**, 43–60 (1921). https://doi.org/10.1037/h0072647

30. Poffenberger, A.T., Barrows, B.E.: The feeling value of lines. J. Appl. Psychol. **8**, 187–205 (1924)

31. Newton, I.: Opticks (1704)

32. Munsell, A.H.: A pigment color system and notation. Am. J. Psychol. **23**, 236 (1912). https://doi.org/10.2307/1412843

33. von Goethe, J.W.: Zur Farbenlehre. Tübingen (1810)

34. Silic, M., Cyr, D., Back, A., Holzer, A.: Effects of color appeal, perceived risk and culture on user's decision in presence of warning banner message. In: Proceedings of the 50th Hawaii International Conference on System Sciences, pp. 527–536 (2017)

35. Pelet, J.É., Papadopoulou, P.: The effect of colors of e-commerce websites on consumer mood, memorization and buying intention. Eur. J. Inf. Syst. **21**, 438–467 (2012). https://doi.org/10.1057/ejis.2012.17

36. Palmer, S.E., Schloss, K.B.: An ecological valence theory of human color preference. Proc. Natl. Acad. Sci. U.S.A. **107**, 8877–8882 (2010). https://doi.org/10.1073/pnas.0906172107

37. Bonnardel, N., Piolat, A., Le Bigot, L.: The impact of colour on website appeal and users' cognitive processes. Displays **32**, 69–80 (2011). https://doi.org/10.1016/j.displa.2010.12.002

38. Fortmann-Roe, S.: Effects of hue, saturation, and brightness on color preference in social networks: Gender-based color preference on the social networking site Twitter. Color. Res. Appl. **38**, 196–202 (2013). https://doi.org/10.1002/col.20734

39. Abegaz, T., Dillon, E., Gilbert, J.E.: Exploring affective reaction during user interaction with colors and shapes. In: Procedia Manufacturing, pp. 5253–5260. Elsevier B.V. (2015)

40. Chang, W., Lin, H.: The impact of color traits on corporate branding. African J. Bus. Manag. **4**, 3344–3355 (2010)

41. Bellizzi, J.A., Hite, R.E.: Environmental color, consumer feelings, and purchase likelihood. Psychol. Mark. **9**, 347–363 (1992). https://doi.org/10.1002/mar.4220090502

42. Westerman, S.J., Sutherland, E.J., Gardner, P.H., et al.: Ecommerce interface colour and consumer decision making: two routes of influence. Color Res. Appl. **37**, 292–301 (2012). https://doi.org/10.1002/col.20690

43. Becker, S.A.: An exploratory study on web usability and the internationalization of US e-businesses. J. Electron Commer. Res. **3**, 265–278 (2002)

44. Seckler, M., Opwis, K., Tuch, A.N.: Linking objective design factors with subjective aesthetics: an experimental study on how structure and color of websites affect the facets of users' visual aesthetic perception. Comput. Hum. Behav. **49**, 375–389 (2015). https://doi.org/10.1016/j.chb.2015.02.056

45. Riegler, A., Riener, A.: Adaptive dark mode: investigating text and transparency of windshield display content for automated driving. In: Mensch und Computer 2019 Workshop on Automotive HMIs, Hamburg, pp. 421–428 (2019)

46. Erickson, A., Kim, K., Bruder, G., Welch, G.F.: Effects of dark mode graphics on visual acuity and fatigue with virtual reality head-mounted displays, pp. 434–442 (2020). https://doi.org/10.1109/vr46266.2020.00064

47. Kim, K., Erickson, A., Lambert, A., et al.: Effects of dark mode on visual fatigue and acuity in optical see-through head-mounted displays. In: Proceedings of SUI 2019 ACM Conference on Spatial User Interaction (2019). https://doi.org/10.1145/3357251.3357584

48. Pastoor, S.: Legibility and subjective preference for color combinations in text. Hum. Factors **32**, 157–171 (1990). https://doi.org/10.1177/001872089003200204

49. Gefen, D.: E-commerce: the role of familiarity and trust. Int. J. Manag. Sci. **28**, 725–737 (2000). https://doi.org/10.1016/S0305-0483(00)00021-9

50. Gefen, D., Straub, D.W.: Consumer trust in B2C e-commerce and the importance of social presence: experiments in e-products and e-services. Omega **32**, 407–424 (2004). https://doi.org/10.1016/j.omega.2004.01.006

51. Huang, M., Ali, R., Liao, J.: The effect of user experience in online games on word of mouth: a pleasure-arousal-dominance (PAD) model perspective. Comput. Hum. Behav. **75**, 329–338 (2017). https://doi.org/10.1016/j.chb.2017.05.015

52. Koo, D.M., Ju, S.H.: The interactional effects of atmospherics and perceptual curiosity on emotions and online shopping intention. Comput. Hum. Behav. **26**, 377–388 (2010). https://doi.org/10.1016/j.chb.2009.11.009

53. Yoon, S.J.: The antecedents and consequences of trust in online-purchase decisions. J. Interact. Mark. **16**, 47–63 (2002). https://doi.org/10.1002/dir.10008

54. Porat, T., Tractinsky, N.: It's a pleasure buying here: the effects of web-store design on consumers' emotions and attitudes. Hum.-Comput. Interact. **27**, 235–276 (2012). https://doi.org/10.1080/07370024.2011.646927

55. Chen, M.Y., Teng, C.I.: A comprehensive model of the effects of online store image on purchase intention in an e-commerce environment. Electron Commer. Res. **13**, 1–23 (2013). https://doi.org/10.1007/s10660-013-9104-5

56. Mardia, K.V.: Measures of multivariate skewness and kurtosis with applications. Biometrika **57**, 519–530 (1970). https://doi.org/10.1093/biomet/57.3.519

57. Finch, H.: Comparison of the performance of nonparametric and parametric MANOVA test statistics when assumptions are violated. Methodology **1**, 27–38 (2005). https://doi.org/10.1027/1614-1881.1.1.27

58. Vartanian, O., Skov, M.: Neural correlates of viewing paintings: evidence from a quantitative meta-analysis of functional magnetic resonance imaging data. Brain Cogn. **87**, 52–56 (2014). https://doi.org/10.1016/j.bandc.2014.03.004

59. Kim, H.Y., Seo, K., Jeon, H.J., Lee, U., Lee, H.: Application of functional near-infrared spectroscopy to the study of brain function in humans and animal models. Mol. Cells **40**(8), 523–532 (2017). https://doi.org/10.14348/molcells.2017.0153

60. Nissen, A., Krampe, C., Kenning, P., Schütte, R.: Utilizing mobile fNIRS to investigate neural correlates of the TAM in eCommerce. In: International Conference on Information Systems (ICIS), Munich, pp. 1–9 (2019)

61. Gefen, D., Ayaz, H., Onaral, B.: Applying functional near infrared (fNIR) spectroscopy to enhance MIS research. AIS Trans. Hum.-Comput. Interact. **6**(3), 55–73 (2014). https://doi.org/10.17705/1thci.00061

62. Hirshfield, L.M., Bobko, P., Barelka, A., et al.: Using noninvasive brain measurement to explore the psychological effects of computer malfunctions on users during human-computer interactions. In: Advances Human-Computer Interaction (2014). https://doi.org/10.1155/2014/101038

63. Krampe, C., Gier, N., Kenning, P.: Beyond traditional neuroimaging: can mobile fNIRS add to NeuroIS? In: Davis, F.D., Riedl, R., vom Brocke, J., Léger, P.-M., Randolph, A.B. (eds.) Information Systems and Neuroscience. LNISO, vol. 25, pp. 151–157. Springer, Cham (2018). https://doi.org/10.1007/978-3-319-67431-5_17

64. Riedl, R., Fischer, T., Léger, P.-M., Davis, F.D.: A decade of NeuroIS research: progress, challenges, and future directions. Data Base Adv. Inf. Syst. **51**, 13–54 (2020)

65. Siok, W.T., Kay, P., Wang, W.S.Y., et al.: Language regions of brain are operative in color perception. Proc. Natl. Acad. Sci. U.S.A. **106**, 8140–8145 (2009). https://doi.org/10.1073/pnas.0903627106

66. Zeki, S., Marini, L.: Three cortical stages of colour processing in the human brain. Brain **121**, 1669–1685 (1998). https://doi.org/10.1093/brain/121.9.1669

67. Hoshi, Y., Kobayashi, N., Tamura, M.: Interpretation of near-infrared spectroscopy signals: a study with a newly developed perfused rat brain model. J. Appl. Physiol. **90**, 1657–1662 (2001). https://doi.org/10.1152/jappl.2001.90.5.1657

68. Huppert, T.J., Hoge, R.D., Diamond, S.G., et al.: A temporal comparison of BOLD, ASL, and NIRS hemodynamic responses to motor stimuli in adult humans. Neuroimage **29**, 368–382 (2006). https://doi.org/10.1016/j.neuroimage.2005.08.065

69. Noah, J.A., Ono, Y., Nomoto, Y., et al.: fMRI validation of fNIRS measurements during a naturalistic task. J. Vis. Exp. 5–9 (2015). https://doi.org/10.3791/52116

70. Sato, H., Yahata, N., Funane, T., et al.: A NIRS-fMRI investigation of prefrontal cortex activity during a working memory task. Neuroimage **83**, 158–173 (2013). https://doi.org/10.1016/j.neuroimage.2013.06.043

71. Riedl, R., Minas, R., Dennis, A., Müller-Putz, G.: Consumer-grade EEG instruments: insights on the measurement quality based on a literature review and implications for NeuroIS research. In: Davis, F.D., Riedl, R., vom Brocke, J., Léger, P.-M., Randolph, A.B., Fischer, T. (eds.) NeuroIS 2020. LNISO, vol. 43, pp. 350–361. Springer, Cham (2020). https://doi.org/10.1007/978-3-030-60073-0_41

72. Müller-Putz, G., Riedl, R., Wriessnegger, S.: Electroencephalography (EEG) as a research tool in the information systems discipline: Foundations, measurement, and applications. Commun. Assoc. Inf. Syst. **37**, 46 (2015). https://doi.org/10.17705/1CAIS.03746

73. Girouard, A., et al.: From brain signals to adaptive interfaces: using fNIRS in HCI. In: Tan, D.S., Nijholt, A. (eds.) Brain-Computer Interfaces, pp. 221–237. Springer, London (2010). https://doi.org/10.1007/978-1-84996-272-8_13

74. Pinti, P., Tachtsidis, I., Hamilton, A., et al.: The present and future use of functional near-infrared spectroscopy (fNIRS) for cognitive neuroscience. Ann. N.Y. Acad. Sci. **1464**, 5–29 (2020). https://doi.org/10.1111/nyas.13948

75. Gazzaley, A., Rissman, J., Cooney, J., et al.: Functional interactions between prefrontal and visual association cortex contribute to top-down modulation of visual processing. Cereb. Cortex **17**, i125–i135 (2007). https://doi.org/10.1093/cercor/bhm113

76. Masquelier, T., Albantakis, L., Deco, G.: The timing of vision - how neural processing links to different temporal dynamics. Front. Psychol. **2**, 1–14 (2011). https://doi.org/10.3389/fpsyg.2011.00151

77. Kornblith, S., Tsao, D.Y.: How thoughts arise from sights: inferotemporal and prefrontal contributions to vision. Curr. Opin. Neurobiol. **46**, 208–218 (2017). https://doi.org/10.1016/j.conb.2017.08.016

78. Delli Pizzi, S., et al.: Functional and neurochemical interactions within the amygdala–medial prefrontal cortex circuit and their relevance to emotional processing. Brain Struct. Funct. **222**(3), 1267–1279 (2016). https://doi.org/10.1007/s00429-016-1276-z

79. Buhle, J.T., Silvers, J.A., Wage, T.D., et al.: Cognitive reappraisal of emotion: a meta-analysis of human neuroimaging studies. Cereb. Cortex **24**, 2981–2990 (2014). https://doi.org/10.1093/cercor/bht154

80. Schienle, A., Wabnegger, A., Schoengassner, F., Scharmüller, W.: Neuronal correlates of three attentional strategies during affective picture processing: an fMRI study. Cogn. Affect. Behav. Neurosci. **14**(4), 1320–1326 (2014). https://doi.org/10.3758/s13415-014-0274-y

81. Dolcos, F., Iordan, A.D., Dolcos, S.: Neural correlates of emotion - cognition interactions: a review of evidence from brain imaging investigations. J. Cogn. Psychol. **23**, 669–694 (2011)

82. Ellard, K.K., Barlow, D.H., Whitfield-Gabrieli, S., et al.: Neural correlates of emotion acceptance vs worry or suppression in generalized anxiety disorder. Soc. Cogn. Affect. Neurosci. **12**, 1009–1021 (2017). https://doi.org/10.1093/scan/nsx025

83. Hirshfield, L.M., et al.: Combining electroencephalograph and functional near infrared spectroscopy to explore users' mental workload. In: Schmorrow, D.D., Estabrooke, I.V., Grootjen, M. (eds.) FAC 2009. LNCS (LNAI), vol. 5638, pp. 239–247. Springer, Heidelberg (2009). https://doi.org/10.1007/978-3-642-02812-0_28

Does Media Richness Influence the User Experience of Chatbots: A Pilot Study

Laurie Carmichael$^{(\boxtimes)}$, Sara-Maude Poirier, Constantinos Coursaris,
Pierre-Majorique Léger, and Sylvain Sénécal

HEC Montréal, Montréal, QC H3T 2A7, Canada
{laurie.carmichael,sara-maude.poirier,constantinos.coursaris,
pierre-majorique.leger,sylvain.senecal}@hec.ca

Abstract. From a user's perspective, this pilot study investigates the contributors
and irritants related to the media content format used by chatbots to assist users
in an online setting. In this study, we use automated facial expression analysis
(AFEA), which analyses users' facial expressions and captures the valence of
their lived experience. A questionnaire and a single-question interview were also
used to measure the users' perceived experience. All measures taken together
allowed us to explore the effects of three media content formats (i.e., an interactive
question and answer (Q&A), a video, and a link referring to a webpage) used in
chatbots on both the lived and perceived experiences of users. In line with Media
Richness Theory (MRT), our results show that an interactive Q&A might be an
optimal chatbot design approach in providing users with sought-after information
or assistance with transactions. Moreover, important avenues for future research
emerge from this study and will be discussed.

Keywords: Chatbot · Media content format · Media richness theory · Task type ·
Automated facial expression analysis

1 Introduction

AI-based software such as chatbots are used by retailers and service providers to offer
24/7 user assistance. It is expected that by 2022, 75–90% of user's queries will be
answered by chatbots [1]. Moreover, human-computer-interaction (HCI) literature has
shown that chatbots have proven to be relevant and effective in assisting consumers
in online settings [2]. To date, most of the research that studied chatbots has tried to
understand the technical aspects of algorithms behind these systems [3]. However, from
a user's perspective, little research has been performed investigating contributors and
irritants related to the media content *format* used by chatbots to provide information to
users. Therefore, there is a need for research that extends our understanding of how the
media content format used by chatbots may affect user experience in e-commerce. This
raises two important research questions: Does the media content format, one that varies
in richness – such as interactive conversation, video, and link to a webpage – used by a
chatbot impact the users' lived and/or perceived experience? Does the task type, whether

F. D. Davis et al. (Eds.): NeuroIS 2021, LNISO 52, pp. 204–213, 2021.
https://doi.org/10.1007/978-3-030-88900-5_23

users ask for information or transactional assistance, moderate the relation between the type of media content format and the users' lived and/or perceived experience?

In this pilot study, we utilize automated facial expression analysis (AFEA), a questionnaire with self-reported measures, and an open-ended question about the users' preference to assess the user's lived and perceived experiences when using chatbots to perform both an informational and a transactional task. Based on Media Richness Theory (MRT) [4], the more a media is synchronized and adapted to the user, the more positive the perceived and lived experience with the technology (here, chatbot) will be at the intersection of HCI and neuroscience. This study contributes to enriching the body of knowledge regarding chatbots and their design. Study results will also shed light on several avenues of future research surrounding the type of task performed with a chatbot and the appropriate type of media content format practitioners would be recommended to use so as to improve their online services and customer support.

2   Theoretical Background and Hypotheses

Chatbots are robots that can maintain a textual or vocal conversation with human users [5]. Due to the growing number of instant messaging services on social media or directly on the retailer or service provider website, in this study we focus on text-based chatbots designed to converse and interact with users [6]. Thus, we defined the chatbot as a "computer program, which simulates human language with the aid of a text-based dialogue system" [7]. The main interest in using chatbots from the user's point of view is that they can increase users' productivity by providing efficient and fast assistance [2].

A positive or negative experience with a chatbot depends on whether the users' expectations are met [8]. Therefore, chatbots must have the ability to correctly interpret user queries and provide accurate answers that are also perceived to be trustworthy and useful [9, 10]. To date, researchers have mainly explored the effect of appearance, personality, and the written style of the chatbot on the overall users' experience, perceived information quality, and satisfaction with it [11]. However, except for question and answer (Q&A) communication [12], the media format used by the chatbot to assist users in accomplishing their task has been overlooked. Next, we present the main media content formats employed by chatbots to redirect users to the right information and explain how these different media can affect users' both lived and perceived experiences.

2.1   Media Content Formats Used by Chatbots

According to MRT, communication media formats are positioned on a continuum from richer to leaner in order to predict their effectiveness during conversation exchanges [4]. To be considered rich, a medium must provide (1) immediate feedback from the receiver to the sender, (2) multiple cues that reduce equivocality of the message, (3) a variety of language (e.g., gesture) and (4) a personal focus [5]. Consequently, face-to-face is ranked as the richness medium while unaddressed documents (e.g., wall posts, flyer, SMS) are classified relatively leaner.

The literature shows that chatbots are communication media for which the richness varies greatly [13]. Therefore, chatbots are built on different technologies and architectures and their answers to user requests vary [1]. In all cases, the chatbot acts as a digital self-service tool that orients users to the right information or that can directly communicate it through textual or verbal exchanges [14]. During this interactive and timely conversation, questions and answers (Q&A) can be provided in an automated format (e.g., frequently ask questions), or in a more natural format (i.e., personalized dialogue) similar to an exchange with a human [14].

In this pilot study, we focus on three ways users may receive information from a chatbot such as a Q&A representing a more natural conversation format and two unaddressed media content formats (i.e., thumbnail or static webpage text). Due to the more personal and adapted exchanges between users and chatbots during Q&A interactions, as well as the elevated level of synchronicity that allows immediate feedback, we posit that this type of communication will result in a more positive lived and perceived experiences than a thumbnail preview of a video or a link leading to a webpage with static text, both previously created for general user consumption. Furthermore, because the video provides visual and auditory cues, we also expect that this media content format will result in a more positive lived and perceived experiences by the user compared to the static text on a webpage. According to the richness of the media content format, less effort during the exchanges with the chatbot should be required from the user to accomplish their query. Moreover, the exchanges with the chatbot should be pleasurable and the information should be perceived as being of quality. Since, the complexity of the communication exchanges also depends on the nature of the task users intend to perform when interacting with a chatbot – ranging from obtaining information to completing a transaction [14] – the next section presents the moderating role of task type.

2.2   Moderating Role of Task Type (Information Versus Transactional)

Chatbots can be divided into two main categories, i.e., whether they are used for conversational or transactional purposes [11]. In the former case, chatbots can provide information to users or act as virtual companions to mainly socialize with them [11]. In the latter, transactional chatbots complete a transaction within the context of the conversation [15]. These chatbots are mainly used in the service sector such as financial, insurance, or telecommunication services to perform tasks usually done by an employee [15]. For instance, chatbots can confirm an outgoing transfer of money or make users sign documents regarding their cell phone contract. Based on the above, we decide to focus on both informational and transactional chatbots so as to explore differences between them in regard to user experience [5].

Through these exchanges, users expect to be provided with the right information in response to a query or to be able to complete a transaction. In the context of informational tasks, the answer provided by the chatbot represents the end of the exchange, whereas for transactional tasks, the user can choose to partially or totally delegate the task to the chatbot. Then, the chatbot may be able to complete the transaction with input from the users or provide users with the right information to let them execute the task. According to [16], engaging users in a collaborative experience with the chatbot improves their experience with chatbots. Therefore, users enter into a dyadic relationship with the chatbot arguably in a more involved way than in the case of an informational task. According to the trust-commitment theory, a transactional chatbot needs to be trusted to accomplish the task correctly so users believe in the benevolence of chatbots to fulfill their needs [17]. Then, expectations are higher toward the chatbot and the final result, because the notion of investment in this exchange is omnipresent. Since it is even more important in a transactional task to directly cooperate with the chatbot to accomplish a task, it is expected that user's lived and perceived experiences will benefit more from media and message richness compared to informational tasks. Thus, we propose the following hypotheses:

H1: Comparing transactional to information tasks, there will be a statistically significant difference in the user's lived experience with the chatbot, such that the *valence* will be more positive (a) for Q&A than for either video or webpage, and (b) for video than for webpage.

H2: Comparing transactional to informational tasks, there will be a statistically signif-icant difference in the user's perceived experience with the chatbot, such that the (a) *pleasure* during exchanges (b) *info quality*, and (c) format *preference* will be higher for Q&A than for either webpage or video. Conversely, (d) the perceived *effort* exerted by the user during the chatbot-user exchange will be lower for the Q&A than for either the webpage or video.

H3: Comparing transactional to informational tasks, there will be a statistically signif-icant difference in the user's perceived experience with the chatbot such that the (H3a) *pleasure*, (H3b) *info quality,* and (H3c) *preference* will be higher for the video than for the webpage. Conversely, (H3d) the perceived level of *effort* deployed by the user during the chatbot-user exchange will be lower for the video than for the webpage.

3   Methodology

3.1   Design and Participants

To test our hypotheses, we have used a 3 (media content formats: Q&A vs. video vs. static webpage text) × 2 (task types: informational and transactional) within-subjects design. For this experiment, 14 participants, aged between 27 to 64 years old, were recruited through a research panel. One participant did not have AFEA data due to a technical problem; we thus excluded their data from the results. Out of the remaining 13 participants, 6 were women and 7 were men. In exchange for their time, the participants were each compensated with $125 CAD.

3.2   Experimental Protocol

Data collection was performed using the procedure proposed by Giroux et al. [18] and Alvavez et al. [19]. Experiment was conducted remotely using Lookback (Lookback Inc, Palo Alto, CA), an online platform that records participants' facial expressions while they are performing tasks. Participants were randomly presented six scenario-based conversations regarding the services of a telecommunications company. These scenarios were provided by the company to test their most often customer-asked informational question and transactional request from customer service. These conversations were between a chatbot and a fictional user, and participants were asked to put themselves in the user's situation.

Moreover, the conversations were presented free of any visual context (e.g., branding, digital environment) to avoid related biases. In the informational task, the fictional user contacted the chatbot for assistance with a technical issue concerning an online service whereas in the transactional task, the fictional user asked the chatbot to subscribe to a new online service. In the Q&A format of the transactional task, the chatbot completed the transaction for the user, whereas in the other two formats, the chatbot only gave the steps on how to proceed to an online video showing its thumbnail as a preview (which was not viewed) or providing a hyperlink to a static webpage containing only textual information (which was viewed), leaving the user to complete the transaction on their own. In the informational task, the fictional user received the same information but in the three different formats. The Q&A format was natural, typing exchanges directly made in the chatbot, the video format was a thumbnail of the video that appeared in the chat with the chatbot and finally, for the webpage, the user received a link in the chat that redirected them toward a webpage that the user visited.

After reading each conversation, participants were presented a questionnaire to assess their perceived experience with the chatbot. Specifically, participants reported their perceived pleasure and effort during exchanges, and also evaluated the information quality of what was provided by the chatbot. Each task type was followed by a question by a moderator to ask the participant's preference between the three media content formats. In total, the experiment lasted about one hour.

3.3 Data Collection, Measurements, and Postprocessing

For the physiological measurement of the lived experience, we exported and analyzed the videos in FaceReader 5 (Noldus Technology Inc, Wageningen), an automated facial expression analysis software [20]. FaceReader is the most widely used AFEA software in NeuroIS research (e.g., [21–24]). *Valence* was calculated using a scale from −1 to 1 that contrasts states of pleasure (e.g., happiness) with states of displeasure (e.g., anger) felt during the reading of each chatbot conversations; specifically, it was measured as the intensity of happiness minus the intensity of the negative emotion with the highest intensity [25].

The perceived experience of participants was measured by using a questionnaire, in which participants self-reported after each of the six conditions their perceptions of our dependent variables. All scales used were adapted from prior research for context, maintaining their original scale answer formats. A 1-item scale was chosen for *pleasure* and *effort* [26]. Thus, participants reported their *Pleasure* during the exchanges with the chatbot on an affective slider from 0 to 100 (0 = low pleasure; 100 = high pleasure) [27]. The perceived level of *Effort* during the task was measured with a single item: "What is the level of effort that you would have deployed if you had this exchange with the chatbot in real life?" on a five-point Likert scale (1 = very low; 5 = very high) [28]. Participants also evaluated the quality of the information (*Info Quality*) provided by the chatbot on a seven-item, seven-point Likert scale (1 = totally disagree; 7 = totally agree) [29]. Finally, after the experiment, a open-ended interview question about users' preference question was asked to participants by a moderator to better understand which condition was their favorite and why (*Preferred*).

Data post-processing and synchronization followed the procedure proposed by Léger et al. [30] and Cube HX (Cube Human Experience Inc, Montréal, Qc) was used prior to analysis [31–33].

4   Results

A linear regression with random intercept model was performed to compare the media formats in terms of *valence*. The 2-tailed p-values are adjusted for multiple comparisons using Holm's method. Wilcoxon Signed Rank tests with significance level of 0.05 were performed to compare the media formats in terms of perceived pleasure, effort, info quality, and preference. Info quality was measured as an index (Cronbach alpha $= 0.9$), taking the average of the responses to the seven items on a scale from 1 to 7 (low to high perceived info quality). The open-ended question's answers were transcribed and its content analyzed to report the preference for each media format per task. The reasons for these preferences were reported by coding and clustering similar answers together.

Results (see Figs. 1 and 2) for the lived experience show that, in the informational task, Q&A format has the highest valence (-0.14), followed by the video (-0.17), and the webpage format (-0.18). Conversely, for the transactional task, the webpage has a higher valence (-0.13), followed by the video (-0.15) and the Q&A (-0.18). Thus, H1a and H1b are supported for the informational task, but not for the transactional task.

For the perceived experience, the pleasure for the informational task is significantly higher for the Q&A (60.53) than for the webpage (50.54), but no statistical difference is observed between the Q&A and the video format (59.92). On the other hand, for the transactional task, the pleasure is statistically significantly higher for the Q&A format (75.92) than both the video (64.54) and the webpage (64.15). Moreover, the perceived info quality in the informational task is higher for the Q&A format (6.08) than both the video (5.60) and the webpage (5.58). The transactional task shows similar results, where the Q&A format (6.48) has a statistically higher info quality (6.48) than the video (5.92) and a marginally significant difference with the webpage format (6.07). Finally, the perceived effort in the informational task is significantly lower for the Q&A (2.23) than the webpage format (3.15), but no statistical difference is found between the Q&A and the video format (2.54). For the transactional task, on its part, the perceived effort is significantly lower for the Q&A (1.69) than both the video (2.77) and the webpage (2.69). Thus, H2a, H2b and H2d are supported.

The Q&A format in the informational task was preferred by more participants (n $= 7$) participants than the video (n $= 5$) and the webpage (n $= 1$), although the difference between the Q&A and video was not statistically significant. Participants preferred the Q&A because the conversation was perceived to be equivalent to a human-to-human interaction (n $= 2$), the answer was directly in the chat (n $= 2$), and that it let user be in control in the problem solving (n $= 1$). Participants (n $= 2$) also preferred the Q&A because it was a clear and fast interaction. For the transactional task, the Q&A format was also preferred by more participants (n $= 12$) than the webpage (n $= 1$) and the video (n $= 0$). Participants preferred the Q&A because it was simple and easy (n $= 11$) and that it avoided a conversation with a human agent (n $= 1$). Thus, H2c is supported.

Moreover, the preference question revealed that 2 participants seemed positively surprised by the chatbot being able to complete the transaction for them:

"Honestly, I was amazed that it was able to add the new service, I found that really pleasant".

Nonetheless, 3 participants also reported discomfort when it came time to disclose personal information to the chatbot, such as their account password, during the transactional task for the Q&A format.

Finally, there is a marginally significant difference in the perceived pleasure between the video (29.92) and the webpage (50.54) in the informational task. Comparatively, no difference is reported between the two formats in the transactional task. There is also no difference in the perceived info quality, effort, and preference between the webpage and the video formats for both task types. Thus, H3 is not supported.

		Valence	Pleasure	Effort	Info quality	Preferred
Info. task	Q&A	-0.14	60.53	2.23	6.08	7/13
	Link to webpage	-0.18	50.54	3.15	5.58	1/13
	Video	-0.17	59.92	2.54	5.60	5/13

Fig. 1. Results of lived and perceived experiences for informational task per media format ([1]Valence calculated excluding the first part of the conversation, which was the same for all conversations. ^Marginal statistically significant difference of 0.10. *Statistically significant difference of 0.05. **Statistically significant difference of 0.01. ***Statistically significant difference of < 0.0001)

		Valence	Pleasure	Effort	Info quality	Preferred
Transac. task	Q&A	-0.18	75.92	1.69	6.48	12/13
	Link to webpage	-0.13	64.15	2.69	6.07	1/13
	Video	-0.15	64.54	2.77	5.92	0/13

Fig. 2. Results of lived and perceived experiences for transactional task per media format ([1]Valence calculated excluding the first part of the conversation, which was the same for all conversations. ^Marginal statistically significant difference of 0.10. *Statistically significant difference of 0.05. **Statistically significant difference of 0.01. ***Statistically significant difference of < 0.0001)

5   Discussion and Conclusion

In the context of informational tasks, findings were in line with what was expected from the MRT in terms of the lived experience, where the Q&A performed the best followed by the video and the webpage text formats. In the context of transactional tasks, however, unexpected results were obtained. To explain this gap, it is plausible that concerns over the (perceived) security of chatbots were at play. Indeed, a few participants indicated they were not comfortable disclosing personal information, such as a username and password, with the chatbot in the Q&A transactional task. This suggests that there may be an interaction effect between task type and perceived security on the lived experience, which would be interesting to investigate further in future research.

For the perceived experience, results were also in line with MRT, such that the Q&A format was associated with a more favorable perceived experience than that of the two other media formats, except for the perceived pleasure and effort between the Q&A and the webpage in the informational task. However, our results did not offer support for the difference in perceived experience between the video and the webpage formats. One factor that could have contributed to this is the fact that the video format was not fully developed. In fact, only a thumbnail of a video was shown to participants. A future experiment could thus study the user experience resulting from different media content formats more in depth by using fully developed chatbot prototypes and the user is actually directed to an online video, which they view, rather having them participate in a review of a scenario-based user-chatbot exchange.

Furthermore, our results indicate there was a difference in the lived experience between the video and webpage formats in the two task types, but participants did not report so in their perceived experience. This suggests the limits of self-reported measurements that rely on recalling an experience. Thus, our results give rise to an important consideration for chatbots developers: evaluating a chatbot solely based on perceived measures is done at the risk of missing key insights that physiological data could help generate.

Moreover, given the differences between the results of the informational vs. the transactional task, our results indicate a plausible moderation of the task type on the relation between media content format and users' lived and perceived experiences. One avenue for future research could be to explore further the difference in user experience resulting from various tasks carried out by chatbots and compare which type produces the best experience overall.

To conclude, we explored the effects of three media content formats used in chatbots on both the lived and perceived experiences of users. Despite the small sample size of this experiment – which limits our ability to generalize our results to the population – it seems that indeed an interactive Q&A format might be an optimal chatbot design approach in providing users with sought-after information or assistance with transactions.

References

1. Kantarci, A.: 84 chatbot/conversational statistics: market size, adoption. AI Multiple (2021). https://research.aimultiple.com/chatbot-stats/
2. Folstad, A., Brandtzaeg, P.B.: Chatbots and the new world of HCI. Interactions **24**(4), 38–42 (2017). https://doi.org/10.1145/3085558
3. Guo, J., Tao, D., Yang, C.: The effects of continuous conversation and task complexity on usability of an AI-based conversational agent in smart home environments. In: Long, S., Dhillon, B.S. (eds.) MMESE 2019. LNEE, vol. 576, pp. 695–703. Springer, Singapore (2020). https://doi.org/10.1007/978-981-13-8779-1_79

4. Daft, R.L., Lengel, R.H.: Organizational information requirements, media richness and structural design. Manage. Sci. **32**(5), 513–644 (1986). https://doi.org/10.1287/mnsc.32.5.554
5. Hussain, S., Ameri Sianaki, O., Ababneh, N.: A survey on conversational agents/chatbots classification and design techniques. In: Barolli, L., Takizawa, M., Xhafa, F., Enokido, T. (eds.) WAINA 2019. AISC, vol. 927, pp. 946–956. Springer, Cham (2019). https://doi.org/10.1007/978-3-030-15035-8_93
6. Insider Intelligence: Chatbot market in 2021: Stats, trends, and companies in the growing AI chatbot industry. Business insider (2021). https://www.businessinsider.com/chatbot-market-stats-trends
7. Zumstein, D., Hundertmark, S.: Chatbots – an interactive technology for personalized communication, transactions and services. IADIS Int. J. WWW/Internet **15**(1), 96–109 (2017). http://www.iadisportal.org/ijwi/papers/2017151107.pdf
8. Jain, M., Kumar, P., Kota, R., Patel, S.N.: Evaluating and informing the design of chatbots. In: Proceedings of the 2018 Designing Interactive Systems Conference (DIS 2018), pp. 895–906. Association for Computing Machinery. New York (2018). https://doi.org/10.1145/3196709.3196735
9. Følstad, A., Nordheim, C.B., Bjørkli, C.A.: What makes users trust a chatbot for customer service? An exploratory interview study. In: Bodrunova, S.S. (ed.) INSCI 2018. LNCS, vol. 11193, pp. 194–208. Springer, Cham (2018). https://doi.org/10.1007/978-3-030-01437-7_16
10. Procter, M., Heller, R., Lin, F.: Classifying interaction behaviors of students and conversational agents through dialog analysis. In: Nkambou, R., Azevedo, R., Vassileva, J. (eds.) ITS 2018. LNCS, vol. 10858, pp. 373–379. Springer, Cham (2018). https://doi.org/10.1007/978-3-319-91464-0_42
11. Rapp, A., Curti, L., Boldi, A.: The human side of human-chatbot interaction: a systematic literature review of ten years of research on text-based chatbots. Int. J. Hum.-Comput. Stud. (2021). https://doi.org/10.1016/j.ijhcs.2021.102630
12. Mori, E., Takeuchi, Y., Tsuchikura, E.: How do humans identify human-likeness from online text-based Q&A communication? In: Kurosu, M. (ed.) HCII 2019. LNCS, vol. 11566, pp. 330–339. Springer, Cham (2019). https://doi.org/10.1007/978-3-030-22646-6_24
13. Sheth, A., Yip, H.Y., Iyengar, A., Tepper, P.: Cognitive services and intelligent chatbots: current perspectives and special issue introduction. IEEE Internet Comput. **23**(2), 6–12 (2019). https://doi.org/10.1109/MIC.2018.2889231
14. Jurafsky, D., Martin, J.H.: Chapter 24: dialog systems and chatbots. In: Speech and Language Processing. 3rd edn., pp. 492–525 (2020). https://web.stanford.edu/~jurafsky/slp3/24.pdf
15. Verani, E.: How to build your own transactional chatbot. Inbenta (2020). https://www.inbenta.com/en/blog/transactional-chatbot/
16. Avula, S., Chadwick, G., Arguello, J., Capra, R.: Searchbots: user engagement with chatbots during collaborative search. In: Proceedings of the 2018 Conference on Human Information Interaction & Retrieval. New York, pp. 52–61 (2018). https://doi.org/10.1145/3176349.3176380
17. Morgan, R.M., Hunt, S.D.: The commitment-trust theory of relationship marketing. J. Mark. **58**(3), 20–38 (1994). https://doi.org/10.2307/1252308
18. Giroux, F., et al.: Guidelines for collecting automatic facial expression detection data synchronized with a dynamic stimulus in remote moderated user tests. In: International Conference on Human-Computer Interaction (2021, forthcoming)
19. Alvavez J., Brieugne, D., Léger, P.M., Sénécal, S., Frédette, M.: Towards agility and speed in enriched UX evaluation projects. In: Human 4.0-From Biology to Cybernetic. IntechOpen (2019). https://doi.org/10.5772/intechopen.89762
20. Riedl, R., Léger, P.M.: Fundamentals of NeuroIS: Studies in Neuroscience, Psychology and Behavioral Economics, 115p. Springer, Heidelberg (2016). https://doi.org/10.1007/978-3-662-45091-8

21. Giroux-Huppé, C., Sénécal, S., Fredette, M., Chen, S.L., Demolin, B., Léger, P.-M.: Identifying psychophysiological pain points in the online user journey: the case of online grocery. In: Marcus, A., Wang, W. (eds.) HCII 2019. LNCS, vol. 11586, pp. 459–473. Springer, Cham (2019). https://doi.org/10.1007/978-3-030-23535-2_34

22. Lamontagne, C., et al.: User test: how many users are needed to find the psychophysiological pain points in a journey map? In: Ahram, T., Taiar, R., Colson, S., Choplin, A. (eds.) IHIET 2019. AISC, vol. 1018, pp. 136–142. Springer, Cham (2020). https://doi.org/10.1007/978-3-030-25629-6_22

23. Beauchesne, A., et al.: User-centered gestures for mobile phones: exploring a method to evaluate user gestures for UX designers. In: Marcus, A., Wang, W. (eds.) HCII 2019. LNCS, vol. 11584, pp. 121–133. Springer, Cham (2019). https://doi.org/10.1007/978-3-030-23541-3_10

24. Giroux, F., Boasen, J., Sénécal, S., Haptic stimulation with high fidelity vibro-kinetic technology psychophysiologically enhances seated active music listening experience. In: 2019 IEEE World Haptics Conference (WHC), Tokyo, Japan, pp. 151–156. (2019). https://doi.org/10.1109/WHC.2019.8816115

25. Noldus Information Technology: FaceReader. Version 6.1. Reference Manual. (2015). https://student.hva.nl/binaries/content/assets/serviceplein-a-z-lemmas/media-creatie-en-informatie/media--communicatie/observatorium/factsheet-facial-coding-reference-manual.pdf?2900513938585

26. De Guinea, A.O., Titah, R., Léger, P.M.: Measure for measure: a two study multi-trait multi-method investigation of construct validity in IS research. Comput. Hum. Behav. 29(3), 833–844 (2013). https://doi.org/10.1016/j.chb.2012.12.009

27. Betella, A., Verschure, P.F.M.J.: The affective slider: a digital self-assessment scale for the measurement of human emotions. PLoS ONE 11(2), e0148037 (2016). https://doi.org/10.1371/journal.pone.0148037

28. Dixon, M., Freeman, K., Toman, N.: Stop trying to delight your customers. Harv. Bus. Rev. July-August issue (2010). https://hbr.org/2010/07/stop-trying-to-delight-your-customers

29. Teo, T.S.H., Srivastava, S.C., Jiang, L.: Trust and electronic government success: empirical study. J. Manag. Inf. Syst. 25(3), 99–132 (2008). https://doi.org/10.2753/MIS0742-1222250303

30. Léger, P.M., et al.: Precision is in the eye of the beholder: application of eye fixation-related potentials to information systems research. Assoc. Inf. Syst. 15(Special Issue), 651–678 (2014). https://doi.org/10.17705/1jais.00376

31. Léger, P.-M., Courtemanche, F., Fredette, M., Sénécal, S.: A cloud-based lab management and analytics software for triangulated human-centered research. In: Davis, F.D., Riedl, R., vom Brocke, J., Léger, P.-M., Randolph, A.B. (eds.) Information Systems and Neuroscience. LNISO, vol. 29, pp. 93–99. Springer, Cham (2019). https://doi.org/10.1007/978-3-030-01087-4_11

32. Courtemanche, F., Léger, P.-M., Dufresne, A., Fredette, M., Labonté-LeMoyne, É., Sénécal, S.: Physiological heatmaps: a tool for visualizing users' emotional reactions. Multimed. Tools Appl. 77(9), 11547–11574 (2017). https://doi.org/10.1007/s11042-017-5091-1

33. Courtemanche, F., et al.: Method of and system for processing signals sensed from a user. Google Patents (2018)

Development of a New Dynamic Personalised Emotional Baselining Protocol for Human-Computer Interaction

Elise Labonté-LeMoyne[✉], François Courtemanche, Constantinos Coursaris, Arielle Hakim, Sylvain Sénécal, and Pierre-Majorique Léger

HEC Montréal, Montréal, QC H3T 2A7, Canada
{elise.labonte-lemoyne,francois.courtemanche,
constantinos.coursaris,arielle.hakim,sylvain.senecal,pml}@hec.ca

Abstract. Measuring emotional responses from users when they interact with a technological artefact is an important aspect of human-computer interaction (HCI) research. People, however, tend to react in an individualized manner to emotional stimuli, thus it is important to compare each user to a personalized baseline. We present, in this paper, the development and preliminary results of a new emotional elicitation protocol that is being validated with our remote psychophysiological measurement ecosystem.

Keywords: Emotional baselining · Emotional elicitation · Remote psychophysiological measurements

1 Introduction

The COVID-19 pandemic and its associated lockdowns and preventative measures have brought about significant challenges for experimental research, most notably the inability to conduct lab-based study. This has forced scholars to adjust existing methods in NeuroIS and innovate in the development of at-home research paradigms to keep participants and researchers safe [1]. In this paper, we present an important piece of this remote psychophysiological ecosystem. Specifically we developed a new dynamic emotional elicitation paradigm to use as a baseline in our remote data collection. As individuals tend to experience emotions in varying ways, research that evaluates emotional reactions needs to employ a differential approach to report emotional reactions relative to a baseline condition prior to the experiment. The experiment includes an emotional baseline that measures how each participant reacts to emotionally loaded stimuli and then compares their responses in the study to this personal baseline.

Emotions have historically been elicited through a variety of methods. Most recently, the presentation of affectively charged images and videos has been most common [2]. For our purposes, we chose to employ dynamic and short social media video stimuli rather than images to provide adequate ecological validity in the context of HCI. Considering the dynamic nature of most artefacts studied in HCI, it is important to be able to evaluate

F. D. Davis et al. (Eds.): NeuroIS 2021, LNISO 52, pp. 214–219, 2021.
https://doi.org/10.1007/978-3-030-88900-5_24

emotional responses using an ecologically valid dynamic IT artefact rather than a static image. With a remote study protocol, it is also important that the videos be free of overly upsetting content in case minors are exposed to them (e.g., by walking into the participant's room) unexpectedly.

We report ongoing work and initial findings of arousal and valence responses to this proposed baseline paradigm for remote NeuroIS research.

2 Phase 1: Development of a Emotionally Balanced Video Database

Methods. To ensure that our sample of videos is current and diverse, we selected the video stimuli from freely available social media platforms (e.g. Youtube, Reddit, Tik Tok). First, we set upon the task of identifying videos with the following criteria: videos should elicit a univocal emotional response that is only positive or negative in nature, should be between 5 and 10 s long, contain no gore or nudity, no politics or cultural elements, have its original audio track, be easily understandable as stand alone and without any need to understand the spoken language. Also, we ensured that there were enough videos in each quadrant of the circumplex model of emotion. Our initial pool contained 115 videos. Next, the videos were reviewed by independent coders with experience in emotional elicitation studies. They rejected any videos that they found did not respect the criteria enumerated above or obtain a consensus. Following this refinement, 64 videos remained.

Psychometric Validation of Videos. The 64 videos were then pre-tested for emotional valence and arousal with 993 participants on Amazon Mechanical Turk using Qualtrics (Qualtrics, Provo, UT). The participants were asked to watch each video in full and with sound before rating their arousal and valence using the affective slider [3]. Each participant watched 15 videos during a single session and they were allowed to complete multiple sessions of different videos if they so wished.

From the 993 responders, we removed 382 participants, who (i) failed an attention check (n = 108), (ii) had a personal standard deviation of less than 10 on a scale of 100, suggesting they most likely selected the same value for all answers (n = 140), or (iii) were consistently more than 1 standard deviation from the center as this indicated they answered randomly (n = 134). The final sample of 611 participants consisted of 346 men, 253 women and 1 non binary, with an average age of 37.6 years. With these pre-test responses, we selected the 42 videos that presented the lowest standard deviations, ensuring the most uniformity in emotional elicitation.

Results. Figure 1 presents the videos plotted on the circumplex model of emotion with self-perceived valence and arousal. In green dots, the responses from the psychometric validation of videos showing that the videos cover the range of the circumplex emotional map, presenting the classic boomerang shape of affective judgement [4].

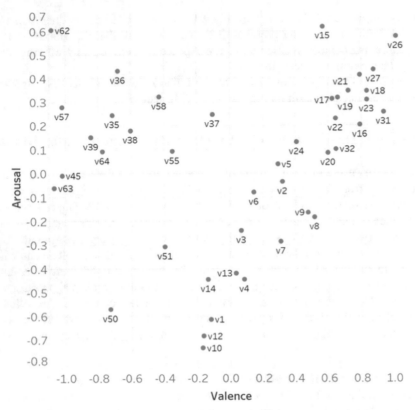

Fig. 1. Plotting of video stimuli on the circumplex emotion model

3   Phase 2: Remote Psychophysiological Evaluation of the Video

Methods. The second phase of this project consists of shipping psychophysiological equipment to participants so they can record their emotional responses to the videos. Building on Vasseur et al. [5] and Giroux et al. [1], this project is ongoing and will be collecting data for the coming months. Each participant receives at home a COBALT system, a proprietary psychophysiological recording wireless system that can be seen in Fig. 2. Figure 2 presents, a depiction of the procedure we've put in place where a moderator instructs a participant in the self-installation of EDA and ECG sensors. Details of this distributed remote psychophysiological data collection protocol can be read in Vasseur et al. [5] and Giroux et al. [1]. The system is built around a BITalino Core, peripheral boards (e.g., microSD, Wifi), and in-house software that records EDA and ECG and allows for post hoc synchronization with our presentation software through the photobooth software [6]. For this project, we developed a presentation software that enabled over the web presentation of videos while still recording precise timing information to allow post hoc synchronisation with the recorded signal [7]. Each week for 4 weeks, participants watch 12 videos presented in a randomized order. Two videos are repeated every week to study the repeated viewing of the same video. The 12 videos are seperated into two blocks of 6 with a 150 s pause between the two blocs and 20 s pauses between each video.

Fig. 2. Moderated remote self-installation of the sensors

Participants. We intend to collect data from a minimum of 100 participants by including this short experiment at the beginning of other studies that use the COBALT system. With approval from our institutional ethics board, participants are recruited through our research lab's study panel and managed per our remote data collection protocol. Participants are instructed and supervised while they self-install the psychophysiological recording equipment. The face of the participant is also recorded using their own webcam during the experiment and the video is post processed through the FaceReader software to decode for emotional valence using automated facial expression analysis (Noldus, Wageningen, the Netherlands). For this paper, we present preliminary data from 12 participants, 4 women, 8 men, with an average age of 25.1 years.

Results. Preliminary results suggest that we are able to detect the evolution of the emotional response of participants while they watch the dynamic stimuli. Figure 3 presents the psychophysiological arousal (y-axis) and valence (colour scale) measures second by second, showing how emotion evolves over time for each video. For illustrative purposes, two videos at the edges of the circumplex model are presented. we can see that the participant responses are in line with the content of the two videos: the first video, v26, is more challenging to understand in the beginning, then something surprising happens and it's fun afterwards. The second video, v62, on the other hand, is scary from the start and remains the same over time, which makes the upward trend in arousal worth investigating further. We also calculated the coefficient of variation to evaluate how much variance there is between participants. We find a coefficient of 31.36 for valence and 12.22 for arousal, showing high variance in this sample.

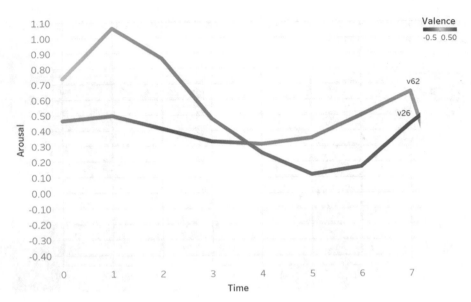

Fig. 3. Illustrative plot of psychophysiological arousal and valence over time

4   Discussion and Conclusion

Our preliminary results suggest that our baseline protocol is effective for emotional elicitation and that it detects, as expected, positive or negative valence on the two polar opposite sample videos. The results also show that there is a high variance in responses to emotional stimulus showing the importance of personalised baselining. Once fully validated, this protocol will be included in the beginning of numerous studies to allow dynamic personalised emotional baselining for each participant. As the variety of videos presented was somewhat limited by considerations of respect for a participant's home environment, we intend to extend the sample with more sensitive videos once in-person data acquisition is safe again to complete the database with the full potential spectrum of emotional reactions. This will ensure that the baselining is able to capture all emotions that can be elicited by HCI, notably in gaming or other subfields of research that could encounter strong emotional responses. As all humans have different backgrounds and experiences, emotional reactions to a given stimulus can vary widely. Emotional baselining can ensure that we as researchers take that into account when evaluating emotional reactions to our artefacts.

References

1. Giroux, F., et al.: Guidelines for collecting automatic facial expression detection data synchronized with a dynamic stimulus in remote moderated user tests. In: Kurosu, M. (ed.) HCII 2021. LNCS, vol. 12762, pp. 243–254. Springer, Cham (2021). https://doi.org/10.1007/978-3-030-78462-1_18
2. Gross, J.J., Levenson, R.W.: Emotion elicitation using films. Cogn. Emot. **9**, 87–108 (1995). https://doi.org/10.1080/02699939508408966

3. Betella, A., Verschure, P.F.M.J.: The affective slider: a digital self-assessment scale for the measurement of human emotions. PLoS ONE **11**, e0148037 (2016). https://doi.org/10.1371/journal.pone.0148037
4. Bradley, M.M., Codispoti, M., Cuthbert, B.N., Lang, P.J.: Emotion and motivation i: defensive and appetitive reactions in picture processing. Emotion **1**, 276–298 (2001). https://doi.org/10.1037/1528-3542.1.3.276
5. Vasseur, A., et al.: Distributed remote psychophysiological data collection for UX evaluation: a pilot project, pp. 1–14 (2014)
6. Léger, P.-M., Courtemanche, F., Fredette, M., Sénécal, S.: A cloud-based lab management and analytics software for triangulated human-centered research. In: Davis, F.D., Riedl, R., vom Brocke, J., Léger, P.-M., Randolph, A.B. (eds.) Information Systems and Neuroscience. LNISO, vol. 29, pp. 93–99. Springer, Cham (2019). https://doi.org/10.1007/978-3-030-01087-4_11
7. Courtemanche, F., et al.: UX Stimuli: an online stimuli presentation software for remote synchronisation of physiological measures (2021)

Mediators of the Relationship Between Self-control and Pathological Technology Use: Negative Affect and Cognitive Failures, but not Self-efficacy

Robert West[✉] and Diana Jiang

Department of Psychology and Neuroscience, DePauw University, Greencastle, USA
robertwest@depauw.edu

Abstract. The widespread adoption of technologies such as smartphones, the Internet, and social media has been associated with the emergence of pathological technology use (e.g., Internet addiction). Prevalence rates of pathological technology use vary widely across age groups, cultures, and medium, although it is not uncommon for rates of mild to moderate pathological use to exceed 20%–30%. These relatively high prevalence rates have motivated researchers to identify the predictors of pathological use. The current study focuses on the relationship between self-control and pathological technology use, and demonstrates that negative affect and cognitive failures, but not self-efficacy, partially mediate the association between self-control and pathological technology use. These findings reveal some of the pathways by which poor self-control could lead to elevated levels of pathological technology use.

Keywords: Self-control · Negative affect · Pathological technology use

1 Introduction

The widespread adoption of smartphones and the technologies that they provide access to (e.g., the Internet and social media) over the last two decades has revolutionized the way we interact with friends and family, conduct business, pursue an education, and spend our leisure time [1]. In contrast to the many benefits these technologies have brought to our lives, there are also significant personal and social costs associated with the abusive (e.g., cyberbullying) and pathological (e.g., Internet addiction) use of technology [2, 3]. Estimates of the prevalence of pathological technology use vary widely across medium, although it is not uncommon for rates to fall between 20%–30% or higher depending upon the sample and form of technology [4–6]. These high prevalence rates have motivated extensive work seeking to identify the causes and consequences of pathological technology use [3, 6, 7]. This research reveals that pathological technology use is consistently associated with diminished self-control [8] and poor mental health outcomes including increased depression, anxiety, stress, and suicidal ideation [9]. The current study sought to extend upon the extant literature by examining whether

© The Author(s), under exclusive license to Springer Nature Switzerland AG 2021
F. D. Davis et al. (Eds.): NeuroIS 2021, LNISO 52, pp. 220–228, 2021.
https://doi.org/10.1007/978-3-030-88900-5_25

the relationships between aspects of self-control and pathological technology use (i.e., smartphone, Internet, social media) are partially or fully mediated through effects on negative affect, cognition, and self-efficacy.

Studies have consistently demonstrated that poor self-control is associated with an increase in pathological technology use [7] and addiction more generally [10]. As an example, Kim et al. [11] found that individual differences in self-control and impulsivity were unique predictors of smartphone pathology. Consistent with this finding, individuals with attention deficit hyperactivity disorder (ADHD) experience higher levels of pathological smartphone use than typically developing individuals [12]. Poor self-control may also account for the association between pathological technology use including video games [13], social media [14], and the Internet [15] and poor decision making in the Iowa Gambling Task (IGT) reported in a number of studies. These findings are consistent with work using electrophysiological and neuroimaging methods demonstrating that pathological technology use is associated with attenuated neural activity related to reward processing [16] and to reduced brain volume in a number of structures within the reward network [17]. The disruption of neural activity and alternations in brain volume and connectivity within the fronto-striatal reward/control network are commonly implicated in disorders of self-control, emotion dysregulation, and addiction [18].

Self-control is a complex construct that represents aspects of impulsivity, risk-taking, and emotion regulation [19, 20]. Poor self-control is associated with increases in negative affect (i.e., depression, anxiety, stress), poor cognitive control, and reduced self-efficacy [19, 21, 22], and each of these are themselves associated with pathological technology use. Therefore, an important question represents the degree to which the influence of self-control on pathological technology use may be mediated through its influence on negative affect, cognitive control, and self-efficacy. To address this question, the current study examined three hypotheses related to possible mediators of the relationship between self-control and pathological technology use:

H1: Poor self-control is associated with increases in negative affect, cognitive failures, pathological technology use, and decreases in self-efficacy.
H2: Negative affect, cognitive failures, and self-efficacy are associated with pathological technology use.
H3: The relationship between self-control and pathological technology use is mediated through negative affect, cognitive failures, and self-efficacy.

To examine these hypotheses, individuals completed measures of pathological technology use (i.e., smartphone, Internet, social media), a multidimensional measure of self-control, a measure of negative affect tapping depression, anxiety and stress, and a measure of cognitive failures. We did not include a measure of pathological Internet gaming as in some previous research an effect of gender has been observed in this domain [12, 23], wherein pathology is much more common in males than in females, an interaction we hoped to avoid in the current dataset. The multidimensional measure of self-control allowed us to distinguish between the influence of aspects of self-control that were differentially related to impulsivity and emotion regulation, and risk-taking on pathological technology use.

2 Method

2.1 Participants

The sample included 210 individuals from the Prolific registry. To participate in the study, individuals had to be 18 years of age or older and have an IP address registered in the United States. The average age of the sample was 31 years (SD = 11 years), with 43% females, 49% males, 4% gender non-conforming, 2% trangender, and 1% other. The sample was predominantly white (63%), followed by Asian (13%), Black or African American (11%), Hispanic or Latinx (10%), Indigenous (1%), and other (2%); 19% had not completed any college, 31% had completed 1–2 years of college, 8% had completed 3 years of college, 28% had completed 4 years of college, and 14% had completed some graduate training. Individuals were compensated $5 US for participation in the study through their Prolific account.

2.2 Materials and Procedure

Cronbach's α was used to estimate the reliability (internal consistency) of the scales for the data obtained for the study, these values are reported in the text. The self-control scale measures six aspects of self- control (Impulsivity $\alpha = .82$, Risk-taking $\alpha = .85$, Self concern $\alpha = .78$, a preference for Simple tasks $\alpha = .87$, a preference for Physical activity over mental activity $\alpha = .78$, Tempe (emotion regulation) $\alpha = .80$) [20]; for the Self-control scale higher scores represent lower self-control. Pathological technology use was measures with the Internet Addiction Test-Short Version ($\alpha = .86$) [24], Smartphone Addiction Scale-Short Version (SAS-SV, $\alpha = .87$) [25], and Bergen Social Media Addiction Scale (BSMAS, $\alpha = .86$) [26]. Negative affect was measured with the Depression, Anxiety and Stress Scale-21 (DASS-21-Depression $\alpha = .93$, Anxiety $\alpha = .86$, Stress $\alpha = .89$) [27]. The Cognitive Failures Questionnaire (CFQ, $\alpha = .94$) [28] was used to assess cognitive control. Self-efficacy was measured with the General Self-Efficacy Scale (GSF, $\alpha = .91$) [29] and Short Grit Scale (perseverance subscale $\alpha = .76$) [30]. Individuals also completed demographic measures including age, gender, education, and race and ethnicity, and other scales not relevant to the current project. There were three attention check questions distributed across the study materials, these included "Which of the words is presented in upper case letters? – cat, RAT, bat, sat; "Which of the words is a color?" – red, grass, sky, water; "Which word is an animal?" – water, pan, cup, fox.

Individuals completed the study through a link posted on the Prolific website. The study materials were presented using Google Forms. At the beginning of the study, individuals read a consent form and checked a box indicating their willingness participant in the study and then continued with the survey materials. Completion of the study materials required 15 to 30 min across participants.

2.3 Data Cleaning and Preparation

All participants answered the three attention check questions correctly and were therefore retaining in the sample. Some participants were missing responses for an item from an individual scale or subscale. These missing values were imputed with the participants mean for the scale or subscale, thereby allowing us to estimate reliability in the entire sample without impacting individual scores for the measures or the pattern of correlation across the measures. For the analyses, scale level measures were formed by calculating the average response across items for each of the scales or sub-scales (Self-control scale). Additionally, composite measures were formed by averaging scores for the three pathology measures (smartphone, internet, social media pathology) and the three negative affect measures (depression, anxiety, stress) given the higher degree of correlation among the relevant measures.

3 Results

Scores for the three pathology measures were highly correlated, as was the case for the measures of negative affect (Table 1). Self-efficacy and perseverance for the Grit scale were also highly correlated ($r = .58$, p < .001); therefore, a composite variable was formed representing the average Z-score for the two measures as they are on different scales. The six subscales of the self-control scale revealed a mixed pattern of correlations (Table 2). A principal component analysis [31] was performed for the Self-control scale in an effort to reduce the number of variables included in the analyses, while capturing the underlying structure within the six subscales. Consistent with previous research [32], the PCA revealed two components with Eigenvalues greater than 1.0 that accounted for 61% of the variance, and were moderately correlated ($r = .24$). The first component (SC1) represented impulsivity (component loading = .76), self-concern (component loading = .64), simple task (component loading = .84), and tempe (component loading = .72), and the second component (SC2) represented risk-taking (component loading = .68) and physical tasks (component loading = .87). Two composite variables were formed that represented the average of impulsivity, self-concern, simple task, and tempe, or risk- taking and physical tasks.

Supporting Hypotheses 1 and 2, the correlations between pathology and the predictors were medium to large, except for the SC2 composite where the correlation was small (Table 3); all correlations were significant. Pathology increased with greater negative affect and cognitive failures, and less self-control, and decreased with greater self-efficacy. The correlations between the SC1 composite and the mediators were also medium to large, and significant. The correlations between the SC2 composite and the mediators were small, and the correlation with negative affect was not significant.

Table 1. Pearson correlations between measures of pathology (left side) and negative affect (right side)

Pathology	1	2	3	Negative Affect	4	5	6
1. Internet				4. Depression			
2. Smartphone	.69			5. Anxiety	.71		
3. Social media	.63	.71		6. Stress	.71	.82	

Note: All correlations are significant at p < .001.

Table 2. Pearson correlations among the subscales of the self-control scale.

	1	2	3	4	5	6
1. Imp.	–					
2. Risk	.40**	–				
3. Self	.29**	.22*	–			
4. Simple	.51**	.05	.35**	–		
5. Physical	.006	.29**	.15#	−.16#	–	
6. Tempe	.40**	.26**	.50**	.35**	.05	–

Note: # p < .05, * p < .01, ** p < .001

Table 3. Pearson correlations between pathology and the predictor variables.

1. Pathology	1	2	3	4	5	6
2. SC 1	.50**					
3. SC 2	.17#	.21*				
4. Negative affect	.52**	.48**	.11			
5. CFQ	.46**	.46**	.19*	.57**		
6. Self-efficacy	−.31**	−.45**	.17*	−.35**	−.31**	

Note: # p < .05, * p < .01, ** p < .001

We tested a set of four mediation models to determine whether the relationship between self-control and pathology was fully or partially mediated by negative affect, cognitive failures, or self-efficacy. The models were fit using the Mediation Model in JASP (0.14, JASP Team 2020) based upon the Lavaan SEM package with R [33]. Maximum likelihood estimation was used. For the direct and indirect effects, we report the z-test for the parameter estimates along with the 95% confidence interval based upon 1000 Bootstrap samples with the Bias-corrected percentile [34]. An initial set of models including both components of self-control revealed that the total effect of SC2 was not significant ($p = .28$). Therefore, this variable was dropped from the analyses to simplify the models. The individual mediator models provided partial support for Hypothesis 2, revealing that both negative affect and cognitive failures were partial mediators of the relationship between self-control and pathology (Table 4). In contrast, although self-efficacy was correlated with both self-control and pathology, the indirect effect was not significant in the self-efficacy model. When negative affect and cognitive failures were included in a combined model, there were significant unique indirect effects for both mediators from self-control to pathology (Table 4, Full model).

Table 4. Standardized path coefficients, z-test, and bootstrap 95% confidence interval for the individual and combined mediator models. The direct effect from SC1 to pathology and the indirect effects through the mediators.

	Direct effect	z	Indirect effect	z	R-squared
Affect	.34 [.22–.46]	5.18**	.18 [.11–.27]	4.58**	.35
Cognition	.36 [.25–.52]	5.73**	.14 [.07–.23]	3.84**	.32
Self efficacy	.47 [.32–.62]	6.80**	.05[−.02–.14]	1.58	.26
Full model	.30 [.17–.44]	4.46**			.37
Affect			.14 [.06–.23]	3.52**	
Cognition			.08 [.006–.17]	2.24#	

Note: #p = .025, * p < .01, ** p < .001. The 95% confidence intervals for the Bootstrap for direct and indirect effects are reported in the [].

4 Discussion

Here we replicated the finding that lower self-control is associated with increased levels of pathological technology use in an adult, community-based sample, with roughly equal numbers of males and females [7]. Our measure of self-control represented two distinct, but correlated, factors. One that was primarily related to impulsivity and emotion regulation (tempe) and one that was primarily related to risk taking. The impulsivity/emotion regulation component was more strongly related to pathology technology use and the three predictors (i.e., negative affect, cognitive failures, and self-efficacy) than the risk-taking component. Additionally, the impulsivity/emotion regulation component was a unique predictor of pathological technology use, while risk- taking was not. These findings provide some refinement of our understanding of the nature of the relationship between pathological technology use and self-control by demonstrating a specific association with impulsivity/emotion regulation. This finding is interesting within the broader literature indicating that pathological technology use may in some instances serve a compensatory role in the context of anxiety and negative affect [35].

We also observed that pathological technology use was associated with elevated negative affect and cognitive failures, and lower self-efficacy. These findings are consistent with previous research [5, 11, 12] and provide the foundation for testing the hypotheses related to mediation of the association between self-control and pathological technology use. The mediation analyses revealed that negative affect and cognitive failures were unique partial mediators of the association between self-control and pathological technology use. In contrast, self-efficacy did not mediate the relationship between self-control and smartphone pathology, although it was significantly correlated with both variables. This finding may indicate that high self-efficacy could provide a buffer against pathological technology that operates outside of the influence of self-control. These findings move beyond the simple observation of associations between self-control and pathological technology use by revealing potential pathways by which this association is manifest. For instance, the mediated path from self-control to negative affect to pathology, may reveal how a reduction in emotion regulation could lead to increased depression, anxiety and stress, that in turn leads individuals to utilize the Internet in an attempt to boost their mood. Likewise, poor impulse control may contribute to lower cognitive control (i.e., increased cognitive failures) making it difficult to resist the temptation to frequently check social media [19, 22].

The current findings may be of interest to Information Systems researchers and professionals alike in demonstrating that there are multiple routes to understanding, and possibly disrupting, the link between self-control and pathological technology use. As an example, based upon our findings, intervention programs designed to reduce pathological technology use may be most successful when focused on both the affective and cognitive aspects of self-control. This suggestion is consistent with some literature examining factors that led to both the emergence of pathological technology use within individuals and also one's ability to overcome pathological use [35]. For instance, in this study individuals reported that pathological use was related to reducing negative affect or creating positive affect consistent with the link for self-control to negative affect to pathological use. Also, others reported that intrinsic and extrinsic motivation that could be related to both self-control and self-efficacy were important in overcoming pathological technology use. Our findings are also interesting within the context of programs designed to disrupt pathological technology use that have a focus on providing resources for monitoring and limiting use [36], possibly serving as a surrogate for endogenous self-control. Together, the current findings and extant literature may demonstrate the important interplay between research designed to both characterize and then reduce pathological technology use.

There are some limitations of the current research worth considering. First, the three measures of pathological technology use were highly correlated, making it difficult to examine unique predictors of different forms of pathology as has been done in some previous research [23]. This might be possible with a larger sample that would be appropriate for estimating what are likely to be small unique effects and would also allow us to examine potential gender differences across the three forms of technology considered in the study. Second, the measures of negative affect were also highly correlated, making it difficult to consider possible unique influences of depression, anxiety, and stress. Again, a larger sample could be useful in addressing this limitation. Third, given the nature of the dataset it is impossible to fully understand the direction or nature of the causal effects reflected in the correlations observed between self-control, negative affect, cognitive failures, and pathological technology use, although we may be able to conclude that the effects of self-control and self-efficacy arise from different sources. The direction of the causal effects could be examined with either longitudinal studies over short or long periods of time, or through targeted intervention studies.

In conclusion, the current study provides a number of insights into the relationship between self-control and pathological technology use. First, we demonstrated that pathological technology use is more strongly related to aspects of impulsivity and emotion regulation rather than risk-taking within the context of self-control. Second, we observed that negative affect and cognitive failures, but not self-efficacy, were partial mediators of the association between self-control and pathological technology use. Together our findings lead to the suggestion that there are multiple pathways by which variation in self-control is associated with pathological technology use.

References

1. Pew Research Center Infographic (2019). https://www.pewresearch.org/internet/fact-sheet/mobile/
2. Bartlett, C., Gentile, D.: Attacking others online: the formation of cyberbullying in late adolescence. Psychol. Pop. Media Cult. **1**, 123–135 (2012)
3. Gentile, D.A., Coyne, S.M., Bricolo, F.: Pathological technology addictions: what is scientifically known and what remains to be learned. In: Dill, K.E. (ed.) Oxford Library of Psychology. The Oxford Handbook of Media Psychology, pp. 382–402. Oxford University Press, UK (2013)
4. Gentile, D.A., Bailey, K., Bavelier, D., Brockmyer, J.F., Cash, H., Coyne, S.M., et al.: Internet gaming disorder in children and adolescents. Pediatrics **140**(Supplement 2), S81–S85 (2017)
5. Gutiérrez, J., de Fonseca, F.R., Gabriel, R.: Cell phone addiction: a review. Front. Psychiatry **7**, 175 (2016)
6. Cash, H., Rae, C.D., Steel, A.H., Winkler, A.: Internet addiction: a brief summary of research and practice. Curr. Psychiatry Rev. **8**, 292–298 (2012)
7. Ning, W., Davis, F., Taraban, R.: Smartphone addiction and cognitive performance of college students. In: Twenty-fourth Americas Conference on Information Systems (2018)
8. Kim, H.-J., Min, J.-Y., Min, K.-B., Lee, T.-J., Yoo, S.: Relationship among family environment, self-control, friendship quality, and adolescents' smartphone addiction in South Korea: findings from nationwide data. PLoS ONE **13**(2), e0199896 (2018)
9. Kim, H.-J., Min, J.-Y., Kim, H.-J., Min, K.-B.: Association between psychological and self-assessed health status and smartphone overuse among Korean college students. J. Ment. Health (2017). https://doi.org/10.1080/09638237.2017.1370641
10. Sayette, M.: Self-regulatory failure and addiction. In: Baumeister, R., Vohs, K. (eds.) Handbook of Self-Regulation: Research, Theory, and Applications, pp. 447–466. The Guilford Press, New York (2004)
11. Kim, Y., et al.: Personality factors predicting smartphone addiction predisposition: behavioral inhibition and activation systems, impulsivity, and self-control. PLoS ONE **11**(8), e0159788 (2016)
12. Kim, J.-H.: Psychological issues and problematic use of smartphone: ADHD's moderating role in the association among loneliness, need for social assurance, need for immediate connection, and problematic use of smartphone. Comput. Hum. Behav. **80**, 390–398 (2018)
13. Bailey, K., West, R., Kuffel, J. What would my avatar do? Gaming, pathology, and risky decision making. Front. Psychol. **4**, 1–10 (2013). Article 609
14. Meshi, D., Elizarova, A., Bender, A., Verdejo-Garcia, A.: Excessive social media users demonstrate impaired decision making in the iowa gambling task. J. Behav. Addict. **8**, 169–173 (2019)
15. Sun, D.-L., Chen, Z.-J., Ma, N., Zhang, X.-C., Fu, X.-M., Zhang, D.-R.: Decision-making and prepotent response inhibition functions in excessive internet users. CNS Spectr. **14**(2), 75–81 (2009)
16. Kirby, B., Dapore, A., Ash, K., Malley, K., West, R.: Smartphone pathology, agency, and reward processing. In: Davis, F.D., et al. (eds.) Information Systems and Neuroscience. Lecture Notes in Information Systems and Organisation, vol. 31, pp. 321–329. Springer, Heidelberg (2020). https://doi.org/10.1007/978-3-030-60073-0_37
17. Horvath, J., et al.: Structural and functional correlates of smartphone addiction. Addict. Behav. **105**, 106334 (2020)
18. Bechara, A.: Decision making, impulse control and loss of willpower to resist drugs: a neurocognitive perspective. Nat. Neurosci. **8**, 1458–1463 (2005)

19. Baumeister, R., Vohs, K., Tice, D.: The strength model of self-control. Curr. Dir. Psychol. Sci. **16**, 351–355 (2007)

20. Hu, Q., West, R., Smarandescu, L.: The role of self-control in information security violations: insights from a cognitive neuroscience perspective. J. Manag. Inf. Syst. **31**, 6–48 (2015)

21. Tangney, J.P., Baumeister, R.F., Boone, A.L.: High self-control predicts good adjustment, less pathology, better grades, and interpersonal success. J. Pers. **72**, 271–322 (2004)

22. Rueda, M., Posner, M., Rothbart, M.: Attentional control and self-regulation. In: Baumeister, R., Vohs, K. (eds.) Handbook of Self-Regulation: Research, Theory, and Applications, pp. 283–300. The Guilford Press, New York (2004)

23. Király, O., et al.: Problematic internet use and problematic online gaming are not the same: findings from a large nationally representative adolescent sample. Cyberpsychol. Behav. Soc. Netw. **17**, 749–754 (2014)

24. Pawlikowski, M., Alstötter-Gleich, C., Brand, M.: Validation and psychometric properties of a short version of young's internet addiction test. Comput. Hum. Behav. **29**, 1212–1223 (2013)

25. Kwon, M., Kim, D., Cho, H., Yang, S.: The smartphone addiction scale: development and validation of a short version for adolescents. PLoS ONE **8**(12), e83558 (2013)

26. Andreassen, C.S., et al.: The relationship between addictive use of social media and video games and symptoms of psychiatric disorder: a large-scale cross-sectional study. Psychol. Addict. Behav. **30**(2), 252–262 (2016)

27. Henry, J., Crawford, J.: The short-form version of the depression anxiety stress scales (DASS-21): construct validity and normative data in a large non-clinical sample. Brit. J. Clin. Psychol. **44**, 227–239 (2005)

28. Broadbent, D., Cooper, P., FitzGerald, P., Parkes, K.: The cognitive failures questionnaire (CFQ) and its correlates. Brit. J. Clin. Psychol. **21**, 1–16 (1982)

29. Schwarzer, R., Jerusalem, M.: Generalized self-efficacy scale. In: Weinman, J., Wright, S., Johnston, M. (eds.) Measures in Health Psychology: A User's Portfolio. Causal and Control Beliefs, pp. 35–37. NFER-NELSON, Windsor (1995)

30. Duckworth, A., Quinn, P.: Development and validation of the short grit scale (Grit-S). J. Pers. Assess. **91**, 166–174 (2009)

31. Svante, W., Esbensen, K., Galadi, P.: Principal component analysis. Chemometr. Intell. Lab. Syst. **2**(1–3), 37–52 (1987)

32. West, R., Budde, E., Hu, Q.: Neural correlates of decision making related to information security: self-control and moral potency. PLoS ONE **14**(9), e0221808 (2019)

33. Yves, R.: lavaan: an r package for structural equation modeling. J. Stat. Softw. **48**(2), 1–36 (2012)

34. Preacher, K.J., Hayes, A.F.: Asymptotic and resampling strategies for assessing and comparing indirect effects in multiple mediator models. Behav. Res. Methods **40**, 879–891 (2008)

35. Maier, C.: Overcoming pathological IT use: how and why IT addicts terminate their use of games and social media. Int. J. Inf. Manage. **51**, 1–9, 102053 (2020)

36. Alrobia, A., Mcalaney, J., Dogan, H., Phalp, K., Ali, R.: Exploring the requirements and design of persuasive intervention technology to combat addiction. In: Bogdan, C., et al. (eds.) HESSD 2016, HCSE 2016. LNCS, vol. 9856, pp. 130–150. Springer, Cham (2016). https://doi.org/10.1007/978-3-319-44902-9_9

High Fidelity Vibrokinetic Stimulation Augments Emotional Reactivity and Interhemispheric Coherence During Passive Multimedia Interaction

Jared Boasen[1,2]([✉]), Felix Giroux[1], Sara-Eve Renaud[1], Sylvain Sénécal[1], Pierre-Majorique Léger[1], and Michel Paquette[1,2]

[1] Tech3Lab, Department of Information Technology, HEC Montréal, Montréal, Canada
jared.boasen@hec.ca
[2] Faculty of Health Sciences, Hokkaido University, Sapporo, Japan

Abstract. Haptic technologies are widely used in multimedia entertainment to psychophysiologically enhance user experience. Psychometric-based research regarding vibrokinetic stimulation during multimedia viewing supports this notion. However, scant neurophysiological evidence exists to verify this effect. Using a between groups design with source-localized electroencephalography, the present study analyzed the effect of high fidelity vibrokinetic (HFVK) stimulation during passive multimedia interaction (i.e. watching a haptically enhanced movie) on self-reported emotional state and intercortical theta coherence. Results indicate that HFVK increases emotional reactivity in association with increased interhemispheric coherence between the right inferiortemporal gyrus and the left insular cortex, thereby conferring neurophysiological support for the efficaciousness of HFVK to enhance emotional response during movie watching.

Keywords: Emotion · Coherence · EEG · Haptics · Multimedia

1 Introduction

Haptic technologies are widespread in modern devices, and in multimedia entertainment such as cinema, music listening, and gaming, where the additional dimension of somatosensory stimulation is used to enhance user experience. Research on high fidelity vibrokinetics (HFVK) supports its psychological, physiological, and cognitive efficaciousness towards this end, with reports of increased emotional reactivity and arousal, and increased immersion in and improved memory of multimedia content [1–8]. However, neurophysiological evidence to support changes in psychophysiological states due to HFVK stimulation remains scant. Such evidence is important to the field of HCI, not just for further validation of HFVK efficaciousness, but also to illuminate the neurophysiological mechanisms upon which HFVK stimulation operates in order to optimize and inform HFVK stimulus design.

One neurophysiological effect that HFVK could have is to moderate connectivity between different cortical regions. Theta (5–7 Hz) oscillatory activity is considered a

© The Author(s), under exclusive license to Springer Nature Switzerland AG 2021
F. D. Davis et al. (Eds.): NeuroIS 2021, LNISO 52, pp. 229–236, 2021.
https://doi.org/10.1007/978-3-030-88900-5_26

functionally relevant frequency band for intercortical communication. In prior research examining the effects of HFVK stimulation during viewing of cinematic opera, we observed a complex yet clear functional effect of HFVK stimulation on theta-band (5–7.5 Hz) brain activity coherence between different cortical regions [9]. This intercortical coherence was particularly notable for connections with the visual and somatosensory processing area of the right inferiortemporal gyrus (ITG) (see Fig. 1 for the cortical location of this and subsequent brain areas mentioned in this paragraph). Specifically the rITG exhibited increased coherence with the auditory and temporal processing areas, right transverse temporal cortex (TTC) and right superiortemporal gyrus (STG). The rITG also exhibited increased coherence with the motor processing areas, right precentral gyrus (PCG) and the right postcentral gyrus (PoCG). Finally, the rITG exhibited decreased coherence with the left insular cortex (IC), which is involved in self-processing, and was proposed as a possibly important mechanism for how HFVK stimulation might increase immersion. However, the HFVK stimuli used in this study were atypical in the sense that they targeted only the musical elements of the audiovisual (AV) stimuli. It remains to be seen whether HFVK cinematic stimuli that target visual aspects of the AV stimuli induce similar changes in intercortical coherence, and whether this can be linked to changes in psychological state.

In the present report, we present a preliminary analysis from a work in progress regarding the neuropsychophysiological effects of HFVK stimulation during a passive multimedia interaction (i.e. watching an haptically enhanced movie). Specifically, we present results on self-reported emotional reactivity and theta-band intercortical coherence, both of which we hypothesized would be increased due to HFVK stimulation. This work broadens neurophysiological evidence regarding the effects of haptic technology which will fundamentally serve investigations of haptics in the field of HCI.

2   Methods and Results

2.1   Subjects

Sixteen right-handed subjects have participated in this study. Subjects were randomly assigned to one of two groups, a HFVK stimulation group or a control group that did not experience HFVK, such that the ratio of sexes were roughly equivalent in each group. Due to technical problems with electrode localization, data from one subject was excluded from the EEG analyses. Thus, data from 15 subjects (M: 4, F: 11) were included in the analyses, six HFVK subjects and nine control subjects. Exclusion criteria were history of neurological disorder, prior viewing of any of the movies more than two times, and for those in the HFVK group, prior experience with HFVK technology. This study was conducted with approval from the Council for Ethics in Research of HEC Montréal and in accordance with the declaration of Helsinki. Written informed consent was obtained from all subjects prior to participation in this study.

2.2   Stimuli

The audiovisual (AV) stimuli of the present study comprised 15 clips, ranging in length from 103 to 288 s in length, from eight different mainstream movies for which commercially available HFVK stimulation existed (D-BOX Technologies Inc., Longueuil,

Canada). The movie clips were played on a Windows operated laptop computer using a proprietary video player designed by D-BOX which permitted coding playback for our targeted clip time periods. The HFVK stimuli were rendered on a specially designed HFVK seat [10]. An accelerometer (BrainProducts, GmbH, Munich, Germany) was attached to the top of the seat and routed into an auxiliary EEG channel in the EEG recording. Meanwhile, the video was displayed on a 70×120 cm high definition Samsung TV. The audio was routed to a Pioneer VSX-324 AV receiver, and played in 5.1 surround sound on Pioneer S-11A-P speakers. Following the method of Boasen et al. [9], the Dolby Theater System (DTS) audio tracks of each film were modified such that the original center audio track was combined with the front left and right audio tracks, and a 5 ms 1000 Hz square wave was put into the new center channel such that its onset would occur precisely $+2000$ ms after launch of the video clip. Accordingly, the center speaker output of the audio receiver was routed to the EEG trigger box where the square waves were registered as triggers in the EEG recording. Meanwhile, the remaining audio channels containing the original audio were routed to their respective speakers.

2.3 Self-reported Emotion

After viewing each movie clip subjects answered a questionnaire which assessed to what extent the extract made them feel the following emotions: happy, sad, angry, disgusted, surprised, and afraid. The emotions were rated on a five-point Likert scale chronologically based on the following ratings: 1) not at all, 2) a little, 3) moderately, 4) a lot, 5) extremely. To avoid confusion and facilitate questionnaire response, the emotions were listed in the same order after each clip. The emotion scores were summed for each movie clip, and then a mean score was calculated across all clips in each subject to provide a mean emotional reactivity score which was used in statistical analyses.

2.4 Procedures

After greeting subjects, a researcher guided them into the experimental room which was blackened with drapes to mimic a cinema environment. Here, the experiment was explained and informed consent was obtained. Then the head dimensions of the subject were measured and 32 electrodes (BrainProducts, GmbH, Munich, Germany) were installed into an appropriately sized EEG cap (EASYCAP, BrainProducts, GmbH, Munich, Germany) according to the 32ch Standard Cap layout for actiCAP. After installing the EEG cap and electrodes on the subject, electrode gel was applied and impedance checked and verified. Then the three-dimensional location of the electrodes were digitized using CapTrak (BrainProducts, GmbH, Munich, Germany). Subjects were sat in the HFVK seat and their feet were elevated just off the ground. Subjects were instructed to sit upright as still as possible with their face directly looking towards the TV during movie playback. Communicating from an adjacent room, a researcher verified that the subject was ready and then initiated playback of the first movie clip. As we used a between groups study design, the order of the clips was the same for all participants. After the clip finished, the subject answered the emotion questionnaire using a tablet device which had been placed on a small table next to the HFVK seat. Once the questionnaire was complete, subjects returned the tablet to the table, and were asked to sit silently without moving for 10 s, after which playback of the subsequent movie clip was initiated. This was repeated until all experimental movie clips were played, at which point the experiment concluded for the HFVK group. However, control subjects were given the option to watch one of the clips of their choice again with HFVK stimulation, the results of which were excluded from our analyses. Then, subjects were guided to the washing facilities and escorted to the building exit whereupon they received $35 compensation for their participation. Total time for the experiment was approximately two hours, including 30 min for preparation and 30 min for cleanup.

2.5   EEG Recording and Pre-processing

EEG was recorded raw at a 500 Hz sampling rate using BrainVision Recorder (Brain-Products, GmbH, Munich, Germany). All EEG processing was performed using Brainstorm[1] running on MATLAB 15a (MathWorks, Natick, MA, USA). First, the EEG recordings were then band-pass filtered from 1–40 Hz. Noisy or dead channels were removed, and then the recordings were cleaned using independent component analysis. For each subject, a 90 s eyes-open and non-movement baseline period recorded at the start of the experiment was used to calculate a noise covariance matrix. An EEG-appropriate forward model was estimated based on the digitized electrode positions using Open-MEEG. Then minimum norm estimation was used to calculate cortical currents without dipole orientation constraints. Ten second segments of the EEG recording at the ending of each of the 15 movie clips were used for analyses of theta-band cortical coherence (see next section). Note that HFVK stimulation was not present during these periods in the HFVK group (verified with the accelerometer recording), and that there was at least one second separating the last HFVK event and the experimental brain activity period so as to avoid inclusion of transient off-responses to cessation of the HFVK stimuli. Analysis periods were marked at 1s intervals and epoched from -1000 ms to 2000 ms, resulting in 150 epochs per subject. All epochs were visually inspected and those with noise or transient artifacts above 100 μV were rejected from further analysis.

2.6   Coherence Processing

Source activity was parcellated into 62 areas according to the Mindboggle cortical atlas (included in Brainstorm). Coherence between each area and all other areas (62×62) was calculated across the time period of 0 to 1000 ms for all epochs at a frequency resolution of 2.5 Hz with a maximum frequency range of 30 Hz using the default pipeline included within Brainstorm. The resulting coherence values were then averaged over all epochs resulting in a single 62×62 coherence matrix with six frequency bins for each subject. Based on brain activity connections previously found to be relevant to HFVK stimulation during cinematic entertainment (Boasen et al., 2020), and as an initial probe into this continuing data collection, we targeted the right ITG and extracted its theta-band frequency bin (5–7.5 Hz) connections to the: right superior temporal gyrus (STG), right tranverse temporal cortex (TTC), right postcentral gyrus (PoCG), right precentral gyrus (PCG), and left insular cortex (IC).

---

[1] http://neuroimage.usc.edu/brainstorm.

2.7 Statistical Analysis

To test the difference in self-perceived emotional reactivity between groups, the mean emotional reactivity scores were compared using an independent t-test. Differences in brain activity coherence between groups for each of the targeted connections were similarly tested with independent t-tests. The threshold for significance was set at p ≤ 0.05. All tests were performed using SPSS software version 26 (IBM, Armonk, NY, USA).

3 Results

As for self-perceived emotional reactivity, the independent t-test revealed that subjects in the HFVK group were significantly more emotionally reactive than those in the control group (mean ± SEM: 12.12 ± 0.614 vs. 9.87 ± 0.632, respectively; p = 0.041). Meanwhile, independent t-tests comparing theta-band coherence between groups were not significantly different for right ITG connections with right STG (p = 0.888), right TTC (p = .594), right PCG (p = 0.929), and right PoCG (p = 0.983). However, the right ITG connection with the left IC was significantly stronger in the HFVK group than the VK group (p = 0.05). A visual summary of the results is shown in Fig. 1.

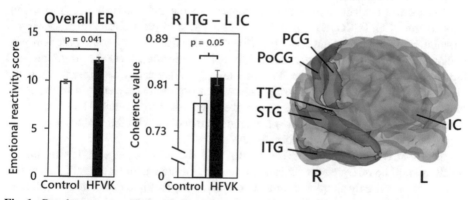

Fig. 1. Results summary. There was higher emotional reactivity (ER) in the HFVK group vs. the control group (left). Of the cortical links to the rITG tested (right cortical map), coherence with the left IC was significantly higher in the HFVK group vs. the control group (middle).

4 Discussion

The present report represents a first look at a work in progress aiming to verify previously observed psychological effects of HFVK stimulation during passive multimedia interaction on self-perceived emotion, and explore the neurophysiological underpinnings of these effects. The present intermediary analysis reveals HFVK stimulation to be associated with both heightened emotional reactivity and increased intercortical theta-band coherence.

That the HFVK stimulus effect on emotional reactivity is discernable with such a small sample size is encouraging. There is always a risk with psychometrics that subjects will self-report in the manner they think is expected of them (i.e. stronger emotion scores). However, the between subjects design employed in the present study should mitigate this risk. Moreover, the present results are congruent with prior psychophysiological evidence from studies involving cinema and music listening [1, 4–6, 8]. Thus, the heightened emotional reactivity of subjects who received HFVK stimulation seems credible, and raises the likelihood of underlying changes in neurophysiology.

This expectation appears supported by our observation of increased theta-band coherence between the right ITG and the left IC. Spatial processing is thought to be right-hemispheric dominant in right-handed people [11], which our subjects were. The ITG is a key visuospatial processing hub [12]. Meanwhile, the functional role of IC is complex, but thought to crucial for self-processing [13]. The IC is furthermore anatomically connected with the limbic system, and is functionally important for accurately recognizing others and our own emotional states [14]. Thus that higher psychological emotional reactivity in subjects stimulated with HFVK would be associated with higher coherence between somatosensory and emotion recognition processing areas seems very logical. The result is also in line with observations that precise integration of the somatosensory system with AV stimulation can enhance task-specific cognitive functions, such as enhanced memory for words when they are written versus typed [15, 16].

Meanwhile, the lack of increased coherence with the motor and auditory processing hubs of the PCG, PoCG, STG and TTC is in contrast to prior observations of HFVK stimulation during cinematic opera [9]. However, this discrepancy could potentially be explained by the fact that the stimuli used with cinematic opera were very rhythmic, and targeted musical elements of the AV stimulus only. Whereas in the present study, the HFVK stimuli predominantly targeted somatosensory elements depicted in the scene that one would expect to feel if they were there with the protagonist. Thus, the HFVK of the present study should arguably have more connection with the limbic system than they would with motor and auditory processing.

That said, the present analysis is naturally limited due to the low sample size, and the results should be interpreted with caution. Indeed, the statistical testing of intercortical coherence was highly simplified, and not corrected for multiple corrections. Nevertheless, the functional correspondence between the neurophysiological and psychological results is promising, and in line with the nature of the HFVK and AV stimuli used in the present study. As the study matures and sample size increases, more in-depth analyses of brain activity will become possible which will likely serve to enrich and add further clarity to the present report.

In conclusion, the presents results suggest that HFVK augments self-reported emotional reactivity in association with increased interhemispheric brain activity coherence. This study expands our knowledge on how HFVK affects neurophysiology, thereby further solidifying a foundation that will be applicable for future research on HFVK during active multimedia interactions. Here, we postulate that the means of HFVK delivery may have differing levels of efficaciousness. Considering that the physical delivery of HFVK tested in the present study was through high fidelity actuators imbedded in a recliner chair, future research should build upon the present methods and results to explore how delivery of HFVK via smaller and wearable vibrotactile mechanisms will affect immersion and other cognitive constructs during IS use contexts.

Acknowledgements. The authors would like to thank David Brieugne, Emma Rucco and all the technical staff at the Tech3Lab for their effort and assistance in executing this study. This study was financially supported by NSERC and Prompt (IRCPJ/514835-16, and 61_Léger-Deloitte 2016.12, respectively).

References

1. Oh, E., Lee, M., Lee, S.: How 4D effects cause different types of presence experience? In: Proceedings of 10th International Conference on Virtual Reality Continuum and Its Applications in Industry, pp. 375–378 (2011)
2. Waltl, M., Timmerer, C., Hellwagner, H.: Improving the quality of multimedia experience through sensory effects. In: 2010 2nd International Workshop on Quality of Multimedia Experience, QoMEX 2010 – Proceedings, pp. 124–129 (2010). https://doi.org/10.1109/QOMEX.2010.5517704
3. Gardé, A., et al.: The effects of a vibro-kinetic multi-sensory experience in passive seated vehicular movement in a virtual reality context. In: Conference on Human Factors in Computing Systems - Proceedings (2018). https://doi.org/10.1145/3170427.3188638
4. Giroux, F., et al.: Haptic stimulation with high fidelity vibro-kinetic technology psychophysiologically enhances seated active music listening experience. In: 2019 IEEE World Haptics Conference, WHC 2019, pp. 151–156 (2019). https://doi.org/10.1109/WHC.2019.8816115
5. Giroux, F., Boasen, J., Sénécal, S., Léger, P.-M.: Hedonic multitasking: the effects of instrumental subtitles during video watching. In: Davis, F.D., Riedl, R., vom Brocke, J., Léger, P.-M., Randolph, A.B., Fischer, T. (eds.) NeuroIS 2020. LNISO, vol. 43, pp. 330–336. Springer, Cham (2020). https://doi.org/10.1007/978-3-030-60073-0_38
6. Pauna, H., et al.: The psychophysiological effect of a vibro-kinetic movie experience: the case of the D-BOX movie seat. Lect. Notes Inf. Syst. Organ. **25**, 1–7 (2017)
7. Pauna, H., et al.: The effects of a high fidelity vibro-kinetic multisensory experience on implicit and explicit brand recognition. J. Neurosci. Psychol. Econ. **12**, 18–33 (2019)
8. Tchanou, A.Q., Léger, P.M., Senecal, S., Giroux, F., Ménard, J.F., Fredette, M.: Multitasking with information technologies: Why not just relax? AIS Trans. Hum.-Comput. Interact. (2021). [Forthcoming]
9. Boasen, J., et al.: High-fidelity vibrokinetic stimulation induces sustained changes in intercortical coherence during a cinematic experience. J. Neural Eng. (2020). https://doi.org/10.1088/1741-2552/abaca2
10. Boulais, S., Lizotte, J. M., Trottier, S., Gagnon, S.: Motion-enabled movie theater seat. United States Patent and Trademark Office (2011)
11. Karnath, H.O., Ferber, S., Himmelbach, M.: Spatial awareness is a function of the temporal not the posterior parietal lobe. Nature (2001). https://doi.org/10.1038/35082075
12. Lobier, M., Palva, J.M., Palva, S.: High-alpha band synchronization across frontal, parietal and visual cortex mediates behavioral and neuronal effects of visuospatial attention. Neuroimage **165**, 222–237 (2018)
13. Evrard, H.C.: The organization of the primate insular cortex. Front. Neuroanat. **13**, 1–21 (2019)
14. Jones, C.L., Ward, J., Critchley, H.D.: The neuropsychological impact of insular cortex lesions. J. Neurol. Neurosurg. Psychiatry **81**, 611–618 (2010)

15. Smoker, T.J., Murphy, C.E., Rockwell, A.K.: Comparing memory for handwriting versus typing. Proc. Hum. Factors Ergon. Soc. **3**, 1744–1747 (2009)
16. Ose Askvik, E., van der Weel, F.R., van der Meer, A.L.H.: The importance of cursive handwriting over typewriting for learning in the classroom: a high-density EEG study of 12-year-old children and young adults. Front. Psychol. **11**, 1–16 (2020)

Explainable Artificial Intelligence (XAI): How the Visualization of AI Predictions Affects User Cognitive Load and Confidence

Antoine Hudon(✉), Théophile Demazure, Alexander Karran, Pierre-Majorique Léger, and Sylvain Sénécal

Tech3Lab, HEC Montréal, Montréal, Canada
antoine.hudon@hec.ca

Abstract. Explainable Artificial Intelligence (XAI) aims to bring transparency to AI systems by translating, simplifying, and visualizing its decisions. While society remains skeptical about AI systems, studies show that transparent and explainable AI systems result in improved confidence between humans and AI. We present preliminary results from a study designed to assess two presentation-order methods and three AI decision visualization attribution models to determine each visualization's impact upon a user's cognitive load and confidence in the system by asking participants to complete a visual decision-making task. The results show that both the presentation order and the morphological clarity impact cognitive load. Furthermore, a negative correlation was revealed between cognitive load and confidence in the AI system. Our findings have implications for future AI systems design, which may facilitate better collaboration between humans and AI.

Keywords: Explainable Artificial Intelligence · XAI · Cognitive load · Confidence in AI · Explanation · Visualization · Cognitive fit theory

1 Introduction

Artificial intelligence (AI) algorithms are growing in complexity, sophistication, and accuracy, aided by increasing data volumes. However, as these algorithms grow in complexity and performance, they become less interpretable and opaquer, making the decisions these machine learning models form hard to comprehend and explain (Rudin, 2019).

Research in explainable AI (XAI) seeks to address the problems associated with a lack of transparency in AI systems. XAI methods attempt to decipher which components of the AI model or system are perturbed to create decisions through visualizations or descriptions of discriminative mechanisms. These methods aim at giving the user access to the AI's decision chain and discriminative variables. Implementing XAI methods within AI systems will help create more transparent and fair systems, thereby helping users become more aware of a system's behavior and support a richer collaboration between humans and AI [1]. Unfortunately, XAI currently focuses more on creating

© The Author(s), under exclusive license to Springer Nature Switzerland AG 2021
F. D. Davis et al. (Eds.): NeuroIS 2021, LNISO 52, pp. 237–246, 2021.
https://doi.org/10.1007/978-3-030-88900-5_27

mathematically interpretable models than on information presentation, neglecting the systems' final users [2], that is, the ones that will benefit from the explanations of those models. Therefore, these interpretability methods may impose a higher cognitive load, making them hard to understand and apply to real-world problems. This study attempts to bridge this gap in XAI by assessing different visualizations of an AI system's explanation, with the aim of providing human-centered explanations without compromising the faithfulness of the AI system's visualizations.

Using the Cognitive Fit theory, we propose to investigate if the visualization of explanations of an AI decision system can affect a user's cognitive load and confidence in the AI system. Specifically, we present in this paper the preliminary results from a study designed to assess two presentation-order methods and three AI decision visualization attribution models, referred to as explanation visualizations (EV), to determine each visualization's impact upon a user's cognitive load and confidence in the system by asking participants to complete a visual decision-making task. For this research, we focus only on identifying the effects of different EVs on a user's cognitive load and confidence in an AI context.

Cognitive Fit theory was initially developed to assess the impact on user's performance when viewing numerical data presented in tabular versus graph format in a symbolic and spatial task [3]. This theory has been applied in several studies within other research domains, adapting it for new information presentation formats (e.g., online rating systems [4], and database structure representations [5]) using numerical, textual, [6] as well as visual data [7]. The data at the core of explanations produced by interpretable models are numerical. However, these explanations are complex for a human to process, requiring translation using shapes and colors to represent their values. This study evaluates which EV type, presentation-order method, and AI decision visualization attribution models result in the best cognitive fit paired with a spatial task. The EV type that results in a better cognitive fit will allow us to make recommendations for future AI systems design, contributing a step on the road towards explainable AI that is more accessible to human mental capacities.

2 Hypothesis Development and Research Model

The research model (Fig. 1) posits that the effect of the EV's adjacency and morphological clarity will affect a user's perceived confidence in the system's judgments. We hypothesize that this effect is mediated by the cognitive load imposed by the visualization on the user's working memory. The rationale behind these relationships is described below.

Cognitive Fit theory proposes that the level of congruence between task and information presentation mediates task performance, such that, while solving a problem, an individual creates a mental representation of the problem based on the information presented [3]. This mental representation's complexity and usefulness are dependent on the user's working memory capacity [6, 8, 9]. Thus, additional cognitive effort is required in the presence of incongruence between task and the format of the information presented to help complete the task [10], requiring the individual to mentally transform that information into a format suitable for accomplishing the task, resulting in reduced performance [6].

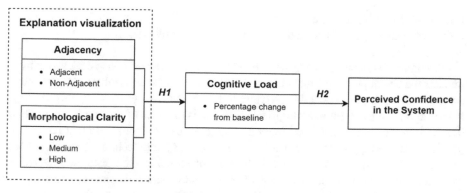

Fig. 1. Research model

We posit that specific EVs will result in a better cognitive fit, reducing cognitive load. Cognitive load is defined as the demand imposed by a task on the user's working memory [11]. Therefore, a task requiring significant mental resources is more likely to prompt more user errors than a task requiring less cognitive resources, resulting in less perceived effort and greater cognitive fit [10, 12].

H1a: Adjacent visualizations will result in a lower cognitive load.

An adjacent EV displays the explanation directly upon the original image by coloring the areas in different colors to indicate their data values [13]. In comparison, a non-adjacent EV is presented with the same explanation data but separated from the image. However, with this latter method, there is a loss in correspondence between the explanation and the original image, requiring significantly greater cognitive effort from the user [14].

Furthermore, we posit that morphological clarity (MC) will play a role in mediating the effect of adjacency upon cognitive workload. Morphological Clarity (MC) represents the degree to which a visualization displays clearly delimited features by adjusting the appearance or removing specific data (e.g., noise) to help make the delimitation clearer for the user. For this research, High MC EVs are faithful to the MC definition by having clear delimited features without noise, providing greater clarity for the user but at the cost of faithfully depicting the model's behavior [14]. On the other hand, Low MC EVs are more faithful to the model's behavior as they precisely illustrate the relevant image's areas, at the cost of being more cluttered. Moreover, Low MC EVs may cause humans to ignore the explanation by giving them too much information to process [14].

H1b: High MC EVs will result in lower cognitive load.
H1c: Adjacent-High MC EVs will result in lower cognitive load.

Providing explanations of system behaviors as a form of transparency positively impacts the development of trust [15] and confidence in new technology [16–20]. However, according to [21], it would be wrong to talk about trust in AI since this implies a loss of human agency. More appropriate, perhaps, is to speak of developing confidence in AI. Confidence in AI has been defined as a measure of risk or surety that AI systems provide correct suggestions and if users consider the system to be reliable [22]. More formally, confidence in this context is the goal of reducing the epistemic uncertainty associated with AI decisions with regard to accuracy, data provenance, and temporal qualities. To investigate this aspect, we formed the following hypothesis:

H2: EVs imposing a lower cognitive load on the user will result in higher perceived confidence in the system.

3 Method

3.1 Experimental Design

We designed a 2×3 within-subject factorial design to investigate the effects of adjacency and MC of AI-EVs on the user's cognitive load and perceived confidence. The first factor considers the representation's adjacency with two levels: EV with (adjacent) or without (non-adjacent) image background. The second factor considers the MC of the EV with three levels: low-cloud of points (CP), medium-heatmap (HM), and high-outline (ON). CP faithfully depicts the model's attributions by highlighting all the image's pixels that positively impact the model's classification toward a result. HM is less precise than CP, showing only the stimuli image's prime focus and does not have precisely delimited features. ON visualization draws only the most essential zones of the image used in the classification but at the cost of pixel-level precision. CP and ON-EVs were both implemented using the Integrated Gradients method [14, 23], and the HM-EV using the Grad-CAM class activation function [24]. See Table 1 for examples of each type of EV used in this study.

3.2 Experimental Procedure

19 participants (24.9 ± 8.3 years old, 10 males) took part in the study, all signed consent following the HEC ERB ethical approval. The task (Fig. 2) consisted of a spatial task repeated over 60 randomized trials. Each trial involves a series of elements displayed on screen in the following order: (1) original image (e.g., image of an elephant), (2) classification of the image given by the AI system[1] (e.g., "Elephant"), (3) the AI EV (e.g., an overlay IA explanation onto the original image) and (4) a perceived confidence question (e.g., confidence in the system). Participants were asked to rate their agreement with

---

[1] The Xception (extreme inception) [30] algorithm which comes with pre-trained weights on the ImageNet dataset was used to classify the images.

Table 1. Examples of EV for the classification "Monkey." Image selected from the ImageNet dataset [25].

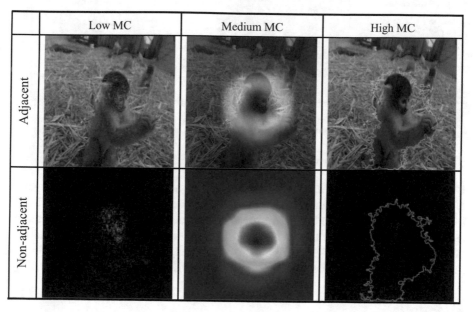

the following statement using a 7-point Likert scale ranging from "Strongly disagree" to "Strongly agree" and "I am confident in the system's ability to classify pictures of similar objects correctly". A baseline image was finally shown at the end of each trial for 1s. Each participant saw all six types of visualizations ten times. To measure pupil dilation, we used the Tobii x60 eye tracker.

3.3 Stimuli Images and Visualizations

Stimuli image categories were chosen based on [26], who defined a standardized set of 260 illustrations of different concepts. From these, 60 concepts were selected as image stimuli. We selected one image for each category from the ImageNet dataset [25], using neutrality (no shocking or disturbing depiction), simplicity, and unambiguity as criteria. Also, no human subjects are present in any of the selected images. These selection criteria allowed us to control the impact of affective mediators on the users-AI confidence. For two rounds of image verification, a panel of 18 judges labeled each image. Tests for inter-rater reliability produced $\kappa = 0.838$ and $\kappa = 0.879$ for the 1st and 2nd verification rounds, respectively, showing a high degree of agreement. Eight images were removed, and new images were added between the two rounds due to a lack of agreement.

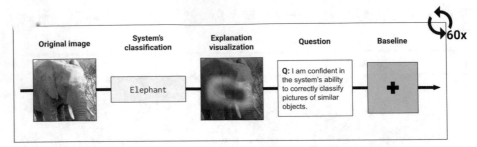

Fig. 2. Task's design. Images selected from the ImageNet dataset [25].

3.4 Calculating Cognitive Effort

Change in Pupil dilation, when a user is faced with a task requiring a high cognitive effort, is referred to as the Task-evoked pupillary response [27]. We used pupil diameter to estimate the user's cognitive effort required to process each EV. We computed the average percentage change from a baseline (PcB) in this preliminary analysis for each participant and each EV. We used the percentage change of pupil diameter rather than the raw pupil size variation due to inter-participant variance [28].

4 Results

We performed a repeated measures ANOVA for the dependent variable PcB, with both Adjacency and MC as factors. The results show a statistically significant main effect of Adjacency ($F(1, 189) = 20.99$, $p < .001$, $\eta^2 = 0.02$), MC ($F(2, 378) = 28.03$, $p < .001$, $\eta^2 = 0.05$) as well as the interaction between both factors ($F(2, 378) = 22.32$, $p < .001$, $\eta^2 = 0.04$). Post hoc comparisons (Bonferroni corrected) reported that Adjacent EV ($M = 1.27$, $SD = 6.36$) results in lower PcB than non-adjacent EV ($M = 3.05$, $SD = 6.52$), providing support for H1a. Concerning MC, post hoc comparisons showed that Medium MC EV ($M = 0.20$, $SD = 6.25$) results in lower PcB than both Low ($M = 2.90$, $SD = 6.25$), and High ($M = 3.40$, $SD = 6.55$), providing no support for H1b. The results (Fig. 3) indicate no significant difference between the three adjacent visualizations. Therefore, H1c is also not supported. Unexpectedly, non-adjacent-Medium MC EV ($M = -0.35$, $SD = 5.63$) resulted in a negative PcB as well as the lowest PcB of all visualizations, which may explain the result of H1b.

With regard to the relationship between PcB and perceived confidence in the system, the results indicate a significant negative correlation between the two variables, $r(1136) = -0.12$, $p < .001$, providing support for H2. However, this requires further analysis. Results of perceived confidence for each EV type can be seen in Fig. 4.

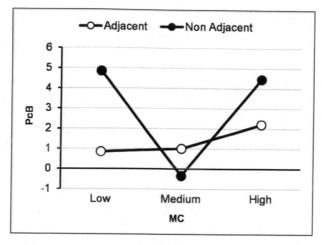

Fig. 3. PcB for each EV type

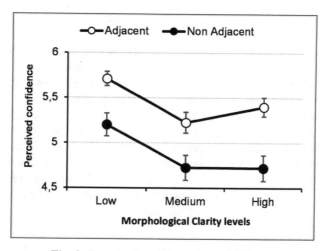

Fig. 4. Perceived confidence for each EV type

5 Discussion and Conclusion

The preliminary results indicate that adjacent EVs result in lower PcB. Potentially, this implies that the cognitive effort required to process and understand the AI's EV through adjacent presentation is significantly lower than non-adjacent EV, given a reduced need to mentally associate the EV to the original image as the association is already made implicit within the EV itself. Indeed, this effect was posited by [13], who stated that combining adjacent presentation-order with a spatial task may lead to faster and more accurate decision making, resulting in a better cognitive fit. Regarding the low PcB reported for non-adjacent-medium MC EV, this type of EV is very abstract, potentially making it difficult to process the original image's target object. We posit that this proved challenging for participants to identify precise forms and associate them with the original image, allowing for a snap judgment and lower cognitive workload through a disengagement effect. This may account for why this EV resulted in one of the lowest scores of perceived confidence.

Moreover, it would appear that a significant negative correlation between cognitive load and perceived confidence does not necessarily result in low confidence when mediated by adjacency. For example, the non-adjacent-low MC EV, which reported the highest PcB, resulted in the highest perceived confidence for non-adjacent visualizations. We propose that the high density of information presented in this EV potentially helped users identify the target object forms by reducing epistemic uncertainty. In that, extraneous but useful information allowed the user to swap from cognitive processing to perceptual processing to understand the model's behavior. Indeed, there is a small body of researchers in human factors investigating this new concept of epistemic uncertainty and how it affects man-machine teaming [29]. Additionally, adjacent-Low MC EV also resulted in the highest perceived confidence, contradicting [14], who stated that users had shown a high over low MC preference. Regarding the negative correlation reported between PcB (inferred cognitive load) and perceived confidence, future work will aim to explain this effect in more detail using further pupillometry measures.

This study investigated the relationships between various types of EVs used to explain an AI system's output and user cognitive load and their effects on a user's confidence in the system. The results indicate that design choices related to EVs can positively impact a user's confidence in AI systems by reducing epistemic uncertainty. In this study, users manifested a preference for visualizations providing precision rather than simplicity, where they can easily associate the target with the explanation. Overall, our results suggest that the careful consideration of cognitive fit theory, information presentation adjacency methods, and explanation visualizations containing low morphological clarity applied to AI interface and task design may help accelerate confidence between a user and an AI decision support system. However, more work is required to tease apart each factor's role to determine how best to increase confidence in users of AI decision support systems.

References

1. Gilpin, L.H., Bau, D., Yuan, B.Z., Bajwa, A., Specter, M., Kagal, L.: Explaining explanations: an overview of interpretability of machine learning. In: 2018 IEEE 5th International Conference on Data Science and Advanced Analytics (DSAA), pp. 80–89, October 2018. https://doi.org/10.1109/DSAA.2018.00018
2. Abdul, A., Vermeulen, J., Wang, D., Lim, B.Y., Kankanhalli, M.: Trends and trajectories for explainable, accountable and intelligible systems. In: Proceedings of the 2018 CHI Conference on Human Factors in Computing Systems - CHI 2018, vol. 2018-April, pp. 1–18 (2018). https://doi.org/10.1145/3173574.3174156
3. Vessey, I.: Cognitive fit : a theory-based analysis of the graphs versus tables literature. Decis. Sci. 22(2), 219–240 (1991). http://dx.doi.org/10.1016/j.jaci.2012.05.050
4. Chen, C.-W.: Five-star or thumbs-up? The influence of rating system types on users' perceptions of information quality, cognitive effort, enjoyment and continuance intention. Internet Res. (2017)

5. Bizarro, P.A.: Effect of different database structure representations, query languages, and task characteristics on information retrieval. J. Manag. Inf. Decis. Sci. **18**(1) (2015)
6. Adipat, B., Zhang, D., Zhou, L.: The effects of tree-view based presentation adaptation on mobile web browsing. MIS Q. **35**(1), 99 (2011). https://doi.org/10.2307/23043491
7. Brunelle, E.: The moderating role of cognitive fit in consumer channel preference. J. Electron. Commer. Res. **10**(3) (2009)
8. Goodhue, D.L., Thompson, R.L.: Task-technology fit and individual performance. MIS Q. 213–236 (1995)
9. Vessey, I., Galletta, D.: Cognitive fit: an empirical study of information acquisition. Inf. Syst. Res. **2**(1), 63–84 (1991)
10. Nuamah, J.K., Seong, Y., Jiang, S., Park, E., Mountjoy, D.: Evaluating effectiveness of information visualizations using cognitive fit theory: a neuroergonomics approach. Appl. Ergon. **88**(June 2019), 103173 (2020). https://doi.org/10.1016/j.apergo.2020.103173
11. Wickens, C.D.: Multiple resources and mental workload. Hum. Factors **50**(3), 449–455 (2008). https://doi.org/10.1518/001872008X288394
12. Palinko, O., Kun, A.L., Shyrokov, A., Heeman, P.: Estimating cognitive load using remote eye tracking in a driving simulator. In: Eye-Tracking Research & Applications Symposium, no. April 2017, pp. 141–144 (2010). https://doi.org/10.1145/1743666.1743701
13. Dennis, A.R., Carte, T.A.: Using geographical information systems for decision making: extending cognitive fit theory to map-based presentations. Inf. Syst. Res. **9**(2), 194–203 (1998). https://doi.org/10.1287/isre.9.2.194
14. Sundararajan, M., Xu, S., Taly, A., Sayres, R., Najmi, A.: Exploring principled visualizations for deep network attributions. In: IUI Workshops, vol. 4 (2019)
15. Bigras, É., Léger, P.-M., Sénécal, S.: Recommendation agent adoption: how recommendation presentation influences employees' perceptions, behaviors, and decision quality. Appl. Sci. **9**(20) (2019). https://doi.org/10.3390/app9204244.
16. Glikson, E., Woolley, A.W.: Human trust in artificial intelligence: review of empirical research. Acad. Manag. Ann. **14**(2), 627–660 (2020). https://doi.org/10.5465/annals.2018.0057
17. Cofta, P.: Designing for trust. In: Handbook of Research on Socio-Technical Design and Social Networking Systems, vol. 731, no. 9985433, pp. 388–401. IGI Global (2009)
18. Eiband, M., Buschek, D., Kremer, A., Hussmann, H.: The impact of placebic explanations on trust in intelligent systems. In: Conference on Human Factors in Computing Systems – Proceedings (2019). https://doi.org/10.1145/3290607.3312787
19. Lee, J.D., See, K.A.: Trust in automation: designing for appropriate reliance. Hum. Factors J. Hum. Factors Ergon. Soc. **46**(1), 50–80 (2004). https://doi.org/10.1518/hfes.46.1.50_30392
20. Meske, C., Bunde, E.: Transparency and trust in human-AI-interaction: the role of model-agnostic explanations in computer vision-based decision support. In: Degen, H., Reinerman-Jones, L. (eds.) HCII 2020. LNCS, vol. 12217, pp. 54–69. Springer, Cham (2020). https://doi.org/10.1007/978-3-030-50334-5_4
21. DeCamp, M., Tilburt, J.C.: Why we cannot trust artificial intelligence in medicine. Lancet Digit. Heal. **1**(8), e390 (2019)
22. Wanner, J., Herm, L.-V., Heinrich, K., Janiesch, C., Zschech, P.: White, grey, black: effects of XAI augmentation on the confidence in AI-based decision support systems. In: Proceedings of Forty-First International Conference on Information Systems, pp. 0–9 (2020)
23. Sundararajan, M., Taly, A., Yan, Q.: Axiomatic attribution for deep networks. In: 34th International Conference on Machine Learning, ICML 2017, vol. 7, pp. 5109–5118, March 2017. http://arxiv.org/abs/1703.01365
24. Selvaraju, R.R., Cogswell, M., Das, A., Vedantam, R., Parikh, D., Batra, D.: Grad-CAM: visual explanations from deep networks via gradient-based localization. Int. J. Comput. Vision **128**(2), 336–359 (2019). https://doi.org/10.1007/s11263-019-01228-7

25. Deng, J., Dong, W., Socher, R., Li, L.-J., Li, K., Fei-Fei, L.: ImageNet: a large-scale hierarchical image database. In: 2009 IEEE Conference on Computer Vision and Pattern Recognition, pp. 248–255, June 2009. https://doi.org/10.1109/CVPRW.2009.5206848
26. Snodgrass, J.G., Vanderwart, M.: A standardized set of 260 pictures: norms for name agreement, image agreement, familiarity, and visual complexity. J. Exp. Psychol. Hum. Learn. Mem. 6(2), 174–215 (1980). https://doi.org/10.1037/0278-7393.6.2.174
27. Beatty, J.: Task-evoked pupillary responses, processing load, and the structure of processing resources. Psychol. Bull. 91(2), 276–292 (1982). https://doi.org/10.1037/0033-2909.91.2.276
28. Attard-Johnson, J., Ó Ciardha, C., Bindemann, M.: Comparing methods for the analysis of pupillary response. Behav. Res. Methods 51(1), 83–95 (2018). https://doi.org/10.3758/s13428-018-1108-6
29. Tomsett, R., et al.: Rapid trust calibration through interpretable and uncertainty-aware AI. Patterns 1(4), 100049 (2020). https://doi.org/10.1016/j.patter.2020.100049
30. Chollet, F.: Xception: deep learning with depthwise separable convolutions. In: 2017 IEEE Conference on Computer Vision and Pattern Recognition (CVPR), vol. 2017-Janua, pp. 1800–1807, July 2017. https://doi.org/10.1109/CVPR.2017.195

Author Index

F. Davis et al. (Eds.): NeuroIS 2021, LNISO 52, pp. 247–248, 2021.
https://doi.org/10.1007/978-3-030-88900-5

Printed in the United States
by Baker & Taylor Publisher Services